ANATOMY and PHYSIOLOGY

Volume 2: Urinary, Respiratory, and Nervous Systems; Sensations and Sense Organs; Endocrine and Reproductive Systems

ANATOMY and PHYSIOLOGY

Volume 2: Urinary, Respiratory, and Nervous Systems; Sensations and Sense Organs; Endocrine and Reproductive Systems

Second Edition

Edwin B. Steen
Ashley Montagu

BARNES & NOBLE BOOKS
A DIVISION OF HARPER & ROW, PUBLISHERS
New York, Cambridge, Philadelphia, San Francisco,
London, Mexico City, São Paulo, Singapore, Sydney

 For information address Harper & Row, Publishers, Inc., 10 East 53rd Street, New York, N.Y. 10022. Published simultaneously in Canada by Fitzhenry & Whiteside Limited, Toronto.

Library of Congress Cataloging in Publication Data
(Revised for Volume 2)

Steen, Edwin Benzel, 1901–
Anatomy and physiology.

(The Barnes & Noble outline series) (COS ; 190–191)
Includes indexes.
Contents: v. 1. Cells, tissues, integument, skeletal, muscular, and digestive systems; blood, lymph, circulatory system—v. 2. Urinary, respiratory, and nervous systems; sensations and sense organs; endocrine and reproductive systems.
1. Human physiology—Outlines, syllabi, etc. 2. Anatomy, Human—Outlines, syllabi, etc. I. Montagu, Ashley, 1905– . II. Title. III. Series: College outline series ; CO/190–191. [DNLM: 1. Anatomy—Outlines. 2. Physiology—Outlines QS 18 S814a]
QP41.S75 1985 612 83-47599
ISBN 0-06-460190-0 (pbk.: v. 1) 85 86 87 88 89 10 9 8 7 6 5 4 3 2 1
ISBN 0-06-460191-9 (pbk.: v. 2) 85 86 87 88 89 10 9 8 7 6 5 4 3 2 1

CONTENTS

PREFACE

This revised edition of *Anatomy and Physiology* is designed to meet the needs of students and others who wish to learn or review the essentials of human anatomy and physiology. It provides a comprehensive summary for students of the biological sciences, nursing, occupational therapy, and physical education. It should be most helpful to medical and dental students at all stages in their education—undergraduate, graduate, and postgraduate—and to students preparing to be physician's assistants, medical technologists, dental hygienists, physical therapists, and others whose activities require a knowledge of the human body and how it works. It should also serve as a useful reference book for laymen who desire information about the human body.

Since the authors have for many years taught students in all the groups mentioned above, they are acutely aware of the fact that in this field of study there is no satisfactory substitute for work in the laboratory. No one can acquire a thorough functional knowledge of anatomy and physiology by merely reading a book. Consequently, this volume is not intended as a substitute for laboratory research but rather as a supplement to it, as a remembrancer following laboratory work, and, finally, as a refresher of the memory long after introductory studies of the subject have been concluded. For laboratory work, the following manual is available: Edwin B. Steen, *Laboratory Manual and Study Guide for Anatomy and Physiology,* 3rd ed., Dubuque, Iowa: William C. Brown Co., 1976.

The Outline constitutes a completely integrated textbook of anatomy and physiology, covering the fundamentals of these inextricably intertwined subjects. The reader is asked to bear in mind always that anatomic structure and physiologic function are but different aspects of the same thing, just as physics and chemistry are but two different ways of looking at matter. Anatomy and physiology are the names we give to two ways in which we look at organic matter, structure being a function of function, and function being a function of structure. There is rarely a structure without function and no function without structure, the two being inseparably associated.

In this new, revised edition, the authors have attempted to bring the subject matter on all topics up to date and to incorporate the latest findings and developments throughout. Many new illustrations have been added to make the text more understandable. The sections at the end of each chapter that list pathological conditions involving the various organ systems have been enlarged and amplified.

The authors wish to express their special thanks and appreciation to Jeanne Flagg, Editor, Barnes & Noble Books, for her work in the preparation of this new revised edition, to Carol Miller for her editorial review of the manuscript, and to Bruce Emmer for his careful work in copy editing. They also wish to thank the publishers who kindly granted them permission to reproduce a number of the figures. Acknowledgment of sources of these figures is made in the legends. To the staff of Harper & Row the authors express their thanks for their work and cooperation in the preparation and production of this revision.

The authors also thank Louise Yorke, librarian at the Princeton Medical Center, and Helen Zimmerberg and Louise Schaffer of the Biology Library, Princeton University, for their always gracious assistance, and Dr. Stanley Rosenberg, of the Princeton Medical Center, for his helpful reading of Chapter 1, "The Urinary System."

While the authors have made every effort to ensure accuracy in their writing, errors and mistakes sometimes occur. They will welcome any information calling attention to possible errors or inaccuracies.

EDWIN B. STEEN
ASHLEY MONTAGU

INTRODUCTORY NOTES

The following special terms should be thoroughly understood inasmuch as they are used repeatedly in the study of anatomy.

TERMS REFERRING TO LOCATION OR POSITION

*Anterior** toward the front of the body
*Posterior** toward the back of the body

Ventral toward the anterior side
Dorsal toward the posterior side

Superior above, upper
Inferior below, lower

Superficial on or near the surface
Deep remote from the surface

Internal† within, inside
External† without, outside

Proximal nearest to the body or to some other point regarded as the center of a system; nearest to the point of attachment
Distal farthest from the body or from some other central point; farthest from the point of attachment
Medial toward the midline or middle plane of the body
Lateral away from the midline or middle plane of the body

Central toward the center of or mid-region; toward the principal axis; inward
Peripheral toward the periphery or outer surface; away from the center; outward

Parietal of, pertaining to, arising from, or located on a wall
Visceral of or pertaining to the viscera or the internal organs located within one of the body cavities

* In quadrupeds, *anterior* means "toward the head end"; *posterior,* "toward the caudal or tail end"; *ventral,* toward the underneath or the belly side; *dorsal,* toward the back or the uppermost side.

† *Inside* or *interior* and *outside* or *exterior* are reserved for reference to body cavities and hollow organs.

TERMS REFERRING TO DIRECTION

Craniad toward the cranium
Cephalad toward the head end
Mesiad toward the median plane
Caudad toward the tail end; away from the head
Laterad toward the side; away from the median plane

TERMS REFERRING TO SECTIONS OF THE BODY

Sagittal a vertical section of the body, or in the same plane as the sagittal suture
Midsagittal a medial sagittal section that divides the body into right and left halves
Transverse, cross, or horizontal a cut made at right angles to the long axis of the body, dividing the body into upper and lower portions
Frontal or coronal a vertical cut made at right angles to the sagittal plane, dividing the body into anterior and posterior portions

The foregoing terms can also be applied to individual organs or structures. In this sense, the axis of the organ (not the axis of the whole body) is the basis of description. Frequently, therefore, in discussions of the body as a whole, *longitudinal* is substituted for *sagittal* and *median longitudinal* for *midsagittal.*

A cut made in any place other than the foregoing is called an *oblique section,* but such a section is not commonly used.

STRUCTURAL ORGANIZATION OF THE BODY

The following is a review of the structural organization of the body.

The Cell. The cell is the basic structural unit of the body. Cells, together with their intercellular material, are organized into *tissues,* tissues into *organs,* and organs into *systems.* An *organ* is a structure of two or more tissues that has a more or less definite form and structure and performs one or more specific functions. A *system* is a group of associated organs that work together in performing a series of related functions.

Tissues. The primary tissues of the body are epithelial, connective, muscular, and nervous. *Epithelial tissues* are those that cover surfaces, line tubes and cavities, and form the ducts and secreting portions of glands. *Connective tissues* form supporting and connecting structures. *Muscular tissues* are found in all structures where movement or change of form occur. They are present in skeletal muscles, the heart and blood vessels, and the walls of visceral structures. *Nervous tissue* is found in the brain, spinal cord, ganglia, nerves, and the sensory portions of sense organs.

Systems. The systems of the body, in which all functional activities occur, are the following:

Integumentary System. This includes the skin and its derivatives (hair, nails, and glands). Its principal functions are protection, prevention of dehydration, and regulation of body temperature. Serving as a base for sensory receptors, it provides information about our environment.

Skeletal System. This includes the bones and cartilage, which form supporting structures, and the ligaments, which bind the bones together at joints. The skeleton functions in support, protects vital organs, and serves for the attachment of muscles by which movement is accomplished. The soft tissue (bone marrow) in the hollow spaces of bones is the seat of the manufacture of blood cells.

Muscular System. This includes the contractile tissues of the body, namely skeletal, cardiac, and smooth muscle. These muscles are responsible for the maintenance of posture and all active movements of the body, including locomotion, change in position of body parts, and the movement of blood, food, and other substances through tubes.

Digestive System. This includes the alimentary canal from mouth to anus and its associated glands. The *alimentary canal* functions in the ingestion, digestion, and absorption of food and water and in the elimination of undigested food and some metabolic wastes. The glands (salivary, gastric, intestinal) are the source of enzymes essential for digestion. Other glands, the pancreas and liver, are essential in the digestion and metabolism of carbohydrates and fats. The pancreas is the source of the hormones insulin and glucagon; the liver is the source of bile.

Circulatory System. This includes the heart and all the blood vessels (arteries, arterioles, veins, venules, and capillaries) that are essential for circulation of the blood to and from the tissues. *Blood* is the medium through which water, nutrients, oxygen, minerals, and other essential substances, such as hormones and vitamins, are carried to all cells and through which the products of cellular activity, such as waste and metabolic products and internal secretions, are carried away.

Also included in the circulatory system is the *lymphatic system,* which includes the lymph vessels (thoracic ducts, lymphatics, capillaries, lacteals) that convey lymph, and the lymphatic organs (lymph nodules, nodes, tonsils, thymus gland, and spleen), which serve various functions. This system functions in the return of tissue fluid to the blood and in the protection of the body from disease-causing agents. The latter is accomplished principally through the action of phagocytes and the production of antibodies.

Urinary System. This includes the kidneys, ureters, urinary bladder, and urethra. In the production of urine, the kidneys remove from the blood toxic waste products of protein metabolism, such as urea and uric acid, and also nontoxic materials, such as water and inorganic salts. The kidneys thus eliminate waste products and, in process, function in the maintenance of water, acid-base, and electrolyte balance. In the male, the urethra functions in the transport of the seminal fluid, which contains spermatozoa.

RESPIRATORY SYSTEM. This system includes the respiratory passageways (nasal cavities, pharynx, larynx, trachea, bronchi, bronchioles, and alveolar sacs) and the lungs. It functions in the exchange of gases (oxygen and carbon dioxide) between the air and the blood and between the blood and the tissues. It also functions in vocalization, in the elimination of excess heat and water, and in the regulation of acid-base balance of the body fluids.

NERVOUS SYSTEM. This includes the brain, spinal cord, ganglia, nerves, and sensory receptors. Impulses originating in sense organs or receptors convey information concerning the nature of the environment, both external and internal, to the spinal cord, the brain, or both, where, either consciously or by reflex action, appropriate action is taken to initiate responses (movement or secretion) by which proper adjustment to the environment is made. Through nervous impulses, the various activities of the body are integrated and coordinated. In addition, the brain is the seat of such higher mental functions as perception, thinking, reasoning, understanding, judgment, and memory; it is also the seat of the emotions and the source of certain hormones.

ENDOCRINE SYSTEM. This includes the glands of internal secretion (hypophysis or pituitary gland, thyroid gland, parathyroid glands, adrenal glands, pancreatic islets of Langerhans, pineal gland, thymus gland, ovaries, testes, and other secreting tissues). These organs secrete *hormones,* chemical agents that are transported by the bloodstream to other organs or tissues (target organs) on which they exert their effects, either stimulating or inhibiting activity. Through hormones, various activities of the body, especially those involved in growth, development, and reproduction, are regulated and coordinated. Hormones are also involved in the regulation of such basic metabolic processes as the utilization of oxygen, sugar, and various minerals.

REPRODUCTIVE SYSTEM. This system is primarily concerned with the production of offspring. The *male reproductive organs* include the testes, which produce spermatozoa; the ducts (ductus deferens, ejaculatory ducts, and urethra), which convey the sperm outside the body; the accessory glands (seminal vesicles, prostate gland, and bulbourethral glands), which contribute to the seminal fluid; and the penis, the copulatory organ. The *female reproductive organs* include the ovaries, which produce ova; the uterine tubes, which serve as a site for fertilization and for conduction of the fertilized egg, or zygote, to the uterus; the uterus, in which the embryo or fetus develops; the vagina, which serves as the female copulatory organ, as a birth canal, and as a passageway for menstrual fluid; and the external genitalia or vulva (the clitoris, major and minor labia, and mons pubis). The mammary glands are sometimes regarded as reproductive organs.

The testes and ovaries also function as endocrine organs producing hormones that regulate the development of secondary sex characteristics and various physiological processes. The penis is also an excretory organ.

SUMMARY. The integrated functioning of all these organ systems is essential for the maintenance of life, for physical and mental health, for adjustment to the environment, and for the continuation of the species. The detailed study of each of these systems constitutes the subject matter of anatomy and physiology. This volume is devoted to the urinary, respiratory, nervous, endocrine, and reproductive systems. Cells; tissues; integument; the skeletal, muscular, and digestive systems; blood; lymph; and the circulatory system are covered in Volume 1.

ANATOMY and PHYSIOLOGY

Volume 2: Urinary, Respiratory, and Nervous Systems; Sensations and Sense Organs; Endocrine and Reproductive Systems

1: THE URINARY SYSTEM

The urinary system (Fig. 1-1) comprises the organs that are concerned with the production and elimination of urine, the fluid vehicle for the discharge of waste products from the body. Consisting of the kidneys, the urinary bladder, and the ureters and the urethra, the urinary system acts in conjunction with the other organs of excretion (lungs, skin, intestines) to *maintain a constant* or *stable internal environment* (*homeostasis*). To accomplish this, the following functions are performed:

1. Excretion of the waste products of metabolism, especially nitrogenous substances. Also, the elimination from the body of excess hormones, vitamins, and foreign substances such as drugs and food additives.
2. Regulation of water balance.
3. Regulation of electrolyte balance and, as a consequence, regulation of extracellular osmolarity.
4. Maintenance of proper acid-base balance accomplished through the regulation of hydrogen ion concentration.
5. Regulation of hemopoiesis and blood pressure accomplished through the secretion of enzymes or hormonelike substances (erythropoietic factor and renin).

STRUCTURE OF THE URINARY SYSTEM

In the following description of the urinary system, the organs are presented in sequence from the kidney, where urine is manufactured, to the urethra, from which it is finally passed out of the body.

Structure of the Kidney. The kidneys (Fig. 1-1) are paired organs, each lying lateral to the vertebral column against the posterior wall of the abdominal cavity at about the level of the last thoracic and the upper two lumbar vertebrae. The right kidney lies slightly lower than the left. Each is approximately 11 to 12 cm long, 5 to 7 cm wide, and 3 cm thick, and each is enclosed in a thin, fibrous *capsule.* The kidneys are *retroperitoneal;* that is, they lie behind the peritoneum. They are embedded in a mass of *perirenal fat.*

On the medial side of each kidney is an indentation, the *hilus.* At this point, blood and lymph vessels and nerves connect and the ureter emerges from the kidney. The hilus opens into a slightly larger space, the *renal sinus,* which is surrounded by kidney tissue and is occupied principally by the *renal pelvis* and *renal calyces* (sing., *calyx*).

In a longitudinal section, the kidney can be seen to consist of two

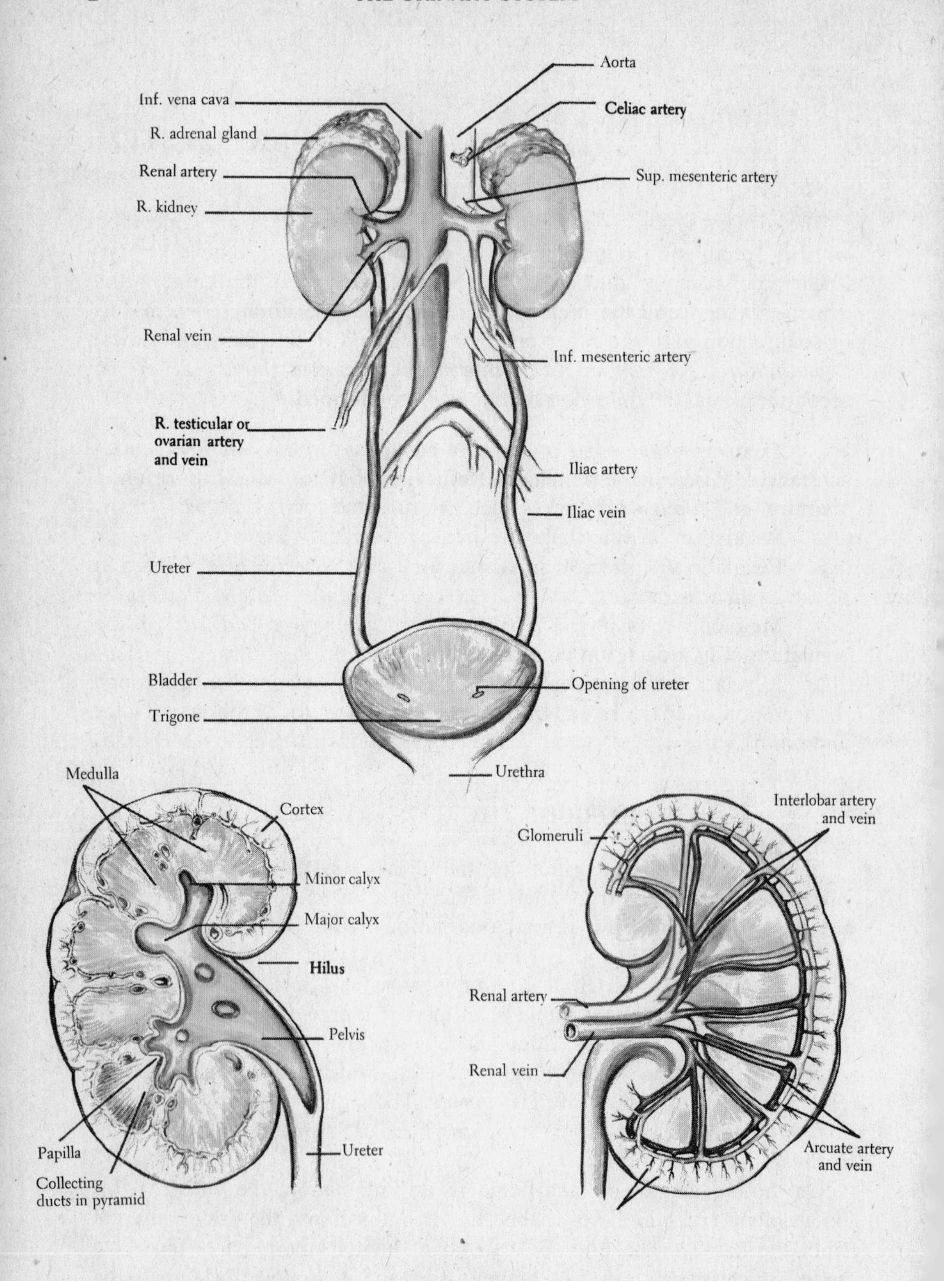

Fig. 1-1. The urinary system and a longitudinal section of the human kidney, showing its internal structure.

general regions: a *cortex* or cortical region and a *medulla.* The cortex forms the outermost layer and at intervals continues into the medulla as *columns of Bertin* or *renal columns.* It consists of regions composed of straight tubules (*medullary rays*) alternating with regions consisting of glomeruli and convoluted tubules. The medulla consists of 8 to 18 cone-shaped *renal pyramids.* The base of each pyramid is adjacent to the cortex; the free end or apex forms a *papilla,* which projects centrally into a minor calyx. Two or three pyramids may terminate in a single papilla. At the tip of each pyramid are 10 to 25 minute openings of the *papillary ducts.*

The human kidney is a *multilobular kidney,* in contrast to the *unilobular kidney* of the rabbit or rat. A *lobe* of the kidney (Fig. 1-2) consists of a medullary pyramid with its overlying cortical tissue. A *lobule* is a subdivision of a lobe. It consists of a medullary or cortical ray and its associated cortical tissue. In the cortex, the lobules are not clearly demarcated from each other.

In the central portion of the kidney is a cavity, the *renal pelvis,* which constitutes the expanded end of the ureter. The pelvis is divided into two or three portions, the *major calyces,* each subdivided into four to six *minor calyces.* Each minor calyx encloses the apical end of one or more pyramids.

NEPHRON. The structural and functional unit of the kidney is the *nephron,* which consists of a *renal corpuscle* and a *renal tubule* (Fig. 1-3). There are a million or more nephrons in each kidney. Nephrons are of two types: cortical and juxtamedullary. In *cortical nephrons,* the renal corpuscles lie in the outer region of the renal cortex, and the tubule penetrates only slightly or not at all into the medulla. In *juxtamedullary nephrons,* the renal corpuscles lie in the inner region of the cortex near the *corticomedullary junction,* and the tubules lie principally in the medulla.

Renal Corpuscle. The renal corpuscle (Fig. 1-4) is a spheroidal body consisting of a glomerulus and the glomerular capsule that encloses it; the capsule is structurally a part of the tubule. The *glomerulus* consists of several looped capillary vessels that connect an *afferent arteriole* through which blood enters with the *efferent arteriole* through which blood leaves. The diameter of the afferent vessel is slightly larger than that of the efferent.

Renal Tubule. Each renal tubule consists of the following parts: a capsule, a proximal convoluted portion, the loop of Henle, and a distal convoluted portion.

The *capsule* of the renal tubule is the expanded end, which encloses the glomerulus. It is a double-walled, cuplike structure, the walls consisting of very thin squamous epithelium. The outer layer of the cup forms the *parietal layer.* At the *vascular pole,* the point where the afferent and efferent vessels enter and leave the glomerulus, this layer is reflected over the glomerulus, which it invests closely to form the *visceral layer* or glomerular epithelium. This layer was formerly thought to be a single layer of simple squamous cells forming a continuous covering of the glomerular capillaries. Electron microscope studies have shown, however, that it is formed of cells

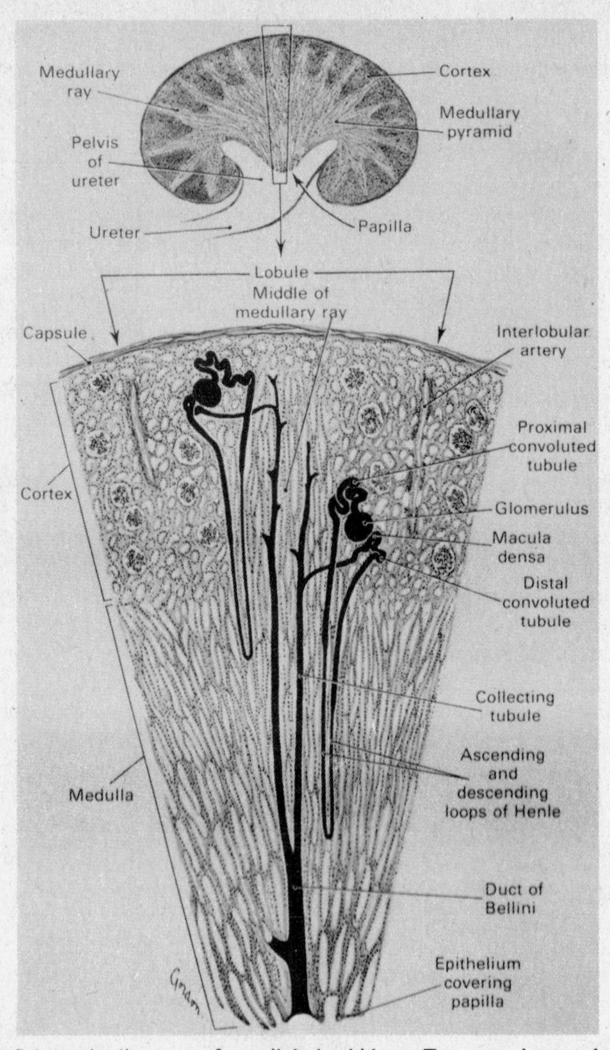

Fig. 1-2. Schematic diagrams of a unilobular kidney. Two complete nephrons and the collecting ducts into which they empty (indicated in black) have been included so that their course may be followed. (Of course, complete nephrons cannot be seen in any single section.) (Reprinted with permission of J. B. Lippincott Co. from A. W. Ham and D. H. Cormack, *Histology,* 8th ed., 1979.)

called *podocytes,* which do not form a continuous layer. Podocytes feature numerous cytoplasmic projections (*major processes*), which possess delicate *minor processes* (*pedicles*) that are closely applied to the basement membrane of the glomerular capillaries.

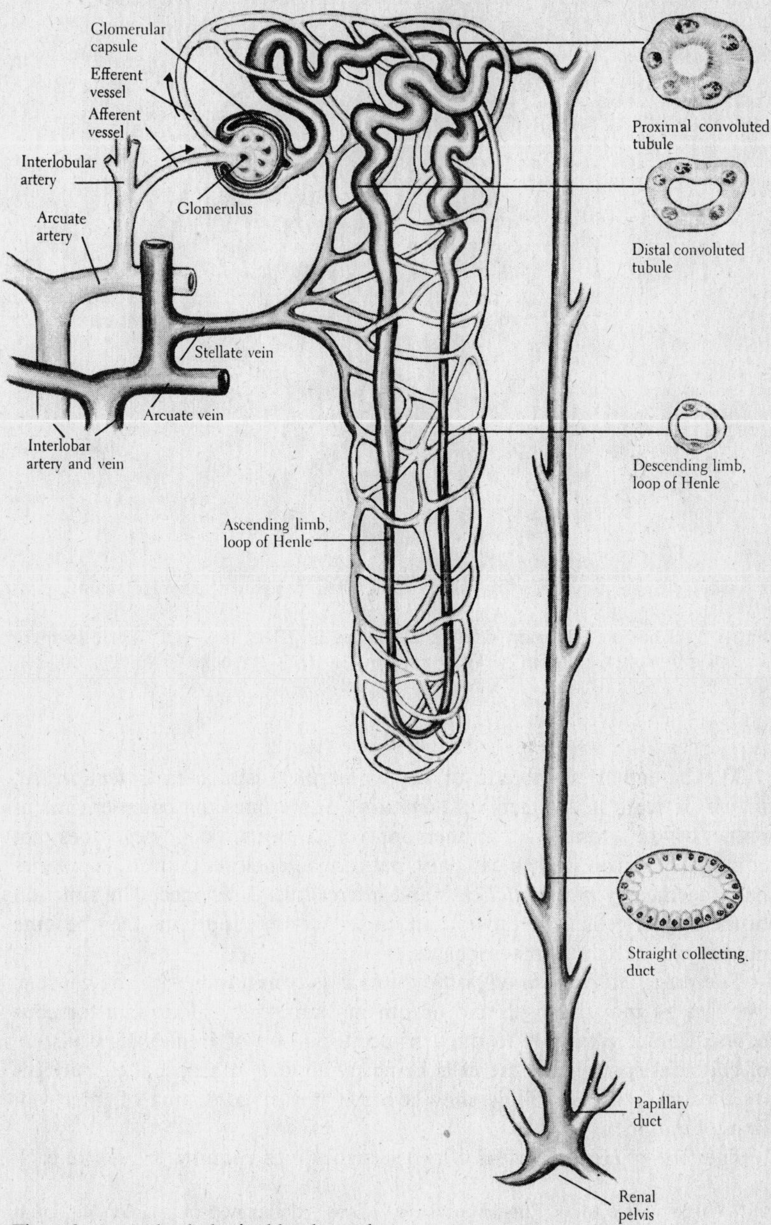

Fig. 1-3. A renal tubule, its blood supply, and its relationship to a collecting duct.

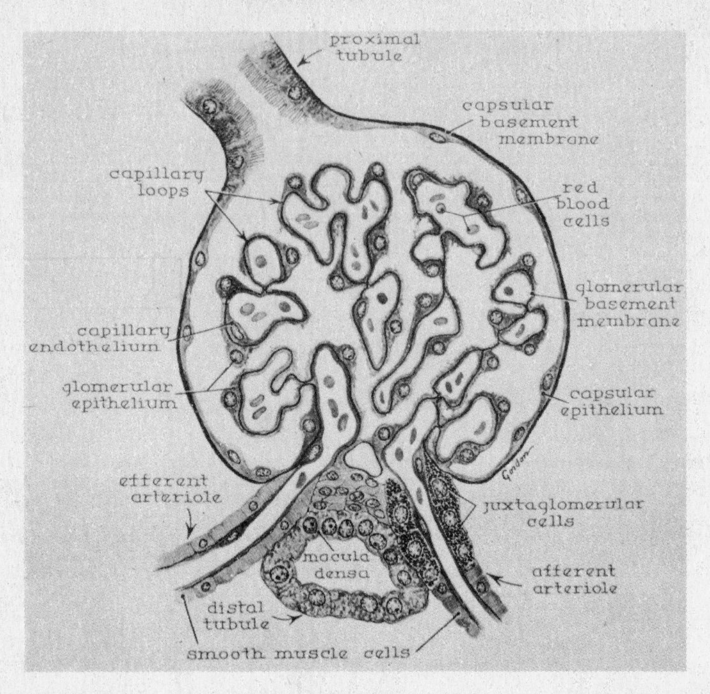

Fig. 1-4. Schematic diagram of a renal corpuscle. (Reprinted with permission of J. B. Lippincott Co. from A. W. Ham and D. H. Cormack, *Histology,* 8th ed., 1979.)

The basement membrane of the glomerular capillaries is *fenestrated;* that is, it bears small openings (*fenestra*). Sometimes the basement membrane of the glomerular epithelium, which bears podocytes, does not completely surround each capillary, but instead encloses a tuft of capillaries held together by *mesanglial cells* and intercellular substance. The function of mesanglial cells is unknown, but under certain conditions they become phagocytic and act as macrophages.

The *proximal convoluted portion,* which is continuous with the capsule, averages 14 mm in length and 60 μm in diameter.* It follows a tortuous course from the capsule to the next portion (loop of Henle). Its walls are of cuboidal epithelium, the cells being pyramidal in shape; their surfaces on the inside of the tubule show a brush border consisting of microvilli lining the lumen.

The *loop of Henle* consists of two straight limbs running parallel to each

* A *micron* is a unit of linear measure variously designated mu, μ, μm. It equals 0.001 millimeter. In the International System (SI) of units, it is called a micrometer and abbreviated μm.

other. The *descending limb* is very narrow (14 to 22 μm in diameter); the *ascending limb* is somewhat thicker (about 33 μm). Most of the loop of Henle lies in the medullary portion of the kidney. The cells comprising the descending limb are of the squamous epithelium type; those in the ascending limb are cuboidal. Near its junction with the distal convoluted portion, the ascending limb makes contact with the capsule from which it arises. It passes between the afferent and efferent arterioles, at which point its cells become specialized and form a region called the *macula densa* (Fig. 1-4). Cells of the macula densa lie adjacent to and in contact with the juxtaglomerular cells of the afferent arteriole.

The *distal convoluted portion* is short (4 to 5 mm) and very convoluted. It leads to a short, arched connecting portion, sometimes called the *junctional tubule,* which empties into a straight *collecting duct.*

COLLECTING OR EXCRETORY DUCT. The nephrons connect with straight excretory ducts in the cortical region. Each collecting duct passes centrally through the outer zone of the medulla to the inner zone, where several may join into a single large collecting tubule, the *papillary duct.* The papillary ducts open on the tip of a renal papilla and empty their contents into a minor calyx of the renal pelvis. The cells of the smaller ducts are cuboidal, those of the larger ducts, columnar.

The collecting ducts were formerly thought to play a passive role of conduction, but it is now known that they are involved in the processes of concentrating urine, conserving sodium ions, secreting potassium ions, and regulating acid-base balance, thus supplementing the action of the tubule.

BLOOD AND LYMPHATIC SUPPLY OF THE KIDNEY. Blood enters the kidney at the hilus through the *renal artery,* which divides into a number of branches that, in turn, lead to the *interlobar arteries.* These extend radially between the pyramids and connect with *arcuate arteries,* which run parallel to the surface in the boundary zone between the cortex and the medulla. From the arcuate arteries, *interlobular arteries* lead peripherally into the cortex. These give rise to the *afferent arteries* leading to the glomeruli.

Near each glomerulus, the smooth muscle cells of the afferent artery become epithelioid in appearance and form a collarlike structure. These cells, called *juxtaglomerular (JG) cells,* instead of containing myofibrils, contain secretory granules and are thought to be the source of *renin,* a substance produced under ischemic conditions. In the bloodstream, renin (an enzyme, not be be confused with rennin) acts on *angiotensinogen,* a plasma protein, converting it to *angiotensin I.* Another enzyme converts angiotensin I into *angiotensin II,* a hypertensive agent causing constriction of arterioles with resulting elevation of blood pressure. Because of the close relationship of JG cells to the macula densa (Fig. 1-4), the two are referred to as the *juxtaglomerular apparatus* or *complex,* and they are thought to be functionally related.

On entering the capsule at the vascular pole, the afferent arteriole divides into four or five branches that subdivide into looped capillaries, which form a compact, globular structure, the *glomerulus.* Anastomoses are infrequent or absent. The capillaries join and leave the glomerulus as the *efferent arteriole,* which continues a short distance and then breaks up into a *second capillary network.* This is a plexus surrounding the convoluted portions of the tubules in the cortex and the loop of Henle in the medulla. Branches of the efferent vessels and straight arteries, branching from the arcuate artery, supply blood to the medullary rays, the medulla, and the pyramids. Those supplying the pyramids are the *arteriolae rectae.*

Capillaries of the cortex converge to form the *stellate veins,* which lead to *interlobular veins,* and these, in turn, lead to the *arcuate veins,* which parallel the arcuate arteries. Blood from the medulla is collected in the *straight veins,* which enter the arcuate veins. Finally, the arcuate veins lead to *interlobar veins* of the medulla. These pass between the pyramids and unite to form the *renal vein,* which emerges from the kidney at the hilus.

The kidney is well supplied with *lymphatic vessels.* Both the capsule and the glandular tissue contain dense networks of lymphatic capillaries, which drain into efferent vessels. These vessels leave at the hilus and enter the lymph nodes lying along the aorta.

NERVE SUPPLY OF THE KIDNEY. Each kidney is innervated by sympathetic and parasympathetic nerve fibers from the *celiac plexus.* Both myelinated and nonmyelinated fibers accompany blood vessels into the kidney substance, where they end principally in the walls of the blood vessels. Both sensory and motor nerve endings are present. The exact relationship between the nerve fibers and the tubules is still in question. Some investigators have described plexuses around the tubules; some have claimed that the nerve fibers penetrate and end in the epithelium of the tubules.

Structure of the Excretory Passages. The urinary excretory passageways include the ureters, the urinary bladder, and the urethra.

URETERS. The renal calyces and the renal pelvis, which lie within the sinus of the kidney, have already been described. They constitute the expanded end of the ureter, a tube that connects the kidney with the urinary bladder. Each ureter consists of an abdominal portion and a pelvic portion. The *abdominal portion* descends behind the parietal peritoneum to the brim of the pelvis, which it crosses. The *pelvic portion* curves laterally, downward, forward, and inward to enter the bladder at the lateral corner of the trigone. The ureter averages 27 cm in length and 4 to 5 mm in diameter.

In cross section, the ureter is seen to possess the following layers:

1. A *mucous coat,* a layer of transitional epithelium with an underlying lamina propria of dense connective tissue. The epithelium lacks a basement membrane. The mucous coat is usually thrown into several longitudinal folds.

2. A *muscular coat,* consisting of an inner longitudinal and an outer circular layer. The layers are not easily distinguishable because fibers from one layer may penetrate the other. In the lower third, another layer of longitudinal fibers lies external to the circular coat.

3. A *fibrous coat* or *adventitia* consisting of fibroelastic connective tissue with bundles of collagenous fibers interspersed with elastic fibers.

URINARY BLADDER. The urinary bladder (Figs. 7-1 and 7-6) is an ovoid muscular sac lying in the anterior portion of the pelvic cavity directly behind the symphysis pubis. In the male, it lies anterior to the rectum; in the female, it lies anterior to the uterus and the vagina. It receives urine from the ureters, which enter the bladder wall obliquely on its lower posterior surface. The urine is stored temporarily and discharged at intervals through the urethra. The triangular area lying between the opening of the urethra and the openings of the two ureters is the *trigone.*

Owing to the oblique course by which the ureters enter the bladder and the presence of a valvelike fold of mucous membrane at each ureteral opening, urine is prevented from reentering the ureters as the bladder fills.

The layers of the bladder wall are the same as those of the ureter. The innermost *mucosa* consists of transitional epithelium with an underlying lamina propria. The *muscular coat* is especially well developed. It consists of three layers of smooth muscle cells somewhat indistinctly separated from each other. The outer layer consists of longitudinal fibers. The fibers of the middle layer run circularly; those of the inner layer, which are especially well developed in the region of the trigone, run longitudinally. Collectively, the three layers constitute the *detrusor muscle,* which, upon contraction, empties the bladder. About the internal opening of the urethra, the smooth muscle forms a dense mass, the *internal sphincter* of the bladder. On the free superior surface, the *fibrous coat* has a layer of peritoneum forming a *serosa.*

The blood supply of the urinary bladder comes from the *hypogastric (internal iliac) arteries.* Veins form several plexuses that empty into the *internal iliac vein.* The bladder is also well supplied with *lymphatics.*

Nerves from both divisions of the autonomic nervous system innervate the urinary bladder. Details of the nerve supply are given in the section on micturition (pages 16–18).

URETHRA. The urethra is the duct that conveys urine from the bladder to outside the body. In the male, it also carries semen and thus serves as a common excretory and reproductive duct. In the female, its function is exclusively excretory.

The *female urethra* (Fig. 7-6) is relatively short, measuring 3 to 5 cm in length. It passes inferiorly along the anterior surface of the vagina, to which it is firmly attached, and opens at the *urethral orifice,* which lies in the vestibule between the vaginal orifice and the glans clitoridis. The *mucous membrane* is lined with stratified squamous epithelium except near the bladder, where it is of the transitional type. It is thrown into longitudinal folds, a prominent fold on the posterior surface forming the *urethral crest.* Urethral glands, the *glands of Littré,* open into lacunae located between these folds. The ducts of the largest urethral glands, the *paraurethral (Skene's) glands,* open just within the urethral orifice. The *lamina propria,* also called the *submucosa,* is rich in elastic fibers and has a cavernous character. The *muscular layer* is thick, with the inner fibers running longitudinally and the outer ones circularly. A fibrous membrane forms the outermost layer. Just within the orifice is a sphincter of circular smooth muscle fibers; at the orifice is a sphincter of striated fibers.

The *male urethra* (Fig. 7-1) extends from the bladder to the tip of the penis. It averages 20 to 25 cm in length and consists of *prostatic, membranous,* and *cavernous portions.*

The *prostatic portion* of the male urethra averages 2 to 3 cm in length and is surrounded by the prostate gland, whose ducts open into it. Its walls possess numerous longitudinal folds, a prominent one on its posterior surface being the *urethral crest.* This crest dilates in the midregion to form the *colliculus seminalis,* which bears on its sides the two small openings of the *ejaculatory ducts.* The prostatic portion passes through the pelvic floor and continues as the membranous portion.

The *membranous portion* averages 1 cm in length; its diameter is considerably less than that of the other two portions. Its walls contain the fibers of a sphincter muscle, sometimes called the *external sphincter* of the bladder.

The *cavernous* or *spongy portion* has two parts: a short proximal portion that is fixed in position and a distal portion that lies in the penis and is mobile. On entering the penis, the urethra dilates to form the *bulb* of the urethra, which contains the openings of the ducts of the bulbourethral glands. At the glans penis the cavernous portion dilates, forming the *navicular fossa,* which opens to the outside through the *urinary orifice* or *meatus.*

The cavernous urethra is surrounded by erectile tissue, the *corpus cavernosum urethrae (corpus spongiosum).* The mucous membrane in the prostatic portion is lined with transitional epithelium like that of the bladder. In the membranous and cavernous portions, the epithelium is stratified columnar or pseudostratified up to the fossa navicularis, where it changes to stratified squamous. In the cavernous portion, the mucous membrane shows many outpocketings called *lacunae.* These receive the openings of numerous small glands, the *glands of Littré.* A definite submucosa is lacking, the mucosa being surrounded by a layer of loose connective tissue containing scattered bundles of smooth muscle cells and numerous veins, the whole being enclosed in a thin layer of connective tissue, the *tunica albuginea.*

PHYSIOLOGY OF THE URINARY SYSTEM

Inasmuch as the urinary system is concerned with the formation, temporary storage, and excretion of urine, it is essential that the nature of this substance be established.

Constituents of Urine. Urine consists of water (95 percent) and solids (5 percent). The solids, which amount to 40 to 50 g per liter of urine, include both organic and inorganic substances. The principal constituents and the average quantities in normal urine are shown in the following table:

Organic Constituents	*g/L*	*Inorganic Constituents*	*g/L*
Urea	23.0	Sodium chloride	9.0
Hippuric acid	0.6	Potassium chloride	2.5
Uric acid	0.6	Sulfuric acid	1.8
Creatinine	1.5	Phosphoric acid	1.8
Other substances	2.0	Ammonia	0.6
		Calcium	0.2
		Magnesium	0.2

Besides those named, other substances, depending on the diet and general health of the individual, may be present in varying amounts, usually in minute quantities or merely as traces. These include carbohydrates, pigments (urochrome, urobilin), fatty acids (acetoacetic, betahydroxybutyric), carbonates, bicarbonates, carbonic acid, mucin and mucinlike substances, enzymes, hormones, and vitamins. Many other substances, the products of normal and abnormal metabolism, may be present in urine, especially in pathological conditions.

Sources of Urinary Constituents. The sources of the principal urinary constituents are as follows:

UREA. Urea, $CO(NH_2)_2$, comprises 60 to 90 percent of all nitrogenous material in urine. It is derived principally from the catabolism of amino acids. The oxidative deamination of amino acids in the liver and other tissues forms ammonia, which combines with carbon dioxide to form urea. The amount of urea in the blood is referred to as *blood urea nitrogen* (*BUN*); it averages 12 to 20 mg/100 ml of blood. The BUN rises in uremia.

AMMONIA. Ammonia, NH_3, occurs in urine in the form of ammonium salts, principally chlorides and urates. It is derived from deamination of amino acids but may also be formed in the kidney itself from glutamine. Ammonia functions to neutralize body acids. In acidosis, the production of ammonium salts by the kidney helps to prevent excessive loss of sodium carbonate from the blood.

URIC ACID. Uric acid is an end product of purine metabolism. It is formed in part from purines ingested in food (exogenous) and from those formed in the body (endogenous). The latter are derived principally from nucleoproteins. Uric acid is markedly insoluble in water, although its salts are more readily soluble. Owing to this insolubility, it tends to crystallize and is a common component of kidney stones. When the uric acid concentration of the blood rises, uric acid and its salts tend to be deposited in and around the joints and in cartilage. Such deposits, called *tophi,* occur in tophaceous *gout.*

CREATININE. A normal, alkaline constituent of the blood, creatinine, $C_4H_7ON_3$, is thought to be derived from *creatine,* a nitrogenous substance present in muscle tissue. Its daily output in urine is remarkably constant, averaging 0.02 mg/kg of body weight.

MINERALS. The amounts of sodium and potassium in urine closely parallel the amounts ingested in foods. Calcium and magnesium are eliminated principally via the large intestine; consequently, they do not appear in significant quantities in the urine.

SALTS. *Chlorides* are derived principally from excess salts in the diet. The concentration of chlorides in the blood remains relatively constant. When salt intake is excessive, the salt content of the urine is high; in individuals on a salt-free diet, it is low. In vomiting, when a considerable amount of chlorides may be lost, the urinary chloride is reduced. *Sulfates* are derived principally from the amino acids of dietary protein. *Phosphates* occur in the urine as sodium compounds (monosodium and disodium phosphate); the former of these is acid, the latter alkaline. Phosphates may also occur in the form of potassium, calcium, and magnesium compounds. Monosodium and disodium phosphates are present in the blood in the ratio of 1:4; they serve as buffers, helping to maintain the normal alkalinity of the blood. In urine their ratio is reversed and is approximately 9:1. By this mechanism, the body's fixed base, chiefly sodium, is conserved.

HORMONES, VITAMINS, AND ENZYMES. Several hormones or their metabolites are excreted in urine, especially those secreted by the adrenal gland (cortex and medulla), reproductive organs (ovaries and testes), and hypophysis. Products excreted include estrogens, pregnanediol from progesterone, and 17-ketosteroids from testosterone. Tests used in the diagnosis of pregnancy are based on the presence in urine of chorionic gonadotropins secreted by the placenta.

Water-soluble vitamins are excreted in urine in variable quantities. Thiamine, nicotinamide and its derivatives, ascorbic acid, and other vitamins are normally present when dietary intake is adequate or excessive. When the amount of thiamine in urine is less than 0.04 mg, a thiamine deficiency is indicated. From 15 to 28 mg of ascorbic acid are excreted each 24 hr.

Enzyme concentrations in urine are usually low and, in general, are of little significance. These enzymes are derived from the blood, from the disintegration of leukocytes and epithelial cells, or from glandular oversecretion in certain pathologic conditions.

Characteristics of Urine. The following are characteristics of urine under normal and abnormal conditions.

COLOR. Urine is normally amber in color due to the presence of a pigment, *urochrome,* formed from a combination of urobilin and urobilinogen with a peptide. Other pigments may be present. Certain drugs, pathological conditions, or the ingestion of certain foods may alter the color.

TRANSPARENCY. Normal urine is usually clear and transparent. On standing, however, it tends to become cloudy, and a sediment may appear, owing to bacterial action that results in the production of phosphates of calcium and magnesium. These substances are insoluble in alkaline urine; consequently, they form precipitates. Bacterial infections of the urogenital tract may produce similar conditions and result in a cloudy urine. The presence of urates, epithelial cells, leukocytes, or pus results in a cloudy urine. Turbidity or smokiness may occur from the presence of red blood cells, semen, mucin, or chyle.

REACTION. Urine is usually *acid,* with a pH around 6.0, the acidity being due chiefly to the presence of sodium acid phosphate, NaH_2PO_4. The reaction varies considerably with the nature of the diet; vegetables and fruits tend to reduce the acidity and to produce an alkaline urine. The acidity of the urine increases in starvation and in certain diseases, such as diabetes mellitus.

SPECIFIC GRAVITY. The specific gravity of urine, in health, varies from 1.015 to 1.025. Excessive water intake results in a large volume of urine of low specific gravity. Low water intake or excessive loss of water through other avenues (lungs, skin, and intestines) decreases the volume of urine formed, with a resultant increase in specific gravity. In pathologic conditions, the kidney may be unable to regulate the specific gravity according to the needs of the body; consequently, it may be high or low, or it may become "fixed."

VOLUME. The average amount of urine excreted daily ranges from 1000 to 1800 ml (average about 1500 ml). Many factors tend to alter the amount, among them *diet* (a high-protein diet increases urinary output because urea, sulfates, and phosphates tend to act as diuretics); *water intake; loss of water* through channels other than the urinary system (in summer, urinary output is decreased owing to increased sweating); and *degree of body activity* (urine flow is decreased during sleep, increasing markedly 2 to 3 hr after arising; strenuous muscular activity results in reduced urine formation).

THE FORMATION OF URINE

The substances the kidney eliminates in the urine are, with a few exceptions, substances that are normally present in the blood. Accordingly, the kidneys act to maintain homeostasis by removing substances present in excess. This is accomplished by the processes of *filtration, reabsorption,* and *tubular secretion.*

Filtration. The glomerulus and glomerular epithelium act as a filter. As the blood passes through the glomerulus, water and dissolved substances pass through the capillary walls and the epithelium into the capsule to form the *glomerular filtrate.* Blood cells and colloidal substances such as proteins, because they consist of particles too large to pass through the epithelium, are retained in the capillaries. Blood pressure (hydrostatic pressure) provides the force for filtration. The reduced size of the efferent vessel of the glomerulus assists in building up filtration pressure.

GLOMERULAR FILTRATE. The glomerular filtrate is similar in composition to the blood plasma, except that it is almost protein-free. Water, salts, and sugar are present in approximately the same concentrations as in the blood. This fluid passes along the tubule and enters the renal pelvis as *urine.* Examination of this urine shows that it contains *less* water, *more* urea, and *no* sugar. The concentrations of salts, too, are altered, some being increased, some decreased.

FILTRATION CAPACITY OF THE KIDNEYS. There are approximately 2 million glomeruli in the two human kidneys, with a filtration surface of about 5000 mm^2/g of kidney. Blood flow through the kidneys is large, averaging 1300 ml/min (about 25 percent of the cardiac output); 180 to 200 L of glomerular filtrate is formed each 24 hr. Through reabsorption this is concentrated and reduced to 1 to 2 L of urine.

Reabsorption. As the glomerular filtrate passes along the tubule, some of the substances are reabsorbed into the blood in the capillaries surrounding the tubule. To be reabsorbed, a substance must pass through a series of membranes. It must pass through the cell membrane of a tubule cell, traverse its cytoplasm, pass through the cell membrane on the opposite side, and enter the interstitial fluid. Then it must repeat the process by passing through the cells of a capillary to reach the bloodstream. This entire process, called *transepithelial transport,* involves both active and passive transport mechanisms. As various substances are absorbed in varying degrees, *selective reabsorption* occurs.

Some substances are *entirely or almost entirely reabsorbed.* These include glucose, chlorides of sodium, potassium, calcium, and magnesium, all important constituents of the body fluids, which are excreted only when their concentrations in the blood are higher than normal. They are present in normal urine only in limited quantities. The concentration of a substance in the blood plasma at which it begins to be excreted in the urine is designated its *renal threshold.* For glucose this is 180 mg/ml. The normal blood glucose level averages 100 mg.

Other substances are *reabsorbed in limited quantities.* These include urea, uric acid, and phosphates, which appear in the urine in considerable amounts. Finally, certain substances are *not reabsorbed.* These include creatinine, sulfates, and ammonia. The ammonia excreted is that synthesized in the kidney, as well as that arising from the deamination of amino acids.

Tubular Secretion. Substances may be transported from the peritubular capillaries into the kidney tubules by the process of *tubular secretion,* which is the reverse of reabsorption. Of the substances normally present in the blood excreted by this process, hydrogen and potassium ions are the most important. It is also by this avenue that foreign substances such as penicillin and para-aminohippuric acid (PAH) are excreted. A salt of the latter is used in testing the effectiveness of the tubular excretory mechanism.

Factors Affecting the Formation of Urine. Although the formation of urine is principally a passive process, the amount of urine produced varies with and is dependent on nervous, chemical, and physical factors. The volume *increases* as a result of an increased rate of filtration, a reduced rate of absorption, or both. The volume of urine *decreases* with a decreased rate of secretion, an increased rate of reabsorption, or both.

FACTORS AFFECTING THE FILTRATION RATE. The primary factors that influence the filtration rate are:

1. *Variations in the extent and nature of the glomerular filtering surface.* The number of glomeruli functioning at any one time is variable, depending on nervous and chemical factors. The number of functional capillary loops within a glomerulus may also vary.

2. *Pressure within the glomeruli.* This depends on systemic blood pressure or on changes within the kidney itself; these factors bring about an increased supply of blood to the kidney or an increased glomerular flow.

3. *Osmotic pressure relationships.* Proteins in the blood exert an osmotic pressure, which acts to draw water from the capsule back into the glomerular capillaries. *Filtration pressure* is the difference between the blood pressure forcing fluids out of the glomerular capillaries and the osmotic pressure drawing the water back in. *Blood pressure* in the kidney capillaries averages about 70 mm Hg; osmotic pressure of the plasma proteins averages about 30 mm Hg. Filtration pressure consequently averages about 40 mm Hg (that is, 70 mm minus 30 mm). The fluids in Bowman's capsule, however, exert a back pressure of about 5 mm; consequently the *effective filtration pressure* is about 35 mm.

FACTORS AFFECTING THE REABSORPTION RATE. The primary factors influencing the rate of reabsorption are:

1. *Rate of flow along a tubule.* When the flow is rapid, there is less opportunity for reabsorption.

2. *Osmotic pressure of the filtrate.* Reabsorption requires the expenditure of energy by the cells of the tubules. As reabsorption proceeds, the filtrate becomes more concentrated, and the osmotic pressure rises. Reabsorption of water ceases when the osmotic pressure exceeds the capacity of cells to absorb water.

3. *Hormones. Vasopressin,* the antidiuretic hormone (ADH), and *aldosterone* play a role in salt and water reabsorption.

Control of Urinary Secretion. Secretion of urine is under nervous and chemical control.

NERVOUS CONTROL. This is accomplished *directly* through the action of nerve impulses on the blood vessels leading to the kidney and on those within the kidney leading to the glomeruli and *indirectly* through the effects of nerve impulses on certain endocrine glands.

CHEMICAL CONTROL. This is accomplished through the effects of chemical substances (hormones) present in the blood. The hypothalamus secretes a hormone, *vasopressin,* that is stored in the posterior lobe of the pituitary. This hormone is also called the *antidiuretic hormone* (*ADH*) because of the role it plays in the reabsorption of water in the distal tubules and collecting ducts. It acts on the cell membranes of the tubular cells, increasing their permeability. In its absence or upon reduced secretion, reabsorption of water is decreased. As a result, a greatly increased volume of a very dilute urine is produced, a condition characteristic of *diabetes insipidus.*

Other hormones of importance in the reabsorption processes are the cortical steroids, especially *aldosterone* from the adrenal gland. Aldosterone facilitates the active transport of sodium ions accompanied by chloride ions out of the kidney tubules. These are exchanged for potassium and hydrogen ions. The transport of sodium changes the osmotic gradient, which also favors the reabsorption of water. In the absence of cortical hormones, sodium and water are excreted in excessive amounts, and potassium is retained. Adrenal cortical insufficiency produces pronounced changes in the salt and water content of body fluids.

MICTURITION

As urine is secreted by the kidneys, it enters the ureters through which it is forced, drop by drop, into the bladder by the peristaltic contractions of the smooth muscles in the ureteral walls. While the urine is slowly accumulating, the *detrusor muscle* in the bladder wall relaxes and adjusts its tone to the increased volume. When 100 to 200 ml of urine has collected and pressure within the bladder is about 4 to 5 cm of H_2O, sensory pressure receptors in the bladder wall are stimulated, and the first urge to urinate is felt. When the volume reaches 400 ml, a marked sense of fullness and urgency is experienced. During micturition, voluntary release of the external sphincter and perineal muscles occurs, and efferent impulses are discharged to the bladder wall, initiating their contraction. As a result, intravesicular pressure is increased, and the discharge of urine occurs, a process called *micturition, urination,* or *voiding.* Involuntary micturition usually occurs if the pressure within the bladder exceeds 10 cm of H_2O.

Innervation of the Urinary Bladder and the Urethra. The urinary bladder and the urethra are innervated by efferent and afferent fibers from the sympathetic and parasympathetic divisions of the autonomic nervous system.

EFFERENT FIBERS INVOLVED IN MICTURITION. Sympathetic efferent fibers originate in the lumbar region of the spinal cord and pass through the inferior mesenteric plexus and superior hypogastric plexus via the presacral nerves to the inferior hypogastric plexus. Postganglionic fibers originating from this ganglion pass to the bladder and the internal sphincter. These fibers carry *inhibitory* impulses to the detrusor muscle and *motor* impulses to the trigone, the internal sphincter, and smooth muscle in the proximal portion of the urethra. Parasympathetic fibers originate in the sacral region of the spinal cord. They pass through the pelvic nerves and synapse with neurons in the bladder wall and in the region of the internal sphincter muscle. These fibers carry *motor* impulses to the detrusor muscle and *inhibitory* impulses to the internal sphincter. These two sets of nerves are antagonistic in their actions, the former operating to fill the bladder and retain urine, the latter to evacuate the bladder. *Somatic efferent* or motor fibers from the spinal cord pass through the *pudendal* nerves to the striated muscle of the external sphincter.

AFFERENT FIBERS INVOLVED IN MICTURITION. Afferent or sensory fibers pass from the bladder through the pelvic and hypogastric nerves to the spinal cord, carrying impulses set up by distention of the bladder wall and impulses of pain, touch, and temperature. Afferent fibers from the urethra pass through the pudendal nerves to the spinal cord, carrying impulses originating from distention of the urethra.

Steps in Micturition. All processes involved in emptying the bladder are reflex in nature, but they are susceptible to voluntary control and, within limits, can be inhibited or initiated at will. The steps in the process are in general as follows:

1. When the volume of urine in the bladder reaches about 300 ml, the resulting distention of the bladder wall initiates afferent impulses that on reaching the brain give the sensation of bladder fullness and the desire to urinate. If voluntary restraint is released, the succeeding steps occur.
2. The smooth muscle in the proximal portion of the urethra is relaxed. Afferent and efferent pathways of this reflex are over the pelvic nerves, with the reflex center located in the spinal cord.
3. The external sphincter is relaxed. The afferent pathway for this is through the pelvic nerves; the efferent pathway is through the pudendal nerves. The reflex center is in the spinal cord.
4. The detrusor muscle of the bladder wall contracts, and this is accompanied by relaxation of the internal sphincter. For this reflex, both afferent and efferent pathways are in the pelvic nerves, with the reflex center in the pons.

(As the urine passes along the urethra, the following two additional reflexes are initiated.)

5. There is continued contraction of the detrusor muscle. The afferent pathway is through the pudendal nerves, and the efferent pathway is through the pelvic nerves, with the reflex center in the medulla. This reflex brings about a continuation of the contraction of the bladder until all the urine has been expelled.

6. There is continued relaxation of the external sphincter. The afferent and efferent pathways are in the pudendal nerves, with the reflex center in the sacral region of the cord.

The expulsion of urine also involves activity by the muscles of the abdominal wall and the perineum. Contraction of the abdominal wall increases intra-abdominal pressure and usually starts the process of micturition. At the same time, the muscles of the perineum are relaxed. Contraction of the bulbocavernous muscle brings about expulsion of the urine that remains in the urethra.

In *infants,* urination is entirely under reflex control; voluntary control is acquired through training, but only after all the neural components have developed. Voluntary control in adults is sometimes lost as a result of spinal cord lesions, disturbances in peripheral innervation, or emotional disturbances.

WATER BALANCE

Water is the most important, as well as the most abundant, compound in the body. Because the kidneys play a primary role in the maintenance of water balance in the body, this topic is included under the urinary system.

Body Fluids. Water is the principal constituent of the body fluids and comprises 65 to 70 percent of body weight. The body fluids are as follows:

INTRACELLULAR FLUID. The fluid enclosed within cell membranes constitutes an integral part of the protoplasm of cells. It averages 28 L and comprises about 40 percent of body weight.

EXTRACELLULAR FLUID. The fluid in spaces outside the cells includes:

1. *Interstitial fluid,* which comprises all intercellular fluid, tissue fluid, lymph, fluid in any of the body cavities (pleural, peritoneal, cranial, spinal, synovial, ocular), or in the secretions of glands. It averages about 11 L and comprises about 16 percent of body weight.
2. *Plasma,* the noncellular portion of the blood, which averages about 3 L and comprises about 4 percent of body weight.

Water in Body Tissues. The distribution of water in tissue varies with the types of tissue; the percentages of water in some representative tissues are given in the following table:

	Percent
Fatty tissues	6–10
Bone (extremities and skull)	14–22
Bone (vertebrae and ribs)	16–44
Tendon	56–58
Brain (white matter)	68–70
Muscle tissue	75–78
Thyroid gland	77–81
Brain (gray matter)	82–85

Water Balance. Balance results when *water intake* is in equilibrium with *water output.* The methods by which the body gains and loses water are as follows:

WATER INTAKE. The total intake of water varies with individuals, but for the average person it is about 2500 ml/day. This amount is acquired in the following ways:

In liquids ingested	1200 ml
In foods ingested	1000 ml
From metabolism of foods	300 ml

WATER OUTPUT. The avenues by which water is lost and the average amount of water lost per day at average temperature and humidity are as follows:

Through the skin	500 ml
Through the lungs	350 ml
Through the kidneys	1500 ml
Through the intestine	150 ml

Water Lost Through the Skin. Naturally, these amounts vary in response to changing conditions. When the body temperature rises, as from muscular exercise, fever, or increased environmental temperature, the amount of water lost through the skin by perspiration is greatly increased. Sweating serves as a means for dissipating excess heat. Excessive sweating is *hyperhidrosis.* With the loss of water, there is also a considerable loss of salts, principally sodium chloride. For normal body functioning, both the water and the sodium chloride must be replaced. This may necessitate an increase in salt intake in hot environments or following increased muscular activity.

Water Lost Through the Lungs. The amount of water lost through the lungs varies with the nature of the air inspired and the rate and depth of breathing. If the inspired air is dry, more water is lost than when the inspired air is laden with moisture. Rapid breathing or breathing through the mouth increases water loss.

Water Lost Through the Kidneys. When a great deal of water is lost through the skin, the amount lost through the kidneys is noticeably reduced. The kidneys are, however, the *primary* organs for the regulation of water balance. When there is an excess of water in the body, the excess is eliminated through the formation and excretion of urine. When the water supply is low, the quantity of water thus eliminated is reduced, and the salts and waste products become concentrated.

Water Lost Through the Intestine. The amount of water in feces is fairly constant, varying little from day to day. It is increased in diarrhea and decreased in constipation.

Other Avenues. Small amounts of water are also lost through the sputum, tears, menstrual fluid, and mother's milk.

Regulation of Water Balance. A number of mechanisms operate to maintain a balance between water intake and water output and thus achieve constancy in the concentration of the body fluids. Water intake is largely regulated by the sensation of *thirst.* Thirst is induced by a reduction in body water and may be due to (1) reduced intake, (2) excessive loss of water through sweating, diarrhea, vomiting, or polyuria, (3) loss of blood, or (4) excessive salt intake or intravenous injections of hypertonic solutions. The sensation of thirst itself comes about from the drying of the mucosa of the mouth and pharynx resulting from reduced secretion of saliva or from general dehydration of the tissues.

Certain cells (*osmoreceptors*) in the hypothalamus are sensitive to osmotic changes in the blood. These cells, which are neurons located in the supraoptic and paraventricular nuclei, act as a *thirst center,* initiating drinking activities and thus increasing water intake. They are the same cells that produce *vasopressin,* the antidiuretic hormone (ADH), which is stored in the posterior pituitary. This hormone acts on the distal portions of kidney tubules, increasing their ability to reabsorb water, thus reducing the loss of water through urine.

Dehydration. Under normal conditions, a considerable amount of water is stored in the body, principally in the loose connective tissue. When water intake is inadequate, the body draws on these reserves. If the imbalance continues for any appreciable length of time, the tissues begin to lose water and dehydration sets in. Loss of water up to 5 to 10 percent of body weight constitutes serious dehydration; a loss of 20 percent is usually fatal. In dehydration, the blood becomes more concentrated (*anhydremia*) and reduced in volume (*oligemia*). Under average conditions, a person may live for 5 to 7 days without water but if also exposed to heat may survive only 2 or 3 days.

Some effects of dehydration are excessive thirst, loss of weight, disturbances in acid-base and electrolyte balance, an increase in nonprotein nitrogen in the blood, elevated body temperature, reduced cardiac output, increased heart rate, drying and wrinkling of the skin, and, in severe cases, exhaustion and collapse.

Some causes of dehydration are (1) reduced fluid intake, which may result from environmental factors as in cases of shipwreck or desert travel, inability to swallow, cachexia, refusal of mental patients to drink, severe dieting; (2) excessive loss of water as occurs in persistent vomiting, protracted diarrhea, excessive sweating, diuresis, or digestive disturbances such as bacillary dysentery, especially in infants; and (3) salt depletion with reduction in electrolytes in body fluids due to adrenal cortical insufficiency, diabetic ketosis, excessive vomiting, or diarrhea.

Water Intoxication. Under normal conditions, when large quantities of water are ingested, the excess is eliminated by the kidneys. In cases of kidney disorder, however, when urinary secretion is reduced, the tissues may become saturated with water, giving rise to *water intoxication.* The symptoms manifested in this condition are headache, dizziness, reduced body temperature, vomiting, convulsions, and coma. Occasionally, it is fatal.

It is a curious fact that water intoxication prevails in the early stages of starvation. It has also been observed that in the course of a restricted dietary regimen, the body may lose little weight during the first week or two; this is explained by the replacement with water from the oxidation of adipose tissue. But if the restricted diet is continued, a sudden loss of weight will eventually occur, owing to the rapid loss of water through the kidneys.

ELECTROLYTE AND ACID-BASE BALANCE

Regulation of Electrolyte Balance. Electrolyte or mineral metabolism is closely related to and dependent on water and acid-base metabolism. A change in the degree of hydration of the body is accompanied by corresponding changes in electrolyte concentrations.

The electrolytes of the body fluids include compounds containing the *positive ions* (cations) sodium, potassium, calcium, and magnesium and the *negative ions* (anions) chloride, bicarbonate, phosphate, and protein. Within the body, these ions are distributed in a specific manner. In the intracellular fluid, the principal cation is *potassium;* in the extracellular fluid, *sodium.* The principal anions within cells are *phosphate* and *protein;* in the extracellular fluid, *chloride.* Differences in the distribution of electrolytes in the two fluids are maintained by active, energy-consuming processes at the cell membrane. Although the amounts of water and electrolytes taken into the body vary from day to day, total body water and electrolyte content and the distribution of ions within the body remain relatively constant under normal conditions.

The kidneys play a primary role in this process. Through a complicated countercurrent exchange between the glomerular fluid within the nephron and collecting ducts and the surrounding interstitial fluid and the blood within capillaries, water and various electrolytes are reabsorbed to the extent required by the body. Any excess is eliminated through the urine. Processes involved include both active and passive diffusion and tubular secretion by the kidney tubules.

Regulation of Acid-Base Balance. In normal bodily metabolism, especially of proteins, an excess quantity of acid is produced that must be removed daily. The kidneys play an important role in maintaining acid-base balance through the formation of an acid urine from a slightly alkaline blood plasma. In the process, the glomerular filtrate, with a pH of 7.4, is converted into urine, which may have a pH as low as 4.8 (average 6.0). A number of reactions are involved in this activity. An important one is the exchange of sodium ions present in the urine of the tubules for hydrogen ions formed in the tubular cells. Here, the transport of sodium across the epithelial membrane is accomplished by a "sodium pump," an active transport mechanism. A bicarbonate anion formed by the dissociation of H_2CO_3 accompanies the sodium ion and, combining with it, forms a part of the alkali reserve of the blood. The hydrogen ions are combined principally with HPO_4 ions to form an acid phosphate or with ammonia (NH_3) to form NH_4, which combines with chlorine and is excreted in the urine. The regulation of acid-base balance is accomplished primarily through the maintenance of proper bicarbonate-ion concentration in the plasma, thus acting to conserve the alkali reserve. The kidneys can, within limits, restore acid-base equilibrium from either an acidic or alkaline condition. In *acidosis,* there is an increase in acid or ammonium salts in the urine; in *alkalosis,* a decrease. In alkalosis, there may also be an increase in the excretion of citric acid.

Supplementing the action of the kidneys in the maintenance of proper acid-base balance are (1) the work of the lungs, which, through a variable respiratory rate, controls the amount of carbon dioxide in the blood, and (2) the presence of buffers in the blood, which permit considerable amounts of acid or alkali to be taken up by the blood without a significant change in reaction (pH).

SECRETORY FUNCTION

In addition to the secretory action of the tubules in which substances are taken from the blood and passed into the urine, the kidney produces two substances that play an important role in circulatory activities. These are *renin* and a *renal erythropoietic factor.* Prostaglandins are also produced by the renal medulla.

Renin. This substance is not to be confused with *rennin,* a digestive enzyme. *Renin,* a kidney enzyme, is produced by juxtaglomerular cells that surround the afferent arteriole just before it enters a glomerulus. When there is a reduction in blood pressure, blood volume, or blood sodium, renin is produced. In the bloodstream, renin acts on a blood globulin, *angiotensinogen,* converting it into *angiotensin I.* A converting enzyme produced in the lung converts angiotensin I into *angiotensin II,* which is biologically active. It increases the heartbeat and is a powerful vasoconstrictor, acting on peripheral arterioles. As a result, systemic blood pressure increases. It also stimulates the production of aldosterone by the adrenal cortex. Aldosterone acts on kidney tubules to increase sodium reabsorption, which tends in turn to increase intravascular volume and blood pressure. Angiotensin II is inactivated by *angiotensinase,* an enzyme found in the kidneys, the blood, and other tissues.

Renal Erythropoietic Factor (REF). This factor is an enzyme or enzymelike substance produced in the kidney. In anemia or conditions of low oxygen tension (hypoxia), it is released into the bloodstream, where it converts a plasma protein into active *erythropoietin.* Erythropoietin acts on stem cells in blood-forming tissues in the bone marrow, stimulating differentiation and proliferation and accelerating the synthesis of hemoglobin. It thus increases the number of erythrocytes in peripheral circulation.

TESTS OF URINE AND OF URINARY FUNCTION

Two of the most commonly employed methods of investigating the urine and urinary function of an individual to diagnose or uncover abnormal conditions are *urine analysis* and the *renal clearance test.*

Urine Analysis. Usually referred to as *urinalysis,* the examination of urine is extremely important in that many pathological conditions or metabolic disturbances can be detected by changes that occur in the urine. Certain substances that are not normal constituents of urine may be present, or the proportions of the normal constituents may be altered. Some of the substances not normally present are glucose, acetone, albumin, blood cells, and bacteria. Such abnormal constituents are seen in the following conditions.

GLYCOSURIA. When glucose (sugar) is found in the urine, it may be a sign of (1) alimentary glycosuria due to excessive intake of sugar; (2) emotional glycosuria, probably the result of excessive activity of the adrenal gland that has caused the breakdown of glycogen and liberation of sugar from the liver; (3) hypoinsulinism from failure of the pancreatic islets of Langerhans to produce insulin, the primary factor in diabetes mellitus; (4) glycosuria of pregnancy or lactation, owing to the presence of lactose; (5) glycosuria due to overactivity of the posterior lobe of the hypophysis; or (6) a low renal threshold for glucose.

KETONURIA. Ketone bodies (acetone, acetoacetic acid, and betahydroxybutyric acid) appear in the urine in metabolic acidosis, in which ketone bodies are incompletely oxidized in the tissues or the liver. This results from a deficiency in intracellular glucose, which may occur from starvation, from a high-fat, low-carbohydrate diet, or from a pathological condition such as diabetes mellitus.

ALBUMINURIA. "Proteinuria" would be a more inclusive term because it would designate more accurately what is meant—the presence in urine of any of the plasma proteins (serum albumin, serum globulin, serum fibrinogen). Albumin is the most common of the proteins to be found in urine, and its presence indicates that the glomerular epithelium is not serving as an effective filter. It is usually regarded as evidence of kidney disease, although it can occur in the absence of renal pathology. Some types of albuminuria are as follows.

Cyclic or Adolescent Albuminuria. In young individuals, small amounts of albumin may appear at certain times of the day.

Postural or Orthostatic Albuminuria. This form of albuminuria appears only when the subject is standing or sitting. It disappears when a reclining position is assumed. It is seen especially in persons afflicted with lordosis and is probably due to mechanical factors that impede circulation in the kidney.

Functional or Benign Albuminuria. This mild albuminuria may result from excessive intake of proteins, strenuous muscular activity, cold baths, pregnancy, or other conditions. It has no pathological significance.

HEMATURIA. The appearance of blood in the urine may be the result of infections involving the kidney or the urinary tract, mechanical injury to the kidney, or a growth within the urinary system.

PYURIA. Pus (white blood cells and bacteria) in urine is taken as evidence of infection involving the kidney or the urinary tract.

CASTS. Casts, usually molded in the form of tubules, are sometimes revealed by microscopic examination of urine. Among the types found are *hyaline casts,* consisting of coagulated protein material; *granular casts,* consisting of blood, epithelial cells, and albumin; and *fatty* or *waxy casts.* Casts in urine are often signs of renal disease or injury, although they may be found in the urine of healthy individuals.

Renal Clearance Tests. Renal plasma clearance of any substance is expressed as the volume of plasma freed (cleared) of that substance by renal activity per unit of time, usually one minute. Depending on the substance, clearance is accomplished by glomerular filtration, cellular transport involving reabsorption or tubular secretion along the tubule, or both. A substance, *inulin,* is filtered by the glomeruli but not reabsorbed. Plasma clearance for inulin is 120 ml/min. Glucose, which is totally reabsorbed, has a clearance of zero; para-aminohippuric acid (PAH), which is filtered, not reabsorbed, and further augmented by tubular secretion, has a clearance of 650. These tests are of value in providing information about kidney function. A reduction in the clearance of urea is indicative of kidney damage. The clearance of inulin gives an accurate measurement of the glomerular filtration rate; plasma clearance of 650 for PAH is a measure of total blood flow to the kidney, which is about 1200 ml/min, approximately 20 percent of cardiac output. Creatine clearance (120 ml/min) is widely used in clinical tests.

The Artificial Kidney. This machine is used by persons whose kidneys have failed to function because of injury or disease. It removes substances from the blood that, if left to accumulate, would prove fatal in a short period of time.

Two types of kidney failure occur, acute and chronic. *Acute renal failure* may occur as a result of injury, infection, shock, or ischemia or from the effects of toxic drugs or poisons. *Chronic renal failure* occurs when the kidneys have been irreversibly damaged, as from prolonged progressive kidney disease. In acute cases, the artificial kidney machine is used until one or both kidneys resume their normal functioning. In chronic cases, the kidney machine must be used as long as the patient lives or until he or she receives an effective kidney transplant.

The artificial kidney machine is a dialysis apparatus in which the blood of the patient passes slowly over a cellophane membrane. The by-products of cellular metabolism (urea, uric acid, potassium, phosphorus, magnesium, creatinine) are removed by diffusion of these substances through the cellophane into the dialysate. The dialysate includes all the electrolytes in the amounts present in extracellular fluids.

Blood from an artery of the patient is conducted by a tube to the machine, where it passes over one side of the cellophane filter. An anticoagulant (heparin) is added before it enters the machine. After it has been cleansed, the blood is conveyed back into a vein of the patient. It takes about 30 min for all the blood of a patient to pass through the machine. Treatments last 4 to 8 hr. Chronic dialysis patients require two or three treatments a week. Dialysis is usually performed at a hospital center, but home dialysers are now available with which a patient with some assistance can, after careful training, carry out the procedure.

Kidney Transplants. Kidney transplants are now an accepted method of therapy in cases of chronic renal failure. If the kidney is from a person other than an identical twin, suppression of the recipient's immune response by drugs is essential. Rejection of the transplant, which is a common reaction, may occur immediately or not for months or even years.

DISORDERS OF THE URINARY SYSTEM

The urinary system reflects many bodily disturbances. The more common of the functional disorders and abnormal distribution of substances associated with this system are described in the following paragraphs. They are indications of a wide range of pathological conditions.

Anuria. This is the cessation of production of urine by the kidneys. Some causes are acute nephritis; obstruction of one or both ureters; blockage of kidney tubules or of blood vessels leading to the glomeruli; poisoning by ether, phosphorus, lead, and turpentine; and the aftereffects of certain diseases (e.g., cholera and typhoid fever).

Diuresis. The secretion of excessive amounts of urine. *Osmotic diuresis* results from increased amounts of such osmotically active substances as urea, glucose, or salt in the glomerular filtrate; *water diuresis* results from reduced reabsorption of water, as in diabetes insipidus due to reduced antidiuretic hormone; *filtration diuresis* results from an increase in glomerular filtration rate such as occurs when cardiac output is increased in hypertension or following renal vasodilatation. *Diuretics,* agents that increase urine production, include caffeine, digitalis, mercury compounds, and xanthines.

Dysuria. Difficult or painful micturition.

Enuresis. In infants, the process of micturition is involuntary. When involuntary discharge of urine occurs after the age of about 2 or 3 years, it is regarded as abnormal. It is called *diurnal enuresis* if it occurs during the day, *nocturnal enuresis* (or *nocturia*) if it occurs at night (usually during sleep). Enuresis may be due to delayed maturation of bladder innervation, psychogenic factors, or (less often) pathological conditions.

Incontinence. The inability to retain urine may be due to structural factors (such as poorly developed sphincter muscles), pathological conditions (lesions involving the reflex centers controlling these functions), or psychogenic factors. Trauma to the sphincter resulting from an operation or accident is a common causative factor.

Oliguria. In this condition, the amount of urine formed is markedly decreased; it occurs in dehydration, cardiac failure, shock, or tubular obstruction, either within or outside the kidney. (See *anuria.*)

Retention. The inability to expel urine from the bladder may be due to loss of muscle tone in the bladder (as from anemia, advanced age, or wasting disease), obstruction of or stricture in the urethra, lesions involving the nervous pathways to and from the bladder or the reflex centers of the brain and the spinal cord, and psychogenic factors.

Uremia. This is a toxic condition in which nitrogenous substances accumulate in the blood. It is a sequela of anuria.

PATHOLOGICAL CONDITIONS

Congenital Anomalies. Developmental defects involving the kidney include *agenesis* (failure of one or both kidneys to develop), *ectopia* (displacement of the kidney), *fusion* (resulting in a horseshoe-shaped kidney), or *hypoplasia* (incomplete development). Uteteral duplication, partial or complete, is a common anomaly.

Cystitis. Inflammation of the bladder.

Cysts. Polycystic kidneys may be present at birth or develop in adults. Cysts of various types may occur; some are simple or solitary, others multiple.

Kidney Stone (Urolith, Renal Calculus). This is an abnormal concretion, usually containing uric acid, calcium oxalate, or phosphates. Small stones may enter and pass through the ureter; this is usually accompanied by intense pain called *renal colic.* Stones in the kidney, ureter, bladder, or urethra interfere with the passage of urine and may give rise to *hydronephrosis* (excessive accumulation of urine in the pelvis), distention of the bladder, and atrophy of the kidney.

Movable Kidney (Nephroptosis). The kidney moves to a limited extent in response to respiratory movements and changes in body position. When the movements are greater than normal, the condition is designated as *movable kidney.* In extreme cases, it is called *wandering* or *floating kidney.* It occurs more frequently on the right side than on the left, more frequently in women than in men.

Nephritis (Bright's Disease). Inflammation of the kidney. It is termed *glomerulonephritis* if it involves primarily the glomeruli, and *pyelonephritis* if it involves the renal pelvis. Each may occur in an acute or chronic form.

Nephrosis. See "nephrotic syndrome."

Nephrotic Syndrome. A condition characterized by a prolonged and severe increase in glomerular permeability to protein, with resulting albuminuria, edema, and hypoalbuminemia. Causes are numerous, including glomerular disease, systemic and immunogenic diseases, infectious diseases, nephrotoxins, allergens and drugs, and various metabolic disorders.

Neurogenic Bladder. This condition involves impaired nervous control of the bladder, which results in difficulties in urination, such as inability to empty the bladder completely or incontinence. It may be due to injuries to the brain, spinal cord, or sacral nerves; infections; tumors; or debilitating diseases.

Pyelitis. Inflammation of the renal pelvis and its calyces.

Stricture. Abnormal narrowing of the ureter or urethra may result from an inflammatory condition or a spasm in the muscles of the passage walls.

2: THE RESPIRATORY SYSTEM

The respiratory system comprises the organs concerned with the exchange of gases between an organism and its environment, specifically the intake of oxygen and the elimination of carbon dioxide. The exchange of gases between the blood and the air taken into the lungs is *external respiration;* the exchange of gases between the circulatory fluids (blood, lymph, tissue fluid) and the cells is *internal respiration.* The significance of respiration lies in the fact that life processes depend primarily on the release of energy from food substances, and the basic reaction for this is:

$$\underset{\text{glucose}}{C_6H_{12}O_6} + \underset{\text{oxygen}}{6O_2} \rightarrow \underset{\text{water}}{6H_2O} + \underset{\text{carbon dioxide}}{6CO_2} \text{ PLUS Release of energy}$$

This is physiologic oxidation or destructive metabolism (*catabolism*). The maintenance of life demands a continual supply of oxygen and continual removal of carbon dioxide. Because the tissues are unable to store any appreciable quantity of oxygen, cessation of respiratory activities or interference with the distribution of oxygen to the organs and tissues results in malfunctioning or deterioration of the organs and tissues and possibly death of the organism.

Accessory functions associated with respiration or accomplished with the aid of respiratory organs are (1) production of sound (vocalization), (2) elimination of volatile waste gases, and (3) elimination of excess heat from the body.

STRUCTURE OF THE RESPIRATORY SYSTEM

The organs included in the respiratory system are the *air passageways* and the *lungs.*

Air Passageways. These include the nasal cavity and its associated paranasal sinuses, pharynx, larynx, trachea, bronchi, bronchioles, alveolar ducts, and alveoli. The mouth and oral cavity, though primarily digestive organs, function in breathing and are of primary importance in speech and the production of sound. The mouth and oral cavity are described in Chapter 7, Vol. 1. The structure of the air passageways is described in the following paragraphs.

NASAL CAVITY. The nasal cavity (Fig. 2-1) is the space between the roof of the mouth and the floor of the cranial cavity. The nasal septum divides this space into the *right* and *left nasal fossae,* the external openings of which are the *external nares* or *nostrils.* Each nostril leads to a *vestibule,*

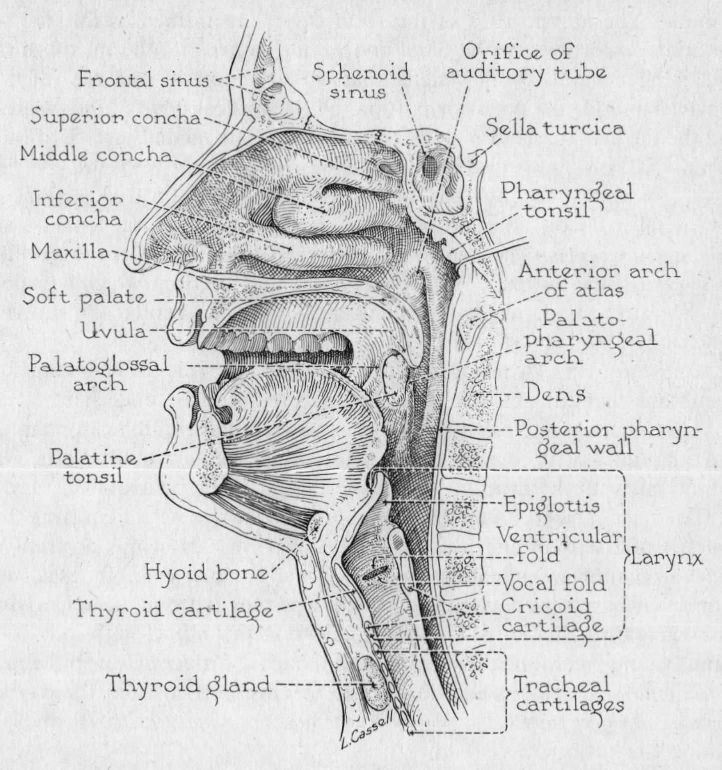

Fig. 2-1. Saggital section through the mouth, larynx, pharynx, and nasal cavity. (Reprinted with permission of W. B. Saunders Co., Philadelphia, from B. G. King and M. J. Showers, *Human Anatomy and Physiology,* 6th ed., 1969.)

the most anterior portion of the fossa, which contains long hairs called *vibrissae.* Posteriorly, the fossae open into the nasal portion of the pharynx through the *choanae* or *posterior nares.* Each fossa is divided into the *respiratory* and *olfactory regions,* differentiated structurally on the basis of the type of epithelium lining them. The respiratory region comprises the major cavity of each fossa and serves as the main air passageway. The olfactory region, a relatively small area in the uppermost portion of each fossa, contains the olfactory receptors for the sense of smell.

Bony Structures of the Nasal Cavity. The *roof* of the nasal cavity is supported as follows: (1) the anterior portion, by the nasal and frontal bones; (2) the middle portion, by the cribriform plate of the ethmoid; and (3) the posterior portion, by parts of the sphenoid, vomer, and palatine. The *floor* is supported by the hard palate, consisting of the palatal processes of the maxillae and the horizontal processes of the palatine bones. The nasal septum is formed by the vomer and the perpendicular plate of the ethmoid. The *lateral walls* of the nasal cavity are formed as follows: (1) the anterior portion, by the nasal and frontal bones and the maxillae; (2) the lateral portion, by the lateral mass of the ethmoid and the inferior conchae; and (3) the posterior portion, by the vertical plate of the palatine and the medial pterygoid plate of the sphenoid. The medial surface of each lateral wall bears three prominent scroll-like projections, the *nasal conchae* or *turbinates;* the superior and middle conchae are inward projections of the lateral mass of the ethmoid bones, whereas the inferior conchae are independent bones. The conchae divide each fossa into three groovelike passageways, the *meatuses* (superior, middle, and inferior), which lie below and lateral to the corresponding conchae. Above and behind the superior conchae is the *sphenoethmoidal recess.*

Nasal Mucous Membrane. The nasal cavity is lined with mucous membrane that adheres closely to the periosteum of the underlying bone. It is continuous posteriorly with the mucous membrane of the nasopharynx and anteriorly with the modified skin lining the vestibules. It is also continuous with the mucous membrane that lines the paranasal sinuses.

The mucous membrane of the *respiratory region* (that covering the conchae and septum and lining the meatuses) consists of pseudostratified ciliated columnar epithelium containing many goblet cells. It rests on a prominent basement membrane, which separates it from the underlying connective tissue (tunica propria). The latter contains numerous seromucous glands whose secretion serves to moisten the nasal surfaces. The epithelium of the *olfactory region* is pseudostratified columnar. It contains three types of cells: (1) *olfactory cells,* (2) *sustentacular* or *supporting cells,* and (3) *basal cells.*

The mucous membrane of the olfactory region contains the receptors for the sense of smell; they are stimulated by molecules of volatile substances present in the air. This membrane lies in the upper portion of the nasal cavity, occupying an area on both sides of the nasal septum and on the superior conchae. The total area in each fossa amounts to about 250 mm^2. The mucosa of this region differs from that of the respiratory region in the following respects: (1) Ciliated cells are absent, (2) a basement membrane is lacking, (3) different kinds of glands are present, and (4) the color is yellowish instead of pinkish.

Blood, Lymph, and Nerve Supply of the Nasal Cavity. The respiratory epithelium throughout the nasal cavity is highly vascular, but on the surfaces of the middle and inferior conchae it is extremely so, containing numerous venous plexuses somewhat resembling erectile tissue. When these plexuses become engorged with blood, as in the nasal congestion that commonly accompanies respiratory infections, the mucosa becomes swollen and turgid, resulting in obstruction to the flow of air through the nasal passageways. Because of this extreme vascularity of the nasal mucosa, *epistaxis* (nosebleed) is common. This may arise from hypertension, trauma, abscesses, or ulcers.

The arteries supplying the nasal cavity are (1) the *sphenopalatine,* a branch of the internal maxillary artery, which supplies the nasal conchae and the lower portion of the septum; (2) the *ethmoidal,* from the ophthalmic artery, which supplies portions of the lateral and medial nasal walls; (3) the *descending palatine,* from the internal maxillary artery, which supplies the floor of the nasal cavity; and (4) the *superior labial,* from the external maxillary artery, which supplies the vestibule. The ethmoidal arteries receive their blood through the internal carotid artery, the others through the external carotid.

The *venous plexuses* of the nasal mucosa are drained by (1) the *sphenopalatine* vein, which leads to the pterygoid plexus and thence into the common facial or exterior jugular vein; (2) the *ethmoidal* veins, which enter the ophthalmic vein and eventually drain into the cavernous sinus; and (3) branches of the *anterior facial* vein.

Lymphatics are numerous, draining into the subdural and subarachnoid plexuses.

The *olfactory nerves,* which pass through the cribriform plate, supply the olfactory epithelium. The respiratory epithelium is supplied principally by sensory fibers of the *trigeminal nerve.*

Functions of the Nasal Cavity. As air passes through the nasal cavity, it is *warmed* and *moistened* through contact with the mucous membrane. It is also *filtered;* larger, coarser particles are caught by nasal hairs, whereas smaller particles (such as dust and bacteria) are trapped by the sticky mucus, which is moved posteriorly by the currents the cilia produce. A lapse of about 10 min is required for mucus passing from the anterior portion of the nose to reach the pharynx, from which it is either expelled through the mouth or swallowed. The nasal cavity also serves as a resonating structure in the *production of sound,* and it contains the receptors for *olfaction,* by which dissolved substances can be smelled.

PARANASAL SINUSES (Fig. 2-2). Air spaces present in bones adjacent to and communicating with the nasal cavity are referred to as the *paranasal sinuses.* The mucosa of these sinuses is continuous with that of the nasal cavity but is thinner and less vascular and hence paler in color. It is also less firmly attached to the adjacent bone structure. Mucous glands are present, and definite ciliated tracts convey the mucus through small channels into the nasal cavity. There are four groups of paranasal sinuses: the maxillary sinus, the frontal sinus, the ethmoidal air cells, and the sphenoidal sinus. The *primary function* of the paranasal sinuses is believed to be that of lightening the bones of the skull. *Secondary functions* are to supply mucus to the nasal cavity and to serve, along with that cavity, as resonating structures in sound production.

Maxillary Sinus. This is the largest of the paranasal sinuses, averaging 15 to 20 cm^3 in capacity. It is paired, being found in the body of each maxilla directly beneath the orbit. Each sinus opens into the middle meatus of the nose. Its opening or *ostium* is disadvantageously placed as a drainage

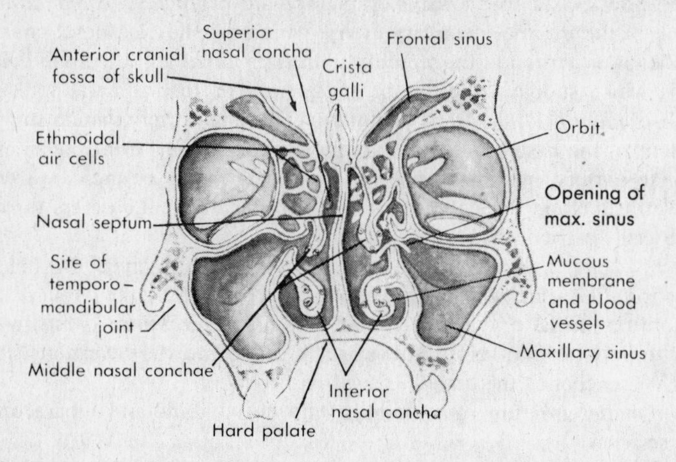

Fig. 2-2. Coronal section through the skull, showing the relationship of paranasal sinuses to the nasal cavity. (From *Anatomy of the Human Body* by Weston D. Gardner, M.D., and William A. Osburn, M.M.A. Copyright © 1978 by W. B. Saunders Co. Reprinted by permission of Holt, Rinehart and Winston, CBS College Publishing.)

opening since it is located at the highest point in the medial wall of the sinus.

Frontal Sinus. This sinus lies in the frontal bone, superior and medial to the orbit. Normally, it is paired, a frontal septum separating the two sinuses. The shape of a frontal sinus is extremely variable. Often the two sinuses are fused to form a single cavity; on the other hand, they may be subdivided to form several cavities. They may be large, small, or even lacking. Their average capacity is 7 cm^3. Each opens into the middle meatus.

Ethmoidal Air Cells. These include numerous irregularly shaped air spaces, which honeycomb the lateral masses of the ethmoid bone. Collectively, they form the *ethmoidal labyrinth,* which occupies the region between the orbit and the nasal cavity but may extend into the nasal conchae. The number of cells in each labyrinth varies from 3 to 18. The cells are arranged in three groups: (1) the *anterior ethmoidal cells,* numbering two to eight; (2) the *middle ethmoidal cells,* numbering one to six; and (3) the *posterior ethmoidal cells,* numbering one to seven. The anterior cells communicate with the middle meatus, the middle cells with the anterior and posterior ethmoidal cells through large ostia, and the posterior cells with the superior meatus.

Sphenoidal Sinus. This sinus, with its mate, lies in the body of the sphenoid bone. It varies greatly in size and shape, sometimes extending into the great wing. Its average capacity is 8 cm^3. Sphenoidal sinuses open into the sphenoethmoidal recess.

Pharynx (Fig. 2-1). The *pharynx* constitutes a common passageway for both air and food. Its detailed structure is described in Chapter 7, Vol. 1, under the Digestive System. Briefly, its three regions are the nasopharynx, the oropharynx, and the laryngopharynx. The *nasopharynx* communicates superiorly with the nasal cavity by the *posterior choanae* or *nares* and with the tympanic cavities of the two ears by the *auditory (pharyngotympanic) tubes.* Through these tubes, a balanced air pressure is maintained on the two sides of the tympanic membranes. The *oropharynx* communicates anteriorly with the oral cavity by the isthmus of the fauces. The *laryngopharynx* communicates inferiorly with the larynx and esophagus. Note that the air and food passageways cross in the oropharynx, but automatic reflex mechanisms close the airway when foods or liquids are being swallowed.

Larynx (Fig. 2-1). Between the pharynx and the trachea lies a tubular structure, the *larynx,* which serves to connect these two organs. It contains a pair of *vocal folds* that regulate the passage of air through the larynx and play a role in the production of sound. For this reason, the larynx is commonly called the *voice box.* Lying in the neck, the larynx is anterior to the esophagus at about the level of the 4th to 6th cervical vertebrae. It is situated directly inferior to the hyoid bone, to which the larynx is firmly bound by the *hyothyroid membrane.* In swallowing movements, the larynx is drawn upward against the base of the tongue. This action causes the *epiglottis,* a cartilaginous flap lying superior to the larynx, to incline backward and downward, directing food into the esophagus and preventing its entry into the larynx.

STRUCTURE OF THE LARYNX (Fig. 2-3). The larynx is made up of a number of cartilages bound together by an elastic membrane; its movement is controlled by muscles. Its lumen is lined with mucous membrane that is continuous with that of the pharynx above and the trachea below. The nine *cartilages* that form the framework of the larynx are:

Single	*Paired*
Cricoid	Arytenoid
Thyroid	Corniculate
Epiglottic	Cuneiform

The *cricoid cartilage* forms the lowermost portion of the larynx and is connected to the first tracheal cartilage. It is shaped somewhat like a signet ring, the broad portion (*lamina*) being located at the back of the larynx, the narrow portion (*arch*) forming the anterior and lateral portions.

The *thyroid cartilage* is the principal cartilage of the larynx. It consists of two broad *laminae* united anteriorly to form a V-shaped structure that in males forms a more or less marked prominence called the "Adam's apple." The angle of fusion in males is about 90°, in females about 120°. Superiorly, the fusion is incomplete, leaving a V-shaped space between the

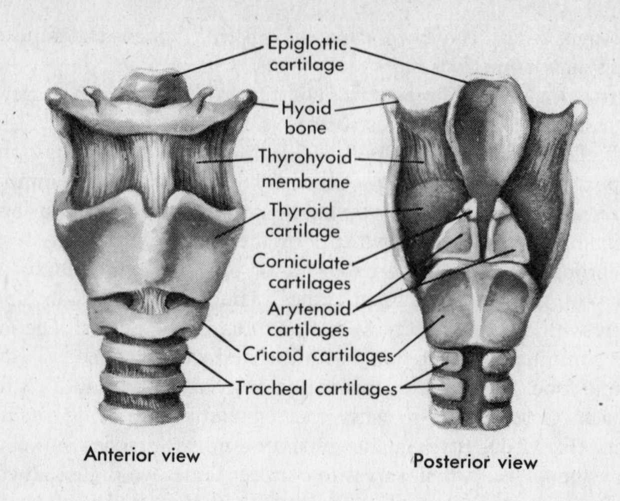

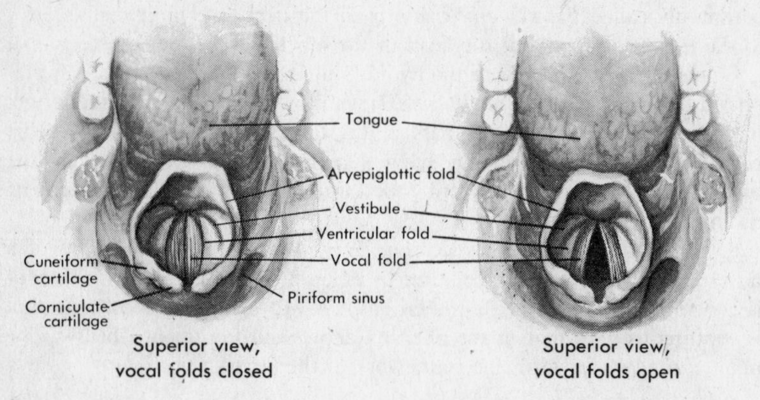

Fig. 2-3. The larynx. (From *Anatomy of the Human Body* by Weston D. Gardner, M.D., and William A. Osburn, M.M.A. Copyright © 1978 by W. B. Saunders Co. Reprinted by permission of Holt, Rinehart and Winston, CBS College Publishing.)

laminae, the *superior thyroid notch.* On the posterior margins of each lamina are two projections, the *superior* and *inferior cornua.* A broad, fibroelastic *thyrohyoid membrane* connects the thyroid cartilage to the hyoid bone.

The *epiglottic cartilage* forms the framework of the epiglottis. It is a thin, leaf-shaped cartilage lying above the thyroid cartilage and between the horns of the hyoid bone. Its upper end, which is slightly notched, is directed upward and posteriorly; its lower end narrows to a slender stalk, the *petiole,* to which the thyroepiglottic ligament is attached. To the lateral margins of the epiglottis are attached folds of mucous membrane, the *aryepiglottic folds.*

The *arytenoid cartilages* are two somewhat triangular-shaped cartilages lying directly superior to the cricoid cartilage, with which they articulate. The vocal ligaments, which lie in the free edges of the vocal folds, are attached to these cartilages. Movement of the arytenoid cartilages is largely responsible for changes in the state of tension of the vocal folds.

The *corniculate cartilages* are two small conical structures located at the apices of the arytenoid cartilages.

The paired *cuneiform cartilages* are small rodlike bodies located in the aryepiglottic folds just anterior to the corniculate cartilages. They are somewhat variable in size and form and may even be absent.

The elastic membrane of the larynx consists of a sheet of elastic fibers forming a part of the walls of the larynx. Its upper portion is called the *quadrangular membrane;* the lower portion, the *elastic cone.*

MUSCLES OF THE LARYNX. The muscles of the larynx comprise two groups: extrinsic muscles and intrinsic muscles. The *extrinsic muscles,* inserted on the larynx, have their origin on structures surrounding the larynx. They are involved in movements of the larynx *as a whole,* as in swallowing. Included in this group are the omohyoid, sternohyoid, sternothyroid, thyrohyoid, and several others. The *intrinsic muscles* lie entirely within the walls of the larynx. Their actions are concerned with movements of the *parts of the larynx in relation to each other.* They include the *cricothyroid, thyroarytenoid* (external and internal), *arytenoid* (transverse and oblique), and *cricoarytenoid* (posterior and lateral) muscles.

The intrinsic muscles perform the following functions: (1) opening and closing the glottis during inspiration and expiration, (2) closing the laryngeal aperture and glottis during swallowing, and (3) regulating the tension of the vocal folds in the production of sound.

Both extrinsic and intrinsic muscles are composed of striated muscle fibers and are innervated by the superior portion of the *accessory nerve* and the laryngeal branches of the *vagus nerve.*

LARYNGEAL CAVITY. The cavity of the larynx is relatively small. It contains within its walls two pairs of folds: the *ventricular folds* (false vocal cords) and the *vocal folds* (true vocal cords) (Fig. 2-3). The *ventricular folds* are two rounded folds, one lying on each side of the laryngeal cavity superior and parallel to the vocal folds. They are not concerned with sound production. The *vocal folds* are two white bands, each forming the edge of a prominence on the lateral wall of the larynx. In length, they measure, when relaxed, 15 mm in the male and 11 mm in the female; when stretched, 20 mm in the male and 15 mm in the female. They extend diagonally from the angle of the thyroid cartilage to the arytenoid cartilages. Each encloses the vocal ligament and vocal muscle, the internal thyroarytenoid. Between the vocal folds is a narrow slit, the *rima glottidis,* through which air passes. The two vocal folds and their intervening space constitute the sound-producing apparatus, the *glottis.*

The ventricular and vocal folds divide the laryngeal cavity into three regions: (1) the *vestibule,* the region superior to the ventricular folds; (2) the *ventricle,* the region between the ventricular and vocal folds; and (3) the *inferior entrance to the glottis,* the region inferior to the vocal folds.

The cavity of the larynx is lined throughout by mucous membrane, which is continuous with that of the pharynx and the trachea. Its surface, except that covering the vocal folds, consists of pseudostratified ciliated epithelium containing many goblet cells. Mucous glands and lymphatic tissue are abundant. The beat of the cilia is directed upward. Over the vocal folds, the epithelium is stratified squamous, and cilia and glands are lacking. In this region there is little submucosa.

BLOOD AND NERVE SUPPLY OF THE LARYNX. The larynx receives blood through the *superior thyroid artery,* a branch of the external carotid, and the *inferior thyroid artery,* a branch of the thyrocervical trunk from the subclavian artery. Innervation is supplied by the *superior laryngeal nerve,* a branch of the vagus nerve, and the *inferior laryngeal,* a branch of the recurrent nerve that comes from the vagus. The larynx also receives sympathetic branches from the *thoracolumbar division* of the autonomic nervous system.

SEX DIFFERENCES IN THE STRUCTURE OF THE LARYNX. Before puberty, the larynx is approximately the same size in boys and girls. After puberty, however, the larynx undergoes noteworthy changes in males: It increases in size, the cavity becomes larger, the thyroid cartilage is more pronounced, and the vocal cords become longer and thicker. The result is a change in the quality and pitch of the voice. These changes are usually completed within 2 years.

LARYNX IN VOCALIZATION. The human voice consists of sounds produced by the vibration of the vocal folds and modified by the resonance chambers (mouth cavity, pharynx, nasal cavity, sinuses). Vocalization is accomplished by the passage of a current of air through the glottis. On the sides of the glottis are the parallel edges of the vocal folds (vocal cords) that, when possessing a certain degree of tension, vibrate and produce a sound. Immediately before speaking or before uttering any sound, one brings the edges of the vocal folds together. Contraction of the abdominal and thoracic muscles increases the air pressure in the lungs. When this pressure reaches a certain level, the glottis opens, and the current of air passes out through the glottis, causing the vocal folds to vibrate, with the resultant production of a sound.

The changes in position and in tension of the vocal folds are brought about principally by the movement of two small triangular cartilages, the *arytenoid cartilages,* to which the posterior ends of the vocal folds are attached. The anterior ends of the folds are attached to the thyroid cartilage;

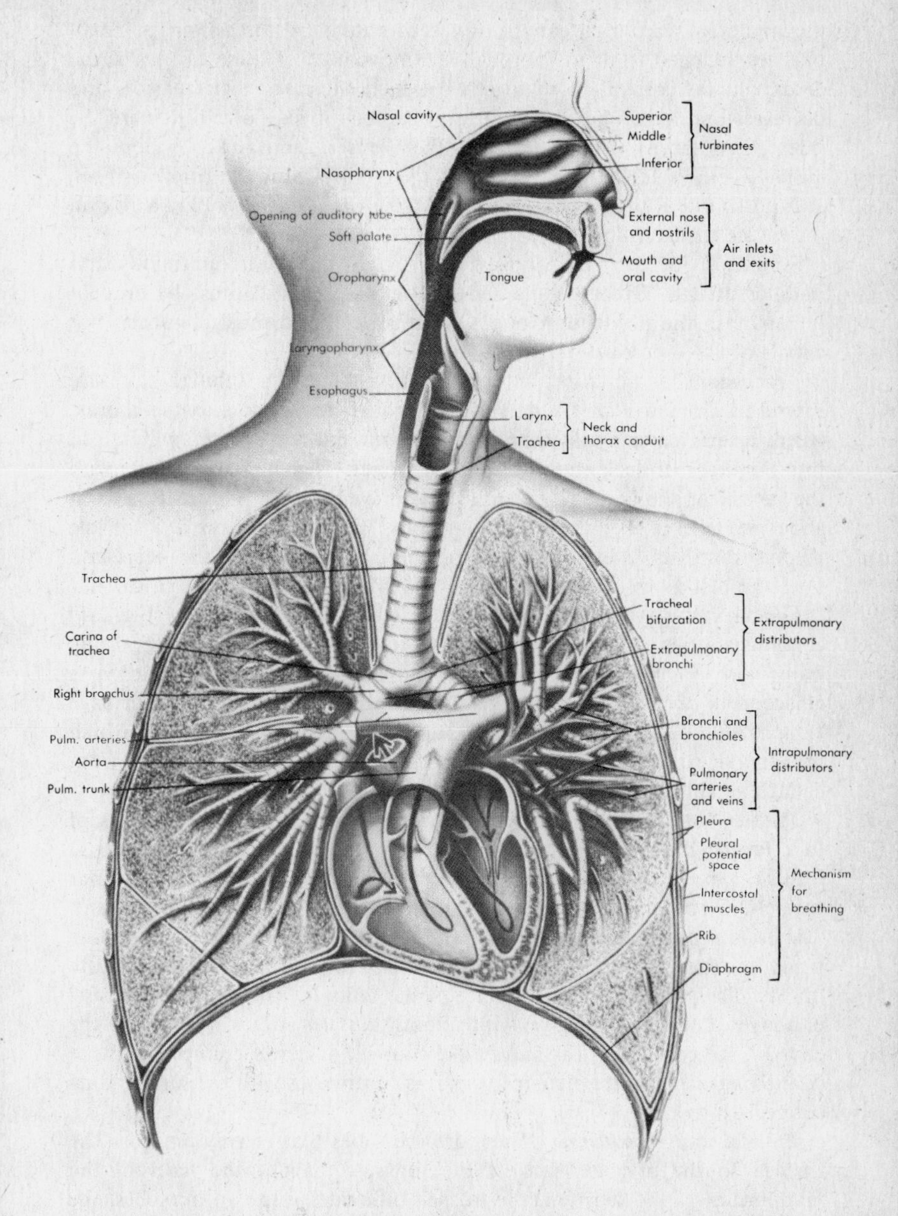

Fig. 2-4. The trachea, bronchi, and bronchial tubes. (From *Anatomy of the Human Body* by Weston D. Gardner, M.D., and William A. Osburn, M.M.A. Copyright © 1978 by W. B. Saunders Co. Reprinted by permission of Holt, Rinehart and Winston, CBS College Publishing.)

the arytenoid cartilages can be made to rotate by the action of small muscles attached to them. Through the movement of these cartilages, the cords can be brought close together (which increases their tension and decreases the size of the glottis), or they can be moved outward toward the sides of the larynx (which decreases their tension and greatly widens the air passageway). In ordinary breathing, the cords occupy the latter position, and no sound is produced. Muscle fibers in the cords also play a role in increasing their tension.

Sound produced by the larynx alone would be crude, monotonous, and undifferentiated. Other organs and cavities of the skull must be brought into play in the production of articulate speech and musical tones. (For details of voice production, see pages 56–58.)

Trachea. The windpipe or *trachea* (Fig. 2-4) is a tubular structure extending from the larynx downward through the neck into the thorax, where it terminates by dividing into the *right* and *left bronchi,* which lead into the lungs. It lies anterior to the esophagus. The trachea connects with the larynx at the level of the 6th cervical vertebra; its bifurcation is at about the level of the 5th thoracic vertebra. Within the thorax it lies in the mediastinum, its lower portion being directly posterior to the heart and the large blood vessels (arch of the aorta and the superior vena cava).

GROSS STRUCTURE OF THE TRACHEA. The wall of the trachea (Fig. 2-5) contains a series of 16 to 20 C-shaped cartilages, which form its supporting framework. Each cartilage is incomplete dorsally, where the trachea is adjacent to the esophagus. Across the open portion of each cartilage extends a fibroelastic membrane containing fibers of smooth muscle, which in general run transversely. The cartilages prevent the trachea from collapsing and thus maintain a free passageway for air.

MICROSCOPIC STRUCTURE OF THE TRACHEA. The trachea consists of four layers: mucosa, submucosa, cartilage, and adventitia. The *mucosa* forms the innermost layer; it is lined with ciliated pseudostratified columnar epithelium containing many goblet cells. The beat of the cilia is upward. The *submucosa* consists principally of loose connective tissue containing many *tracheal glands,* compound tubuloalveolar glands, which secrete mucus. These open onto the surface of the mucosa. The *cartilage* consists of a layer of hyaline cartilage, with smooth muscles stretching between the ends of the cartilages. The *adventitia* consists of dense connective tissue containing elastic and reticular fibers; it is continuous with the surrounding connective tissue.

Bronchi and Bronchioles. These are the tubes that convey air from the trachea to the alveolar sacs of the lungs. At about the level of the manubrium of the sternum, the trachea bifurcates into two branches, the *right* and *left primary bronchi,* each of which follows a diagonal course to the lungs. The right primary bronchus is the shorter and wider of the two (it is about 2.5 cm in length); the left primary bronchus measures about 5 cm in length and lies more horizontally than does the right. For this

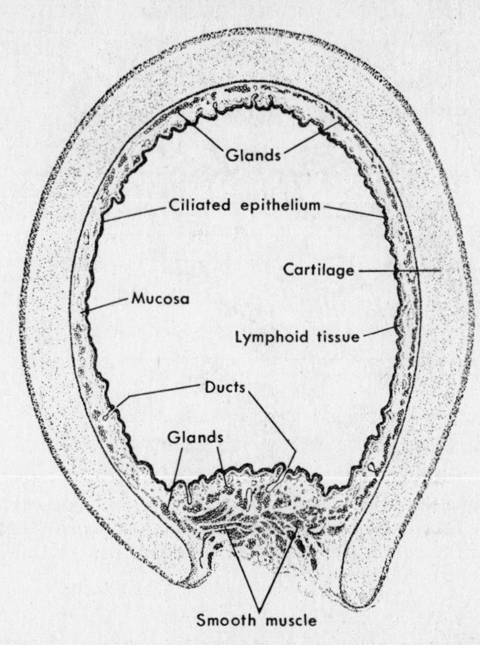

Fig. 2-5. Cross section through the trachea. (Redrawn and modified from Kölliker–von Ebner.) (Reprinted with permission of W. B. Saunders Co., Philadelphia, from W. Bloom and D. W. Fawcett, *Histology*, 10th ed., 1975.)

reason, foreign bodies drawn into the trachea enter the right bronchus more frequently. Each of the bronchi is similar in microscopic structure to the trachea; the right bronchus possesses 6 to 8 rings of cartilage, the left, 9 to 12.

On entering the hilus of the lung, each bronchus divides into smaller bronchi (two in the left lung, three in the right), which enter the primary lobes of the lung. Within the lung, these smaller bronchi divide and subdivide into still smaller bronchi, the diameter decreasing with each subdivision. The cartilage rings of the bronchi become progressively smaller and are eventually replaced by irregularly distributed cartilage plates. When the air tubes acquire a diameter of about 1 mm, the cartilage disappears and the tubes are known as *bronchioles.* The branch entering a lobule is called an *intralobular bronchiole.* This branches into *terminal bronchioles,* (Fig. 2-6), each of which gives rise to two or more *respiratory bronchioles.* These branch into *alveolar ducts,* which lead to the *alveolar sacs.* Each of these sacs (also called *air sacs*) has numerous outpocketings, the *alveoli.* A bronchus and all its branches comprise a *bronchial tree.*

As the cartilage in the bronchial tube grows less abundant, the amount of smooth muscle tissue increases, becoming most prominent in the

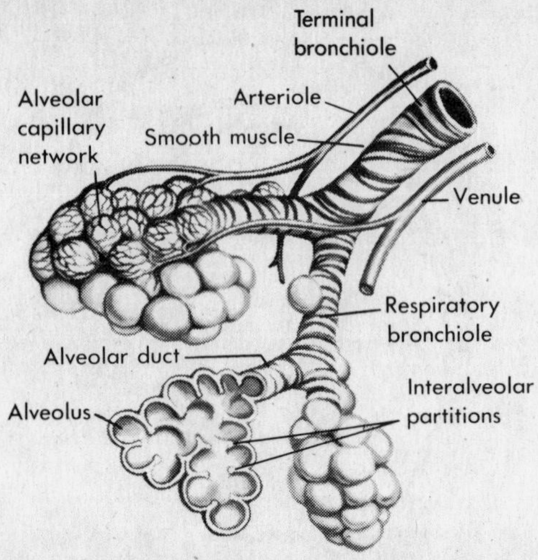

Fig. 2-6. A terminal bronchiole and its branches. (From *Anatomy of the Human Body* by Weston D. Gardner, M.D., and William A. Osburn, M.M.A. Copyright © 1978 by W. B. Saunders Co. Reprinted by permission of Holt, Rinehart and Winston, CBS College Publishing.)

terminal bronchioles, where it is intimately associated with the elastic layer, forming a *myoelastic layer.* In the respiratory bronchioles, the ciliated columnar epithelium of the conducting tubes undergoes transition to the cuboidal type, and goblet cells become fewer. Distally the cilia disappear, and in the alveolar sacs, the epithelium becomes of the simple squamous type, adapted for ready passage of gases to and from the surrounding capillaries. The alveoli contain two types of epithelial cells: small, flat *alveolar (Type I) cells* and larger *septal (Type II) cells.* The latter are secretory, producing a lecithin-containing surfactant that contributes to the elastic properties of pulmonary tissues. Also present are *alveolar macrophages* or *dust cells,* which are highly phagocytic.

The smooth muscles of the bronchioles are innervated by parasympathetic constrictor fibers from the vagus nerve and dilator fibers from the sympathetic trunk. The diameters of the bronchioles adjust reflexly to the respiratory needs of the body.

Lungs. The lungs are the primary organs of respiration. They serve essentially as structures that permit the interchange of gases (oxygen and carbon dioxide) between the blood and the air.

The lungs lie in the thoracic cavity, one on each side of the mediastinum. Each is roughly cone-shaped, with its *base* resting upon the diaphragm, its tip or *apex* extending superiorly into the root of the neck. On the medial surface of each lung is a depression, the *hilus.* At this point several large structures enter or leave the lung, among them the bronchi, the pulmonary artery and veins, the bronchial arteries and veins, lymphatic vessels, and nerves. Collectively, these structures constitute the *root of the lung.* Each lung lies free in the thoracic cavity, attached only by its root and by a pulmonary ligament.

SURFACES OF THE LUNGS. Each lung has three surfaces: costal, diaphragmatic, and mediastinal. The *costal* surface lies against the thoracic wall, the *diaphragmatic* (the base) rests against the diaphragm, and the *mediastinal* is directed medially and bears a concavity, the *cardiac fossa,* which accommodates the heart.

LOBES OF THE LUNGS. Each lung is divided by a deep *interlobar fissure* into two lobes, a *superior lobe* and a larger *inferior lobe.* In the right lung, however, a secondary fissure, located below the interlobar fissure, extends horizontally and gives rise to a third or *middle lobe.*

GROSS STRUCTURE OF THE LUNGS. The right and left lungs are slightly *asymmetrical,* the right being shorter and broader owing to its position above the liver, the left narrower and longer owing to its position in relation to the heart. In *texture,* the lungs are soft and spongy. They are elastic and are constantly changing their form with inspiratory and expiratory movements, assuming the outline of the thoracic cavity. The lungs of a fetus, which contain no air, will sink when placed in water. Following birth, after respiration has been established, some air is retained in the lungs, and they will float. The *color* of the lungs varies, depending on the amount of air, blood, pigment, and dust particles present in the tissue. In general, they are red in fetuses, pinkish in children, a mottled gray in adults. At times the lungs may become almost entirely black from inhalation of carbon particles.

Each lung consists of a connective tissue *framework* or *stroma* that supports a system of branching air tubes (bronchi, bronchioles, alveolar ducts, air sacs, and alveoli) and associated structures, such as blood vessels, lymph vessels, and nerves.

The surface of each lobe is divided into small irregular areas demarcated by dark lines. These areas are the bases of *secondary lobules.* The dark lines indicate the position of the *interlobular septa,* which separate these lobules. The structural and functional unit of the lung is the *primary lobule,* consisting of a terminal bronchiole and the air ducts connected with it (respiratory bronchioles, alveolar ducts, alveolar sacs, and alveoli) together with its associated blood vessels, lymphatics, and nerves. In addition to the lobes, each lung is further subdivided into segments (apical, posterior, superior, anterior, and lateral) as portions of the bronchial tree.

BLOOD AND LYMPH SUPPLY OF THE LUNGS. The principal blood vessels supplying the lungs are the bronchial and pulmonary arteries. The *bronchial arteries* arise from the aorta or from an intercostal artery. There are two for the left lung, one for the right lung. They enter at the hilus and follow the bronchi and their branches, terminating at the distal ends of the alveolar ducts. These arteries supply oxygen and nourishment for the lung tissue. Bronchial veins from the lung tissue terminate in the innominate, azygos, intercostal, or pulmonary veins. *Pulmonary arteries* also enter each lung at the hilus and follow the main bronchus, with branches accompanying the successive divisions of the bronchial tubes. On entering a pulmonary lobule, a branch accompanies each alveolar duct and terminates in a capillary net in the walls of the alveoli. Here the deoxygenated blood from the tissues is separated from the alveolar air only by the capillary wall and the thin wall of the alveolus. This arrangement permits the ready diffusion of gases between the alveolar air and the blood in the capillaries. Venules from the alveolar capillaries and from those of the pleura lead to larger veins, which in turn lead to the *pulmonary veins* (two for each lung), which make their exit from the hilus carrying oxygenated blood, to the left atrium.

Lymph vessels are present in the walls of the bronchi, pleura, and pulmonary arteries and veins. Lymph nodes are found at the branching of the larger bronchial tubes. Nodules of lymphoid tissue are found along the pulmonary vessels. These frequently contain considerable quantities of carbonaceous material, especially in persons of advanced age.

Thoracic Cavity. The thoracic cavity is the space lying within the walls of the thorax, its floor being formed by the diaphragm, which separates it from the abdominal cavity. Superiorly, the thoracic cavity extends a short distance into the base of the neck.

DIVISIONS OF THE THORACIC CAVITY. Two *pleural cavities,* right and left, contain the lungs. The region between these pleural cavities is the *mediastinum* or *mediastinal septum.*

Pleurae. Enclosing the lungs is a serous membrane called the *pleura,* which forms a closed doubled-walled sac consisting of two layers. The *pulmonary pleura* or *visceral layer* closely invests the lungs, covering almost their entire surface. The *parietal layer* forms the lining of the thoracic wall, lying against the endothoracic fascia. It may be differentiated into costal, mediastinal, diaphragmatic, and cervical regions. The *isthmus* is the portion connecting the visceral and parietal pleurae. It forms a tubelike structure investing the structures that constitute the root of the lung.

The space between the visceral and parietal layers is a potential rather than a real cavity, for expansion of the lungs causes the visceral layer to come into contact with the parietal layer from which it is separated by a very thin space filled with serous fluid. This fluid serves as a lubricant permitting free movement of the lungs within the thoracic cavity.

Mediastinum. Between the two pleural cavities lies the *thoracic mediastinum* or *mediastinal septum,* which contains the organs and structures occupying the interpleural space. These are:

1. The heart and its pericardial covering.
2. The great vessels entering and leaving the heart (thoracic aorta and branches from the aortic arch, pulmonary artery, superior vena cava, and thoracic portion of the inferior vena cava).

3. The thymus gland.
4. The esophagus and the lower portion of the trachea.
5. The thoracic duct and thoracic lymph nodes.
6. Nerves passing into or through the thorax, including the vagi and the phrenic, recurrent laryngeal, and cardiac nerves.

PHYSIOLOGY OF RESPIRATION

External respiration or *breathing,* also called *ventilation,* consists of two phases: *inspiration,* the taking in of air, and *expiration,* the expulsion of air. It is accomplished through changes in size of the thoracic cavity brought about by the contraction and relaxation of respiratory muscles. The thoracic cavity is a closed space with no external openings. Contained within it are the lungs, elastic baglike structures that communicate with the outside by means of the respiratory passageways. When the thoracic cavity enlarges, a reduction in pressure causes air to rush in and fill the lungs. When the inspiratory movements cease, the size of the thoracic cavity is reduced, and the resultant increase of pressure on the lungs causes air to be expelled.

Pressure Changes. The pressure within the thoracic cavity, that is, between the two layers of pleura, is known as *intrapleural* or *intrathoracic pressure.* It is always slightly less than atmospheric pressure (a "negative pressure"), due primarily to the elasticity of the lungs. If an opening is made in the thoracic wall, air will enter the pleural cavity, and the lung will collapse (*pneumothorax*).

The pressure within the lungs and the respiratory passageways is the *intrapulmonary pressure.* When it is below atmospheric pressure, inspiration occurs; when it is above atmospheric pressure, expiration occurs. The movement of air into and out of the lungs is due to changes in intrathoracic pressure, which in turn affects the intrapulmonary pressure.

Respiratory Movements and Rate of Respiration. The rhythmic movements of the thorax by which air is drawn into and expelled from the lungs constitute the respiratory movements. With inspiration and expiration counted as one combined movement, the normal rate in quiet breathing is about 14 to 18 a minute. This rate is variable. It is increased by muscular activity, by higher body temperatures (as in fever), and in certain pathological conditions (such as hyperthyroidism). It varies with *sex,* females having the more rapid rate (16 to 20/min), and with *age,* the approximate rates being as follows: at birth, 40 to 60; at 5 years, 24 to 26; at 15 years, 20 to 22; finally, at 25 years, 14 to 18. The *position of the body* has an effect on the rate of respiration, which drops to 12 to 14 when the body is prone or during sleep. With the body in a semierect or sitting position, the rate increases to about 18; with the body in a standing position, the rate is 20 to 22. *Emotional conditions,* too, may slow down or speed up the rate.

Types of Respiration. The following terms describe various types and conditions of breathing.

Apnea. Temporary cessation of breathing movements.

Dyspnea. Difficult or labored breathing, usually accompanied by discomfort and the sensation of breathlessness.

Eupnea. Normal breathing, with usual quiet inspirations and expirations.

Hyperpnea. Increase in rate and depth of breathing; abnormal exaggeration of respiratory movements.

Orthopnea. Dyspnea when the body is in a horizontal position. It is usually relieved upon the assumption of an upright position.

Polypnea. Rapid breathing such as results from increased activity or from emotional states; panting.

Tachypnea. Excessively rapid and shallow breathing.

Muscular Activity in Breathing. As previously stated, changes in pulmonary pressure are directly responsible for determining whether air will flow into or out of the lungs, and these changes are brought about by changes in the size of the thorax. These latter changes are accomplished through the action of the *respiratory muscles,* which include the diaphragm, abdominal muscles, and various muscles that act on the ribs.

INSPIRATION. The *diaphragm* is a dome-shaped muscle that forms the floor of the pleural cavities, separating them from the abdominal cavity. Contraction of the radial muscle fibers causes the diaphragm to become slightly flattened and to descend. This increases the vertical diameter of the thorax and reduces intrathoracic and intrapulmonary pressures, so air from the outside, having a higher pressure than that in the lungs, rushes into these organs. In its descent, the diaphragm exerts a pressure on the abdominal contents, causing them to press against the abdominal muscles. Being generally relaxed, these muscles give way, and the abdominal wall moves outward. This process is known as *diaphragmatic* or *abdominal breathing.*

The chest cavity can also be enlarged by raising the sternum and the ribs. This process is called *costal* or *thoracic breathing.* The ribs extend laterally from the vertebral column with their curved anterior ends extending downward and inward. Contraction of the *external intercostal muscles* elevates the anterior ends of the ribs, especially the third to sixth pairs, and these ends move upward and outward. Since they are attached to the sternum through the costal cartilages, the sternum also moves forward and upward. These actions increase the anteroposterior and lateral diameters of the thorax. In forced respiration, additional muscles may assist the external intercostals in elevation of the ribs and sternum; these are the *scaleni, levatores costorum, pectoralis major,* and *serratus posterior superior.*

EXPIRATION. Normal, quiet respiration is a *passive* process. No active muscular effort is expended. At the end of inspiration, the diaphragm and the external intercostals relax. The recoil of the stretched costal cartilages and the weight of the walls of the thorax, together with the recoil of the elastic lungs, act to bring the chest wall back to its original position. The relaxed diaphragm moves upward as a result of the pressure of the abdominal contents and the negative pressure within the thoracic cavity. Intrathoracic pressure is increased, and air is expelled from the lungs.

In forced or labored respiration such as occurs during strenuous muscular activity or during voluntary deep breathing, expiration is an *active* process involving the activity of a number of muscles. The muscles primarily involved are the *abdominal muscles* (rectus abdominis, external and internal oblique, transverse abdominis), which constrict the abdominal cavity, causing pressure to be exerted on the underside of the diaphragm, and the *internal intercostals, serratus posterior inferior,* and *quadratus lumborum,* which depress the ribs. Also in deep breathing, a number of muscles may assist the diaphragm in inspiration. These include the *sternocleidomastoid, serratus anterior, pectoralis minor,* and *scalenes.*

Respiratory Sounds. Passage of air through the air passageways to the alveoli produces sounds that can be heard by use of a stethoscope or by applying the ear to various regions of the thorax.

NORMAL SOUNDS. The normal sounds are the *bronchial* or *tubular sound,* a high-pitched sound heard principally over the trachea, bronchi, and larger bronchioles, occurring in both inspiration and expiration, and the *vesicular murmur,* a soft, rustling sound heard only during inspiration and at the beginning of expiration and thought to be caused by the distension of the alveoli with air.

ABNORMAL SOUNDS. Obstruction of the respiratory passageways, such as occurs in pathological conditions, results in marked changes in respiratory sounds; hence these sounds play an important role in the diagnosis of pulmonary disorders. The term *rale* is applied to an abnormal sound arising in the lungs or air passageways and heard on auscultation of the chest. It results from the passage of air through bronchial tubes containing a secretion of exudate or narrowed by a spasm or a swelling of their walls. Various terms are applied to different types of rales, such as "crepitant," "fine," "medium," "coarse," "moist," and "dry." *Wheezing* refers to a whistling or sighing sound. Alterations in normal breathing sounds or the appearance of sounds not normally heard occur in diseases such as pneumonia, tuberculosis, and asthma. Other sounds of importance in the diagnosis of respiratory disorders are the *pleural friction sound, voice sound,* and *percussion sounds.* The pleural friction sound occurs when the roughened surfaces of the pleura rub against each other, as in pleurisy. The whispered voice sets up sounds in the chest that are useful in detecting the presence of pulmonary infiltration. Percussion sounds (those resulting from tapping the chest) indicate various chest conditions by their degree of resonance. If the air of the lung is replaced by fluid or a solid substance, as in pneumonia, a short, feeble high-pitched sound known as *dull* or *flat* is produced instead of the normal sound, known as *normal* or *vesicular resonance.*

Respiratory Volumes. The amount of air that can be taken into and expelled from the lungs can be measured with a *spirometer.* The volumes determined are classified as *tidal volume, expiratory reserve volume,* and *inspiratory reserve volume.*

TIDAL VOLUME. The amount of air inhaled and exhaled during normal quiet breathing averages about 500 ml for the adult male. Of this amount, about 150 ml remains in the respiratory passageways; this is *dead space air,* which is unavailable for respiratory exchange. The remainder (350 ml) constitutes *alveolar air.*

INSPIRATORY RESERVE VOLUME. This is the maximum amount of air that, in addition to tidal air, can be inhaled after an ordinary expiration that has followed a normal inspiration. For an adult male, this averages about 3000 ml.

EXPIRATORY RESERVE VOLUME. This is the amount of air that can be expelled by the most forceful effort after an ordinary expiration that has followed a normal inspiration. For an adult male, this averages about 1000 ml.

VITAL CAPACITY. The total of tidal, inspiratory reserve, and expiratory reserve volumes constitutes *vital capacity.* This is the amount of air that can be expelled by the greatest effort following the deepest possible inspiration. It averages 4500 ml for the average adult male. This measurement serves as an index to the general physical fitness of an individual; it is higher in athletes and others engaged in strenuous physical activity, lower in persons leading a sedentary life. Improper posture, lung disease, obesity, and deformities of the thorax or abdomen are among the conditions that decrease vital capacity.

RESIDUAL AIR. This term refers to the air that remains in the lungs after the most forcible expiratory effort. It averages about 1500 ml. Even in case of a collapse of a lung, which occurs when the thorax is opened, some air remains trapped in the alveoli, and the lung will float in water. This is also known as *minimal air.* It is of importance in medicolegal cases, since its presence or absence is the criterion for determining whether or not a baby is stillborn (fetal lung tissue sinks when placed in water).

The following table shows graphically the relationships just described as applied to the *average adult male:*

Total lung capacity 6000 ml (6 L)	Tidal volume	Alveolar air	350 ml	500 ml	Vital capacity 4500 ml
		Dead space air	150 ml		
	Inspiratory reserve volume			3000 ml	
	Expiratory reserve volume			1000 ml	
	Residual air				1500 ml

Exchange and Transport of Gases in Respiration. In the lungs, oxygen is taken up by the blood, and carbon dioxide is given off. In its passage throughout the body, the blood surrenders oxygen to the tissues and takes up carbon dioxide.

CHEMICAL CHANGES. The following table shows the chemical changes that take place in the air *within the lungs:*

Gases	*Inspired Air*	*Expired Air* (percent volume)	*Alveolar Air*
Oxygen	20.96	16.3	14.2
Carbon dioxide	0.04	4.0	5.5
Nitrogen and other rare gases	79.0	79.7	80.3
Water vapor	low	high	high

MECHANICS OF EXCHANGE OF GASES. Within the body the movement of gases is governed by the same physical laws that govern the movement of gases outside the body. The principal process involved is *diffusion.* Briefly, this is the tendency of molecules of a gas to become uniformly distributed. When there are differences in pressure or tension, that is, differences in the number of molecules per unit of space, the molecules of a gas tend to move from a region of higher tension to one of lower tension. In a mixture of gases, each gas acts independently of the others.

When air is drawn into the lungs in a normal inspiration, the lungs are only partly filled with fresh air since tidal air amounts to only one-eighth of vital capacity. But diffusion occurs, oxygen diffusing from the alveoli into the blood and carbon dioxide in the reverse direction. In respiration, with fresh air being repeatedly drawn into the lungs, this process is taking place continuously.

Diffusion in the Alveolar Walls. In the alveolar walls, diffusion of gases takes place between the alveolar air and the blood. About 5 L of blood is pumped through the lungs each minute. In the capillaries of the lungs this blood is brought into close relationship with the alveolar air, from which it is separated by the extremely thin membranes of the capillaries and the alveolar walls. It is estimated that the capillary surface within the lungs (that is, the surface over which the blood is exposed to the alveolar air) exceeds 1000 ft^2. The differences in pressure of oxygen and carbon dioxide in blood leaving the lungs and blood being brought into them are as follows:

	Arterial Blood	*Mixed Venous Blood* (mm Hg)	*Difference*
Oxygen	100	40	60
Carbon dioxide	40	46	6

As a result of the foregoing pressure differences, oxygen diffuses from the alveoli of the lungs into the blood, and carbon dioxide diffuses from the blood into the alveoli. *In the tissues,* the process is reversed. Oxygen tension in the tissues is low owing to the continuous use of oxygen in the oxidative processes of metabolism; carbon dioxide tension is high owing to the continuous production of that compound in the same process. As a result of these differences in tension, oxygen diffuses from the blood into the tissues, and carbon dioxide diffuses from the tissues into the blood.

Transport of Oxygen by the Blood. Oxygen is carried in the blood in two forms: (1) in solution in the plasma, a negligible amount (only about 3 percent) being carried in this form, and (2) in combination with hemoglobin, as oxyhemoglobin.

When the blood is fully oxygenated, each 100 ml of blood contains approximately 20 ml of oxygen. Under normal conditions, arterial blood is 95 to 97 percent saturated and contains (per 100 ml) 19 ml of oxygen (18.7 ml in the red blood cells in combination with hemoglobin; 0.30 ml in the plasma in solution).

Oxygen combined with hemoglobin forms an unstable chemical compound called *oxyhemoglobin.* Such hemoglobin is said to be "oxygenated" rather than oxidized, for no oxide is formed. When oxyhemoglobin gives up its oxygen, it is referred to as *reduced hemoglobin.*

At rest, the body uses about 250 ml of oxygen per minute. The total oxygen capacity of *all* the blood is roughly 1 L, which amount is consumed in about 4 min when at rest or within 1 min during strenuous activity. There is *no mechanism for storage of oxygen.*

The quantity of oxygen that can be held as oxyhemoglobin in red blood cells depends on the partial pressure of the oxygen being held in solution in the plasma, which in turn depends on the pressure of oxygen in the alveolar air of the lungs or in the tissue fluids or tissue cells. In the lungs the partial pressure of oxygen is approximately 100 mm Hg. At this pressure the blood takes on oxygen and the hemoglobin becomes about 95 percent saturated. In the tissues and the tissue fluids, the oxygen pressures are much lower (30 mm Hg); consequently, in the tissues, oxygen is given up by the hemoglobin.

Factors that favor the dissociation of hemoglobin (i.e., the giving up of oxygen) are low oxygen pressures, relatively high carbon dioxide pressures, a rise in temperature, and an increase in hydrogen ion concentration. The last two reduce the quantity of oxygen the blood can hold at any given pressure. In muscular activity, a local rise in temperature (resulting from increased oxidations) and an increased hydrogen ion concentration (resulting from production of carbon dioxide and lactic acid) favor the greater release of oxygen to the tissues.

Hemoglobin can transport oxygen only when it is contained within the red blood cells. If hemoglobin is released into the circulating fluid, as occurs in hemolysis, it is quickly lost through three processes: excretion by the kidneys, destruction by the cells of the reticuloendothelial system, and conversion into a brown pigment, *methemoglobin.* This last occurs after poisoning by certain substances such as acetanilid and phenacetin.

Transport of Carbon Dioxide by the Blood. The amount of carbon dioxide in arterial blood varies from 44 to 52 percent by volume, the average being 49 ml/100 ml of blood. The carbon dioxide content of venous blood ranges from 50 to 80 percent. The principal processes involved in the transport of carbon dioxide are as follows:

About 5 percent of the CO_2 is dissolved in the plasma. The remainder enters the red blood cells and the following reactions occur:

1. About 90 percent of the CO_2 combines with water to form carbonic acid. The reaction is speeded up by the presence of an enzyme, *carbonic anhydrase,* which is absent from the plasma.

$$CO_2 + H_2O \xrightleftharpoons{\text{carbonic anhydrase}} \underset{\text{carbonic acid}}{H_2CO_3}$$

$$H_2CO_3 \leftrightarrows HCO_3^- + H^+$$

The bicarbonate ions diffuse into the plasma, and the hydrogen ions react with hemoglobin. Hemoglobin thus acts as a buffering agent preventing the accumulation of an excess of carbonic acid, which would alter the pH of the blood.

2. About 5 percent (2 to 10) of the CO_2 reacts directly with hemoglobin to form carbamino compounds according to the following reaction:

$$\underset{\text{hemoglobin}}{HbNH_2} + CO_2 \leftrightarrows \underset{\text{carbaminohemoglobin}}{HbNH-COOH}$$

Only a small part of the carbon dioxide carried by the blood is actually combined with hemoglobin, but through the release of its alkali, the hemoglobin is involved indirectly in the carriage of over 85 percent of the blood carbon dioxide.

Chloride Shift. An increase in the carbon dioxide content of the blood brings about an increase in the bicarbonates, which increase the alkali reserve. For this reason, even though large quantities of carbon dioxide can be taken up by the blood, the blood reaction (pH) will change very little, if at all. As carbon dioxide increases in the blood, the concentration of sodium bicarbonate ($NaHCO_3$) increases. The source of the Na^+ ions is the sodium chloride (NaCl) of the plasma; the bicarbonate ions (HCO_3^-) come from the carbonic acid and carbonates within the red blood cells. The chloride ions (Cl^-), left free from the sodium chloride's giving up its Na^+ ions, shift from the plasma to the cells, where they combine with potassium. This mechanism is known as the *chloride shift.* In the lungs, the reverse process occurs. When carbon dioxide is given off, HCO_3^- ions move into the cells, and Cl^- ions move out into the plasma.

Control of Respiratory Muscles. Respiratory movements are essentially involuntary. This is somewhat anomalous, since the respiratory muscles are principally skeletal and therefore subject to voluntary control. Consequently, although respirations occur automatically, one can at will speed up, slow down, or even stop respiration for a limited time (generally about 45 sec).

Respiratory Center of the Brain. Control of the rate and depth of

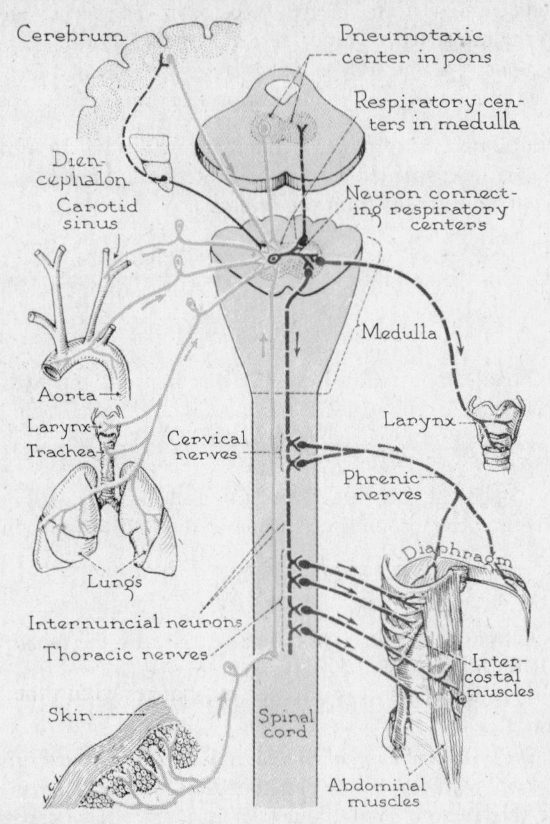

Fig. 2-7. Diagram illustrating nervous control of respiration. Nerves on the left carry impulses to the respiratory centers; those on the right carry impulses from the respiratory centers to the muscles of respiration. (Reprinted with permission of W. B. Saunders Co., Philadelphia, from B. G. King and M. J. Showers, *Human Anatomy and Physiology,* 6th ed., 1969.)

breathing movements resides in a respiratory center consisting of three paired groups of neurons located in the brain stem. This center (Fig. 2-7) consists of three regions: (1) a *medullary rhythmicity center* in the medulla that contains inspiratory and expiratory neurons that control basic respiratory movements, (2) an *apneustic center* in the pons that arrests respiration in inspiration, and (3) a *pneumotaxic center* in the pons that also prevents apneusis. Automatic, involuntary respiratory movements are due to the rhythmic discharge of nerve impulses from these centers. When neurons supplying impulses to inspiratory muscles are active, impulses to expiratory muscles are inhibited, and vice versa.

Inspiration results from the discharge of impulses from the inspiratory center. These impulses pass down the spinal cord to motor neurons in the 3rd, 4th, and 5th cervical segments, whose axons conduct the impulses out of the spinal cord and through the *phrenic nerves* to the diaphragm. Each phrenic nerve, a branch of the cervical plexus, passes downward through the thorax to reach the diaphragm. Efferent impulses from the respiratory center may also descend to the motor neurons of the 3rd to 6th thoracic segments, where they pass through peripheral nerves to the intercostal muscles. Others may pass to other respiratory muscles, such as those in the nostrils, pharynx, and larynx. Contraction of the diaphragm and the accessory respiratory muscles enlarges the thorax and increases the volume of the lungs. Inspiration occurs as a result.

Expiration results from the discharge of impulses from the expiratory center. These are inhibitory impulses that pass to the inspiratory center and bring about the cessation of the outflow of inspiratory impulses. When this occurs, inspiration ceases and expiration follows automatically, since it is passive in nature. In forced expiration, impulses from the expiratory center may be sent to expiratory muscles.

FACTORS INFLUENCING THE RESPIRATORY CENTER. Although respiratory movements are rhythmic and automatic, their rate and degree (that is, whether deep or shallow) can be altered readily. These changes are brought about by certain factors that influence the respiratory center. The automaticity of this center is thought to be due to its capacity to send out impulses as a result of changes occurring within the center itself in the same way that the heartbeat is initiated by changes within the heart tissue itself. The factors referred to are chemical, physical, and nervous.

Chemical Factors. The roles of carbon dioxide, hydrogen ions, and oxygen in the activity of the respiratory center are described in the paragraphs that follow.

Carbon dioxide plays a dominant role in the regulation of respiratory movements. It stimulates the respiratory center, increasing the rate by (1) acting directly on the center and (2) stimulating sensory chemoreceptors in the carotid body and the aortic arch. Any condition that only slightly increases the carbon dioxide tension of alveolar air, which in turn increases carbon dioxide tension in the blood, brings about increased respiratory activity.

Voluntary forced breathing for 3 or 4 min (*hyperventilation*) is followed by a period of apnea, which lasts 40 to 60 sec. This is due to depletion of carbon dioxide, which is demonstrated by the fact that if a subject performs this same experiment but breathes into a paper sack, thereby rebreathing the same air, apnea does not occur. It is also known that forced breathing, if continued for a prolonged period of time, will bring on dizziness and possibly "blackout" (loss of consciousness). All these are the consequences of depletion of the carbon dioxide content of the blood, a condition known as *hypocapnia* or *acapnia.*

Hydrogen ions act in the same way and have the same effect on respiration as carbon dioxide. There is a close relationship between the pH of the blood and the carbon dioxide content: An increase in the latter increases the carbonic acid content.

A decrease in arterial oxygen stimulates breathing to a slight extent, but its role is secondary to that of an increase in carbon dioxide and a rise in blood pH. The effect is brought about through stimulation of chemoreceptors. However, when oxygen lack becomes pronounced, as at high altitudes (above 3000 m), the chemoreceptors of the carotid body and the aortic arch are stimulated, and the respiratory rate is increased reflexly. Oxygen content of the air may be reduced by half without affecting the rate of respiration. Oxygen excess has little effect. Breathing pure oxygen may, however, have a detrimental effect on the body, causing lesions of the lungs. An oxygen concentration of 60 to 70 percent is the maximum that can be breathed for any length of time with safety.

Physical Factors. Physical factors influencing the activity of the respiratory center include temperature, blood pressure, and air pressure within the lungs.

An *increase in body temperature,* as in fever or during muscular activity, increases the respiratory rate; a *reduction in body temperature* is accompanied by a reduced rate of respiration. A *fall in blood pressure* increases the respiratory rate; a *rise in blood pressure* decreases the rate. Blood pressure acts on the pressure receptors of the carotid sinus and the aortic arch, affecting the respiratory center reflexly. *Inflation of the lungs* is a primary factor in the regulation of normal breathing movements. As the lungs become inflated during inspiration, the expansion of the alveoli stretches the lung tissue and stimulates sensory receptors (proprioceptors). This initiates a stream of impulses that pass over afferent nerves (the vagi) to the respiratory center in the brain. These impulses have an inhibitory effect, which decreases the activity of the respiratory center, bringing about cessation of inspiration. Expiration, as already stated, follows automatically. The action just described is known as the *Hering-Breuer reflex.*

Nervous Factors. Nervous control of respiratory movements falls into two categories: voluntary control and reflex control.

Voluntary control is accomplished through impulses discharged from the higher centers of the brain, that is, the cerebral cortex. Normal rhythmic respiratory movements can be altered at will. They may be speeded up, slowed down, or stopped completely for a limited time. Voice control, as in speaking or singing, and straining movements, as in defecation, involve alterations in respiratory movements, and these are principally under voluntary control. Emotions, too, have a marked effect on the rate and depth of respiration. In fright or excitement, respirations are accelerated; under conditions of suspense, apprehension, or close attention, they may become slow and shallow. The sigh of disappointment, the spasmodic breathing of laughter, the gasp of astonishment, and the sobbing that accompanies grief are manifestations of the effects of emotions on respiration.

TABLE 2-1 RESPIRATORY REFLEXES

Receptor	*Stimulus*	*Effect on Respiration*
Proprioceptor endings in:		
Lungs	Stretching of alveoli and bronchioles	Inhibits or ends inspiration
Carotid sinus and aortic arch	Rise of blood pressure	Decreases rate
	Fall of blood pressure	Increases rate
Pharynx	Presence of food or saliva	Stops inspiration and closes glottis
Joints	Bending	Increases rate
Chemoreceptor endings in:		
Carotid body and aortic arch	CO_2 excess, extreme O_2 lack, H ion increase	Increases inspiratory rate
Olfactory epithelium and respiratory passageways	Presence of gases, dust, infectious organisms	Stops inspiration, initiates coughing reflex
Somatic afferent endings in:		
Skin	Cutaneous stimuli (including cold or pain)	May inhibit or stimulate inspiration
Any sense organ	Usual stimuli intensified	May inhibit or stimulate inspiration
Visceral afferent endings in:		
Internal organs	Pain or pressure	May inhibit or stimulate inspiration

Involuntary control is accomplished reflexly. It is responsible not only for normal, quiet breathing but also for the responsiveness of the respiratory mechanism in adjusting to the varying needs of the body. These reflexes are summarized in Table 2-1.

Asphyxia. Because the primary function of respiration is the conversion of deoxygenated blood into oxygenated blood, the mechanisms that control it are very sensitive to changes in the chemical nature of the blood. In any condition that interferes with gaseous exchanges in the lungs (for example, strangulation, drowning, or an obstruction in the respiratory passageways), the oxygen content of the blood is diminished and its carbon dioxide content is increased. When such a condition prevails, the respiratory center is stimulated both directly and indirectly, and respiratory movements become more rapid and forceful. This condition is referred to as *dyspnea* or "labored" breathing. If the cause of the dyspnea is not removed, the respiratory center is stimulated to greater activity. The respiratory muscles contract more vigorously, and the accessory muscles, especially those involved in expiration, become more active. Soon nearly all the muscles of the body are thrown into a state of convulsive activity. With the onset of these convulsions, a state of *asphyxia* sets in. In a short time the muscles become exhausted, inspirations are fewer in number and become progressively weaker, and, finally, a last inspiration is taken, and breathing ends with an expiratory gasp.

Coma or death from asphyxia results from two conditions: *hypoxia* (reduced oxygen content of the blood) and *hypercapnia* (increased carbon dioxide content of the blood). Oxygen starvation and carbonic acid poisoning are both injurious to tissues. Cyanosis resulting from an increased amount of reduced hemoglobin is characteristic of asphyxia.

INTERNAL RESPIRATION

The exchange of oxygen and carbon dioxide between the blood and the body cells and the utilization of oxygen by the cells constitute *internal* or *cellular respiration.* Oxygen combines in the tissues with many substances; the final products are carbon dioxide and water. Through oxidative processes, the major portion of the energy needed for bodily activities is released. The rate of oxidation in any tissue is an index of *vital activity.* Cessation of oxidations results in death of the tissue or the organism. The amount of oxygen used by the tissues depends on the type of tissue and the extent of its activity. Muscles and glands consume the greatest quantities of oxygen. In a resting state, oxygen consumption is low; during activity, it is greatly increased.

Nature of Cellular Oxidation. Fundamentally, oxidations occurring within the body are similar to those occurring outside the body, in that oxygen combines with a substance, which is consumed, and waste products are produced. However, when a substance is burned in air, the rate and extent of combustion are determined by the supply of oxygen and the rate of removal of waste products. In living tissues, the rate of combustion is carefully controlled to meet the needs of the organism. Besides, in the body cells, foodstuffs do not unite directly with oxygen; instead, they are broken down, step by step, to simpler products (metabolites), and intermediate products become progressively richer in oxygen. The final products are carbon dioxide (CO_2) and water (H_2O). In this long series of chemical reactions, all of which take place at a relatively low temperature (about 37°C), catalysts (enzymes and coenzymes) are essential, each step requiring a specific catalyst.

The following processes are regarded as oxidations: (1) addition of oxygen, (2) withdrawal of hydrogen (dehydrogenation), and (3) loss of electrons. The reverses of these processes are regarded as *reductions.* All three types of oxidations are alike in that there is loss of one or more electrons from the substance oxidized.

The oxidation of glucose, a primary source of energy for the body, involves a long series of reactions in which 12 oxygen atoms are utilized. When glucose, the *donor,* is oxidized, it loses electrons. These electrons are transferred to *acceptors,* which in turn transfer the electrons to other acceptors. Each reaction is catalyzed by an enzyme. Finally, the electrons are transferred to oxygen, which is reduced to water. In the step-by-step transfer of electrons in this *electron-transport chain,* energy lost in oxidation is used to convert inorganic phosphate into a high-energy phosphate, the most important of which is *adenosine triphosphate* (*ATP*). ATP is the primary source of energy for most physiological processes, especially secretion and muscle contraction. All the processes involved in the release of energy take place within the mitochondria of cells.

ENZYME SYSTEMS. The enzymes that catalyze oxidations of foodstuffs, all of which are intracellular enzymes, fall into two groups: dehydrogenases and oxidases. *Dehydrogenases* are enzymes that act to withdraw hydrogen from food molecules; they catalyze reactions in which hydrogen is given up. *Oxidases* are enzymes that serve as oxygen activators; they catalyze reactions in which hydrogen is oxidized to water by molecular oxygen.

Some of these enzymes consist of a protein component and a nonprotein component, both of which are essential to the reaction being catalyzed. The protein portion is usually referred to as the "enzyme"; the nonprotein portion is called the "coenzyme." When both these components are working together, they constitute an "enzyme system."

CYTOCHROME SYSTEM. In cellular oxidations, an important enzyme system is the *cytochrome system,* which consists of a cytochrome oxidase and three pigments: cytochromes *a, b,* and *c.* These red iron-containing compounds, present in cells, serve to make oxygen available for the oxidation of hydrogen liberated from cellular metabolites by the dehydrogenases.

VITAMINS. An important group of enzymes are the *flavoproteins,* which are combinations of *riboflavin* with proteins. Flavins are yellowish pigments found in practically all cells. Flavoproteins are components in the electron-transport system and play important roles in oxidation-reduction reactions. In addition to riboflavin, other vitamins play significant roles in the oxidation processes. These include *thiamin* (B_1), *niacin* (*nicotinic acid*), and *pantothenic acid.*

Inhibition of Cellular Respiration. A number of substances have been found that inhibit the oxidation processes within cells. Rotenone, barbiturates, and cyanide affect the electron-transport system, preventing the utilization of oxygen. *Cyanide* blocks the action of cytochrome oxidase, an essential enzyme, by binding to the metal ion. This accounts for its extreme deadliness. Other inhibitors act in certain uncoupling reactions.

PHYSIOLOGY OF VOICE PRODUCTION

Language is the mechanism by which humans communicate ideas and express feelings; *speech* is a highly developed form of language in which thoughts and feelings are expressed by articulate sounds. *Voice* is the purposive production of sound by means of the respiratory organs. Paradoxically, the voice may consist of "voiceless" sounds (such as whispers, produced merely by the movement of a column of air through the air passageways) or "voiced" sounds produced by the vibrations of the vocal folds.

The human voice has certain fundamental properties that are characteristic of all sounds, namely, intensity or loudness, timbre or quality, and pitch (highness or lowness). *Intensity* depends on the amplitude of the vibrations, which in turn depend on the force of the column of air that moves past the vocal cords. *Timbre* depends on the number and intensity of the overtones or harmonics, which are determined principally by the shape and size of the resonating chambers. *Pitch* depends primarily on the length, tension, and thickness of the vocal cords; these factors affect the frequency of the vibrations of the cords. Since women and children have shorter vocal cords than men, their voices are more highly pitched. Although innate structure largely predetermines the nature of the foregoing properties for individuals, all are subject to variation through voluntary control of the muscles of the respiratory organs.

Mechanics of Voice Production. The human voice mechanism is similar in many respects to most tone-producing musical instruments. It consists of a *vibrator* or source of sound, a *motor* or force for setting the vibrator in motion, and a *resonator* or amplifier, which reinforces certain vibrations. In the human voice mechanism, the vocal folds (true vocal cords), located in the larynx, serve as the vibrator. The force that sets them to vibrating is the breath exhaled from the lungs, which strikes the tense vocal cords as it makes its exit from the chest cavity. The resonator consists of a number of parts (the pharynx, mouth, and nasal cavities) that function in the amplification of the tone initiated by the vocal folds.

The human voice mechanism differs from most tone-producing instruments in that it has, in addition, a series of *articulators* that modify the tone into specific speech sounds; these articulators are the lips, teeth, tongue, and hard and soft palates and the parts that form the walls of the resonating cavities.

The production of a speech sound is accomplished in the following manner: After inhalation, the vocal folds of the larynx are approximated and made tense by the action of the intrinsic muscles of the larynx. Upon exhalation, the air under pressure passes through the narrow glottis. When it strikes the tense vocal folds, these are set into motion, giving rise to a sound wave and the production of a more or less undifferentiated vocal tone. The sound wave set in motion by the vocal cords passes upward and enters the resonating chambers (pharynx, mouth and nasal cavities, sinuses), where it is amplified and, finally, by the action of the walls of these chambers and by the articulators, is modified and refined in quality. The amplified and modified sound wave then assumes the characteristics of a specific speech sound.

The resonating cavities, along with the articulating parts, serve (1) to reinforce the sound that was initiated by the vocal cords, (2) to give quality to it by the selection of harmonic overtones, and (3) to produce the differentiated vowels and consonants. Some of the parts are *static* or fixed; others, such as the soft palate and uvula, tongue, lips, teeth, and cheeks, are *mobile.* Malformations of any of these structures may cause marked modifications in the quality of the voice, as is indicated in speech defects caused by harelip, cleft palate, absence of teeth, or the presence of adenoids.

In the production of vowels, the resonating tube (especially the lips) assumes a specific form for each vowel sound. Consonants are produced by interrupting the passage of the air column at some point in the expiratory pathway; they are called "labial," "dental," "palatal," "stopped," "nasal," "open," or "guttural," depending on the site and manner of their formation.

Nervous Mechanism of Speech. The development of speech depends on the association of sounds (words) with sensations (visual, tactile, etc.) aroused by objects in the external environment. Impulses arising from these stimuli pass to the association regions of the brain, where they are "stored" as memories. For example, the word *cat,* when heard, brings to mind a mental image or picture of that animal. Talking or *verbal expression of an idea* consists essentially of (1) coordination of sensory impulses in the association centers of the cerebrum and (2) transmission of impulses from these centers to the muscles of respiration and the muscles of the larynx and other structures involved in the production of speech. Learning to read consists of associating the visual symbols of speech (words) with auditory symbols (spoken words). Learning to write consists of expressing auditory and visual impressions by means of the coordinating action of the digital muscles with cerebral impulses. Association centers for the understanding of spoken or written symbols have not been localized; they seem to be rather widely distributed over the cortex of the left cerebral hemisphere. In left-handed persons, these centers are in the right hemisphere.

Defects of Speech. Speech is considered to be defective when it deviates from the normal to the extent that it is noticeable and interferes with ordinary communication. Defective speech often leads to severe personality maladjustment. Speech defects fall into four general classes:

DEFECTS OF RHYTHM. Stuttering, stammering, and cluttering are defects of rhythm in speech that are characterized by repetitions of words or phrases or the first sounds or syllables of words. The speech becomes hesitant, and stoppages occur. Such defects may be accompanied by facial contortions and spasms and by abnormal respiratory movements.

DEFECTS OF ARTICULATION. These are characterized by distortions of speech sounds. Included are "baby talk," lalling (pronouncing *r* so that it sounds like *l*), lisping (*th* for *s* or *z, w* for *l*), and delayed speech.

DEFECTS OF PHONATION. These include aphonia (lack of voice), pitch disorders (monotone, too high pitch, or too low pitch), and disorders in voice quality (hypo- or hypernasality).

DEFECTS OF SYMBOLIC FORMULATION AND EXPRESSION (DYSPHASIA). Rather than being defects of the organs that produce speech, these defects arise from disorders in the higher brain centers. *Aphasia* is the inability to use words as symbols of ideas. Aphasia may be motor, sensory, or both. In *motor aphasia* (*Broca's aphasia*), the subject, although unable to speak or write normally, is still capable of understanding what is said and is able to read; there is no paralysis of the muscles of articulation. In *sensory* or *receptive aphasia,* the subject, though able to hear and see, is unable to understand spoken words (*word deafness*) or written words (*word blindness*) but retains the ability to speak and to write. In *total aphasia,* disturbances in both the sensory and the motor spheres preclude all these activities. *Anarthria* (loss of speech) and *dysarthria* (difficult speech) are terms applied to disorders of speech due to paralysis of the muscles involved in articulation. These are usually the result of lesions of the brain or the muscles.

PRACTICAL CONSIDERATIONS

Respiratory Emergencies: Artificial Respiration. A respiratory emergency is a condition in which breathing stops or is reduced to the extent that life is threatened. The objective of all methods of artificial respiration is to keep air moving into and out of the lungs until the respiratory center recovers and becomes capable of initiating respiratory movements. If the oxygen supply to the tissues is cut off for a period of more than 4 to 6 min, irreparable damage is done to the central nervous system, with little hope of recovery. But if the heart is still beating and the blood circulating, resuscitation is possible.

Causes of respiratory failure include the following:

1. *Anatomic obstruction,* as when the tongue drops back into the pharynx, or when air passageways are constricted, as in acute asthma, croup, or diphtheria.
2. *Mechanical obstruction,* as by a foreign object within the air passages; the presence of a large food mass within the esophagus, which lies behind the trachea; or the accumulation of substances, such as mucus, blood, or vomitus, in the pharynx or larynx.
3. *Failure of the respiratory center* as may occur in drowning, severe electric shock, circulatory collapse (shock), or drug overdose.

If respiratory failure is due to an obstruction, such should be removed if possible. Often a person chokes when a mass of food becomes lodged in the pharynx or esophagus. As this frequently happens in restaurants and is generally misdiagnosed as a heart attack, it is called a "café coronary." An effective treatment if applied *immediately* is the *Heimlich technique* or *maneuver,* which is performed with the victim standing or sitting. The procedure is as follows: From the rear, place your arms around the victim and grasp the clenched fist of one hand with the other. With locked hands placed below the rib cage and slightly above the navel, exert a quick, firm thrust inward. Repeat several times if necessary. The resulting increase in air pressure within the thoracic cavity will usually bring about the forceful expulsion of the object. If the victim is in a supine position, pressure exerted by the palm of the hand on the abdomen below the rib cage will usually bring about the same result.

METHODS EMPLOYED IN ARTIFICIAL RESPIRATION. These include mouth-to-mouth or mouth-to-nose resuscitation, manual methods, and the use of mechanical devices.

Mouth-to-Mouth and Mouth-to-Nose Resuscitation. This method is superior to any of the manual methods formerly used extensively. The operator first clears the victim's mouth of any foreign matter and sees that the air passageways are clear. Then, with one hand under the victim's neck and the victim's head tilted backward with the chin pointing upward, the operator closes the nostrils between the thumb and index finger, takes a deep breath, and then, applying the mouth closely to the victim's mouth, blows forcibly. If the airway is clear, only moderate resistance will be felt, and the victim's chest will rise. On removal of the mouth, exhalation will occur. This action is repeated at a rate of about 12/min (one breath every 5 sec). For children, the rate should be 20/min. In mouth-to-nose resuscitation, the victim's mouth is held shut and the air is blown into the nostrils.

For aesthetic reasons, certain adjunctive devices have been developed by which direct mouth-to-mouth contact can be avoided. These include a face mask, an oral airway or breathing tube, and a self-inflating bag with mask. These devices require training and experience in their use, and they are not always available when needed. None is as effective as direct mouth-to-mouth contact.

Manual Methods. If it is impossible to employ mouth-to-mouth or mouth-to-nose resuscitation procedures, as in the case of severe facial injuries, a manual method may be resorted to. Two methods that are employed are the following:

1. *Arm lift and back pressure (Holger-Nielsen method).* The operator kneels at the head of the prone subject (lying face down) and alternately raises the patient's arms and applies pressure to the upper part of the thorax.

2. *Arm lift and chest pressure (Sylvester method).* The operator kneels at the head of the supine (face-upward) subject, whose arms are then alternately extended above the head and folded across the chest with pressure applied.

Mechanical Devices. If artificial respiration is required for long periods of time, perhaps days, weeks, or months, mechanical devices must be employed. These are utilized in cases where the respiratory center fails to function, in spinal cord injuries, or in the paralysis of respiratory muscles as in poliomyelitis. An apparatus often used is the *Drinker respirator,* commonly called the *iron lung.* The patient's body, except for the head, is placed in a hermetically sealed cabinet, and air pressure within the cabinet is alternately increased and decreased by a motor-driven pump.

Other mechanical devices, variously called *pulmotors, pulsators, respirators,* and *barospirators,* have been developed to enable persons with seriously impaired breathing problems to survive.

Cardiopulmonary Resuscitation (CPR). This is a combination of artificial respiration and artificial circulation, emergency measures that should be employed when respiratory failure occurs and symptoms of cardiac arrest are apparent. These include sudden unconsciousness, absence of pulse, apnea, and absence of heart sounds.

Cardiac arrest includes three conditions in which blood flow is either absent or inadequate to sustain vital activities. These conditions are (1) *cardiovascular collapse,* in which the heart may still be beating but its beat is so weak that the flow of blood to vital tissues is inadequate; (2) *ventricular fibrillation,* in which the muscles of the heart chambers do not beat in a synchronized fashion; and (3) *cardiac standstill,* in which the heart has stopped beating.

In emergency situations, especially those involving heart attacks, the application of artificial respiration and artificial circulation may restore respiration and heartbeat to the extent that the patient can be transported to a medical center where more advanced life-support measures can be employed.

Artificial circulation involves the application of *external cardiac compression* to a patient in a horizontal position. This technique, also called *external cardiac massage,* involves the placing of both hands, one over the other, on the lower portion of a patient's sternum. With the heel of the underneath hand on the sternum slightly above the xiphoid process, the operator brings his or her shoulders directly over the victim's sternum and applies a downward pressure followed immediately by relaxation. Maintaining the hands in position, the procedure is repeated at the rate of about 60/min. Checking the pupils of a patient's eyes will indicate whether blood is reaching the brain. Constriction of the pupils upon exposure to light indicates restored circulation.

Precordial Chest Thump. In some cases of cardiac arrest, a single blow to the chest administered by a clenched fist directly over the heart (*precordial chest thump*) may be effective in restarting the heartbeat. It must be administered immediately after stoppage, while the heart muscle is still oxygenated.

Modified Forms of Respiration. Forms of inspiration or expiration that are spasmodic, exaggerated, voiced, or otherwise radically different from normal breathing are essentially modified forms of respiration. Examples are coughing, sneezing, hiccoughing, snoring, yawning, sighing, crying, sobbing, and laughing.

COUGHING. A cough is a violent expiration preceded by an inspiration of greater than normal depth. The muscles of expiration, especially the abdominal muscles, contract suddenly, and the closed glottis increases the intrathoracic pressure. The partial opening of the vocal cords allows air to rush out with hurricane force, and the resultant action on the vocal cords produces a sharp sound. Coughing serves a protective function in that it provides a means for dislodging foreign particles from the respiratory passageways and ejecting them from the body. Coughing is commonly initiated by any irritation of the respiratory mucosa, such as occurs in respiratory infections. However, stimulation of other parts of the body or psychic factors may also induce a cough.

SNEEZING. A sneeze (*sternutation*) is a sudden violent expiration, with the air being discharged through the oral and nasal cavities. It is a reflex action usually resulting from irritation of the nasal mucosa, but it may be induced by other stimuli, such as suddenly directing a bright light into the eye.

HICCOUGHING. A hiccough (hiccup) results from a sudden, involuntary contraction of the diaphragm, the inspiration being cut off by a quick closure of the glottis. It is reflex in nature, initiated by stimulation of afferent nerve endings in almost any part of the body. Disorders of the stomach and esophagus or abdominal disturbances are frequent causes. It is common in infants, probably due to distention of the stomach. Hiccoughing may also result from lesions or inflammatory conditions involving the respiratory center or its afferent or efferent pathways or from psychogenic factors.

SNORING. The production of coarse breathing noises during sleep or while in coma is due to the vibration of the soft palate or the uvula.

YAWNING. A yawn is a long, deep inspiration with the mouth open wide, followed by a slow expiration. It is induced by weariness, drowsiness, fatigue, or boredom.

SIGHING, CRYING, AND SOBBING. These forms of respiration are automatic, rhythmic reactions to physical or psychic stimuli, such as disappointment, sorrow, grief, and pain.

LAUGHING. Laughter is strictly a human phenomenon, characterized by spasmodic and involuntary expirations and inarticulate sounds indicative of amusement, mirth, joy, or merriment. Laughter can be reflexly produced by tickling; sometimes it is a manifestation of hysteria. Its causes are numerous, its physiologic functions obscure. In general, it is a social phenomenon involving other people.

Principles of Room Ventilation. Prolonged breathing of air in a poorly ventilated room eventually brings on discomfort; a feeling of stuffiness and oppression develops, vitality is reduced, and headache may ensue. Contrary to popular belief, these symptoms are not due to lessened oxygen supply or increased carbon dioxide consumption. The former may be reduced to 14 percent (normal amount, 21 percent) and the latter increased to 6 percent (normal amount, 0.04 percent) before noticeable effects develop. Even in the most poorly ventilated room, changes of this kind do not occur. Investigations have disclosed that the ill effects of poor ventilation are due principally to changes in the temperature and moisture content of the air, with a consequent loss of cooling power. Proper ventilation of a room requires that the air (1) be clean, that is, relatively free of dust, smoke, and microorganisms; (2) possess the proper amount of moisture; (3) have the proper temperature; and (4) be kept in constant circulation.

Formerly it was thought that in crowded rooms, poisonous substances discharged from the lungs and skin exerted certain toxic effects. Although occasionally volatile substances, such as acetone, may be present in the breath, there is no evidence that these or other gaseous substances discharged by expiration have any harmful effects on other persons who breathe the same air. Infectious organisms may, however, be present on the minute droplets of moisture that are discharged by coughing or sneezing. It is believed that some communicable diseases are transmitted in this way.

Effects of Increased Air Pressure. Persons who work under abnormally high air pressure (divers, workers in diving bells or caissons) sometimes experience a condition known as *caisson disease* or *decompression sickness.* Under high air pressure, the blood and tissues absorb excessive quantities of the gases in the air (nitrogen, oxygen, carbon dioxide). This does not prove harmful if the internal and external pressures remain constant. But if the individual passes too rapidly from a region of high pressure to one of low pressure, serious effects ensue. The excess nitrogen in the blood is released in the form of small bubbles that accumulate in the tissues and the blood vessels, giving rise to such symptoms as nausea, dizziness, muscular pains, abdominal cramps, and pains in the joints. The last of these symptoms has given the condition its popular name—the "bends." The symptoms, which in some cases may be fatal, can be avoided or relieved by passing the patient through a series of "decompression chambers" by means of which the outside pressure is reduced gradually. This permits blood gases to come slowly to an equilibrium with the atmospheric gases.

Hypoxia (Anoxia). Any condition in which the tissues fail to receive an adequate supply of oxygen leads to *oxygen want* or *hypoxia.* This may be *local,* as occurs when an organ or tissue is deprived of its blood supply, or it may be *general,* as occurs when the oxygen in the blood is reduced *(hypoxemia).*

The four types of hypoxia and the primary causative factors involved are listed in the following table:

Type	*Causative Factor*
Hypoxic hypoxia	Low oxygen tension in inspired air resulting in low oxygen tension in the blood
Stagnant hypoxia	Slow circulation of the blood, either local or general
Anemic hypoxia	Oxygen-carrying capacity of the blood is below normal
Histotoxic hypoxia	Inability of tissue cells to utilize oxygen due to impairment in the oxidative-enzyme mechanism

HYPOXIC HYPOXIA. This is the condition in which the blood, in passing through the lungs, fails to acquire its normal oxygen saturation (95 percent). It may be due to low oxygen tension of the atmospheric air, as in high altitudes; vitiated air, as in mines; interference with the air supply to the lungs, as from obstruction of the respiratory passageways by a foreign object; or reduced alveolar surface, as in pneumonia or emphysema. Hypoxic hypoxia that occurs at high altitudes is called *mountain sickness.* The first symptoms of this begin to appear at about 3000 m, at which point oxygen saturation of hemoglobin is about 85 percent. Dyspnea, nausea, vomiting, and cyanosis develop. There may also be emotional disturbances; instances of faulty judgment are common. At extremely high altitudes, consciousness may be lost. The symptoms disappear on return to lower altitudes. The body is capable, however, of adjusting in time to the rarefied atmosphere of high altitudes (up to 6000 m). Constant exposure results in an increased chest volume and an increased number of red blood cells, which may reach 6 to 8 million/mm^3.

STAGNANT HYPOXIA. This condition results from a reduced flow of blood through the tissues. It may be *general,* as in congestive heart failure, shock, or impaired venous return, or *local* following interference in blood flow, as from an embolus, a thrombus, or the application of a tourniquet.

ANEMIC HYPOXIA. When the oxygen-carrying capacity of the blood is reduced, anemic hypoxia results. It may be caused by a reduction in the number of red cells or a reduced amount of hemoglobin (as occurs following a severe hemorrhage or in anemia) or by the inability of hemoglobin to carry oxygen. The last of these occurs in carbon monoxide poisoning. Hemoglobin has 200 times the affinity for carbon monoxide as for oxygen; consequently, small quantities of carbon monoxide in the atmosphere (as little as 0.1 percent), when breathed for some time, may cause death. When this gas combines with hemoglobin, a relatively stable compound, *carboxyhemoglobin* (COHb), is formed; this prevents hemoglobin from uniting with oxygen. Death results from oxygen deprivation.

A number of other substances, such as nitrites, nitrates, acetanilid, chlorates, and sulfonamides, may combine with hemoglobin and reduce its oxygen-carrying capacity.

HISTOTOXIC HYPOXIA. In this condition, the oxidation processes within tissue cells are depressed or abolished due to impairment of the oxidative-enzyme systems. It occurs in cyanide poisoning, in which cyanide combines with cytochrome oxidase, blocking its action and thus inhibiting cellular oxidation.

Oxygen Therapy. In the treatment of certain types of hypoxia, administration of oxygen has proved to be of considerable value. It is employed in cases of pneumonia or pulmonary edema, when the alveolar surfaces for diffusion of oxygen are much reduced. It is also used in the treatment of pulmonary tuberculosis, certain types of asphyxia, poisoning by narcotics, and congestive heart failure. It is not effective in the treatment of anemia. The oxygen may be administered through an open cone or a nasal catheter or in an *oxygen chamber* or *oxygen tent.* The concentration of oxygen in the oxygen chamber or tent is maintained at between 40 and 60 percent. Sometimes carbon dioxide in a concentration of 5 to 10 percent is administered along with the oxygen to stimulate the respiratory center.

DISEASES AND DISORDERS OF THE RESPIRATORY SYSTEM

Adenoidal Hypertrophy. Enlargement of the nasopharyngeal tonsils, located in the upper portion of the pharynx immediately posterior to the nasal cavity. The condition is common in children and frequently congenital. Adenoids may obstruct the posterior nares and cause mouth breathing, nasal speech, and sometimes altered facial expression (*adenoid facies*). They may also obstruct the auditory tubes, with resulting middle-ear disorders.

Asthma. Paroxysmal attacks of dyspnea accompanied by abnormal breathing sounds. This condition is due to a narrowing of the lumens in the bronchial tubes resulting from a spasm in the muscles in their walls, usually accompanied by edema and the excessive secretion of mucus. The reactions are an allergic response to allergenic agents, which may include pollens, mold spores, animal dander, and other substances present in the environment, or to bacterial proteins resulting from bacterial infections of the respiratory tract. Asthma is treated with such drugs as epinephrine, ephedrine, and atropine, which act through the autonomic nervous system to relax the smooth muscles of the bronchioles. In some cases asthma may have a psychogenic origin.

Atelectasis. An airless condition of the lung or a part of it in which the lung is shrunken and collapsed, usually due to bronchial obstruction. In fetal atelectasis, the lungs fail to expand at birth.

Bronchiectasis. Chronic dilatation of bronchi, usually accompanied by accumulation of pus, giving rise to paroxysmal coughing and expectoration of mucopurulent matter.

Bronchitis. Inflammation of the bronchial tree caused by infectious organisms or by physical or chemical agents.

Consolidation. Accumulation of matter in the air spaces of the lungs.

Croup. A condition frequently seen in children, characterized by dyspnea, laryngeal spasm, and sometimes formation of a false membrane that often blocks the air passageways. Also called *laryngotracheobronchitis.* It usually results from various viral infections.

Diphtheria. An acute infectious disease, a toxemia caused by the Klebs-Loeffler bacillus. It is characterized by the formation of a false membrane on the mucous surfaces of the throat and by extreme prostration. The latter symptom is due to the effects of a toxin secreted by the bacillus.

Dysbarism. Decompression illness and related disorders.

Emphysema. Enlargement and overdistension of the alveoli by gas or air, with destructive changes in the alveolar walls.

Empyema. Presence of pus in the body cavity, especially the pleural cavity.

Halitosis. Offensive breath. Foul odors may result from oral or respiratory infections or bacterial decomposition taking place in the mouth or the respiratory passageways, from the ingestion of certain odoriferous foods, or from the use of alcohol, tobacco, or certain drugs. *True* or *essential halitosis* is caused by faulty metabolism in which noxious volatile substances enter the circulating blood and are excreted through the lungs, as in liver failure, uremia, or diabetes.

Hay Fever. An allergic disease affecting the mucous membranes of the upper respiratory passageways and the conjunctiva of the eye. It may be caused by inhalation of foreign substances, such as pollen or dust, and is characterized by sneezing, rhinorrhea, and nasal congestion.

Hemoptypsis. The expectoration of blood or blood-stained mucus, especially from respiratory organs.

Hyperventilation. A condition in which there is an increased amount of air entering the lungs brought on by prolonged rapid and deep breathing. It reduces carbon dioxide tension in the blood and may lead to alkalosis.

Lung Cancer. Bronchiogenic carcinoma (lung cancer), which includes the majority of malignant tumors of the lower respiratory tract, causes more deaths in males than any other form of cancer. Statistical evidence that cigarette smoking is the primary causative factor has led the Surgeon General's Office to require the following warning on all packages of cigarettes and in all cigarette advertising: "Warning: The Surgeon General Has Determined That Cigarette Smoking Is Dangerous To Your Health."

While there is strong evidence that the increase in lung cancer is the result of increased cigarette smoking, other factors may be involved, such as *urban air pollution.* A number of carcinogenic agents have been found in smoke discharged from industrial plants and in exhaust fumes from automobile engines. Lung cancer is also an occupational disease of workers who handle radioactive substances and those in industries using asbestos and certain other metals.

Pleurisy. Inflammation of the pleura, the membrane that lines the thoracic cavity and covers the lungs.

Pneumonia. Inflammation of the lungs characterized by consolidation. It is usually an acute infection involving the bronchi, bronchioles, and alveoli. There are many causative agents including viruses, mycoplasmas, bacteria, and fungi. Noninfectious pneumonia, as that resulting from inhalation of oil droplets, may occur.

Respiratory Distress Syndrome (RDS; Hyaline Membrane Disease). A disorder affecting primarily premature infants, characterized by the formation of a hyalinelike membrane within the alveoli and collapse of the alveoli (*atelectasis*). It is due to failure of the surfactant system to function.

Rhinorrhea. The discharge of thin, watery matter from the nose.

Tonsillitis. Inflammation of the tonsils, also referred to as *quinsy.* Peritonsillar tissue may become involved, with resultant abscess formation.

Tuberculosis. An infectious disease caused by an acid-fast bacterium, *Mycobacterium tuberculosis.* It is characterized by the formation of tubercles in the tissues. Any organ may be affected: lungs, bones and joints, kidneys and bladder, or lymph glands. It may be localized or general. *Pulmonary tuberculosis* is the most common form. The *primary* or first-infection type (childhood tuberculosis) results in enlargement and calcification of lymph nodes in the region of the hilus of the lung. Constitutional symptoms are few and vague. The disease is considered to be noninfectious in this stage. The *secondary* or reinfection type (adult tuberculosis) involves the lungs, usually the apices, and results in the formation of cavities. This is the so-called *active tuberculosis,* the infective organisms of which can be readily transmitted. Chemotherapy is a very effective method of treatment in most cases.

Whooping Cough (Pertussis). An infectious disease characterized by recurrent attacks of coughing that end in a "whooping" respiration. It is caused by the Bordet-Gengou bacillus. Whooping cough is frequently fatal to very young infants.

SPECIAL TERMS

Cyanosis. Bluish coloring of the mucous membranes and skin owing to the presence in the capillaries of excessive amounts of reduced hemoglobin (hemoglobin from which oxygen has been removed). It is associated with the hypoxic and stagnant types of hypoxia previously described. Cyanosis may be due to (1) incomplete oxygenation of blood in the lungs (*arterial* or *central cyanosis*), (2) slow circulation through the capillaries (*peripheral cyanosis*), or (3) alterations in the hemoglobin, such as the production of methemoglobin or sulfhemoglobin resulting from intoxication by sulfanilamide, aniline, and other substances.

Pneumothorax. The accumulation of air or gas within the pleural cavity. This condition may be induced deliberately (*artificial pneumothorax*) by puncture of the chest wall (*thoracocentesis*).

Tonsillectomy. Removal of the palatine or pharyngeal tonsils (adenoids). The radical method is by surgery; conservative method is electrocoagulation by a high-frequency electric current (diathermy).

Tracheotomy. Surgical procedure in which an incision is made into the trachea and a metal tube is inserted into the opening to serve as an air passageway. It is resorted to in cases of an obstruction of the air passageway (as in cancer of the larynx or in diphtheria) or the presence of a foreign body. The opening made into the trachea is called a *stoma.*

3: THE NERVOUS SYSTEM

The regulation and control of bodily activities and their integration resides in two systems, the *nervous system* and its associated sense organs, and the *endocrine system.* Together these two systems coordinate the activities of the internal organs and maintain a state of homeostasis in the body fluids.

The *nervous system* functions through a complicated system of cells by which information is collected and analyzed and appropriate responses are brought about. Through the *sense organs,* information is obtained about the external environment that enables the body to adjust to changes in temperature, moisture, and chemical and physical conditions that affect all living things. Through *receptors* within the body, information is collected on the level of carbon dioxide in the blood, the heart rate, changes in blood pressure, and the presence of food within the digestive tract and fluid within the bladder. This information is transmitted in the form of *nerve impulses* along nerve fibers to the central nervous system, where it is integrated and where nerve impulses are discharged to effector organs (muscles or glands), which respond. Thus organs of the body perform the activities essential to life (digestion, respiration, circulation, excretion) in a coordinated manner. The nervous system is essentially a *stimulus-response mechanism* through which quick accommodations to changing environmental conditions can be accomplished.

The *endocrine system* functions through the action of chemical substances that act locally or are transported through the bloodstream. These substances, which include *hormones, neurohormones,* and *neurohumors* (*neurotransmitters*), act on the cells of specific organs and tissues, controlling their activities through regulation of the rate of specific biochemical reactions occurring within the cells. These substances regulate growth, development, and reproductive processes and have profound effects on physical and mental development.

The two systems are closely interrelated and are sometimes regarded as a single *neurohumoral system.* All nerve cells secrete at the terminations of their axons substances that act at synapses with other neurons or at their connections with effector organs. Some neurons, especially those of the hypothalamus, secrete hormones that act on distant organs, and some reflex responses involve an afferent neural and an efferent endocrine pathway. However, because of anatomical and functional differences, the two systems will be considered separately.

ORGANIZATION OF THE NERVOUS SYSTEM

The nervous system consists of the *brain, spinal cord, ganglia,* and *nerves.* The structural unit is a specialized type of cell, a *neuron.* The functional unit is a group of two or more neurons, constituting a *reflex arc.* This system has two divisions, the central and the peripheral. The *central nervous system,* made up of the brain and the spinal cord, is contained within the cranial cavity of the skull and the vertebral canal of the spinal column. The *peripheral nervous system* includes all nervous structures (ganglia and nerves) that lie outside the cranial cavity and the vertebral canal. The following is a schema of the nervous sytem:

- *Central Nervous System*
  - Brain (Encephalon)
    - Cerebral hemispheres
    - Cerebellum
    - Brain stem
      - Diencephalon
      - Midbrain
      - Pons
      - Medulla oblongata
  - Spinal cord
- *Peripheral Nervous System*
  - Craniospinal nerves
    - Cranial nerves (12 pairs)
    - Spinal nerves (31 pairs)
  - Sympathetic division of the *autonomic nervous system**
    - Sympathetic trunks (2)
    - Sympathetic ganglia
    - Nerves

* The term *autonomic nervous system* is a functional rather than an anatomic one. This system includes all structures (ganglia and nerves) that innervate the involuntary organs (smooth muscles, cardiac muscles, glands).

FUNCTIONS OF THE NERVOUS SYSTEM

The primary function of the nervous system is *regulation, integration,* and *coordination of body activities.* This system is the seat of consciousness, memory, and intelligence. It provides the basis for such higher mental processes as reasoning, thinking, and judgment. The central nervous system is also the center of emotional responses.

All activities of the body are adjustments of the organism to a changing environment. Changes are constantly occurring outside the body in the *external environment* and within the body in the *internal environment.* Nerve cells (receptors) respond to these changes (stimuli), and a wave of irritability sweeps over the nerve cell. This constitutes a *nerve impulse.* Because all parts of the organism are connected through the nervous system, an impulse may be conducted to any or all parts of the body. Upon reaching such organs as muscles or glands, the impulse may induce activity (contraction of the muscle or secretion by the gland). In some cases, inhibition of activity results. In either case, an effect is produced; hence these organs are called *effectors* or *effector organs.* Through the responses of effectors, an organism adjusts to a changing environment.

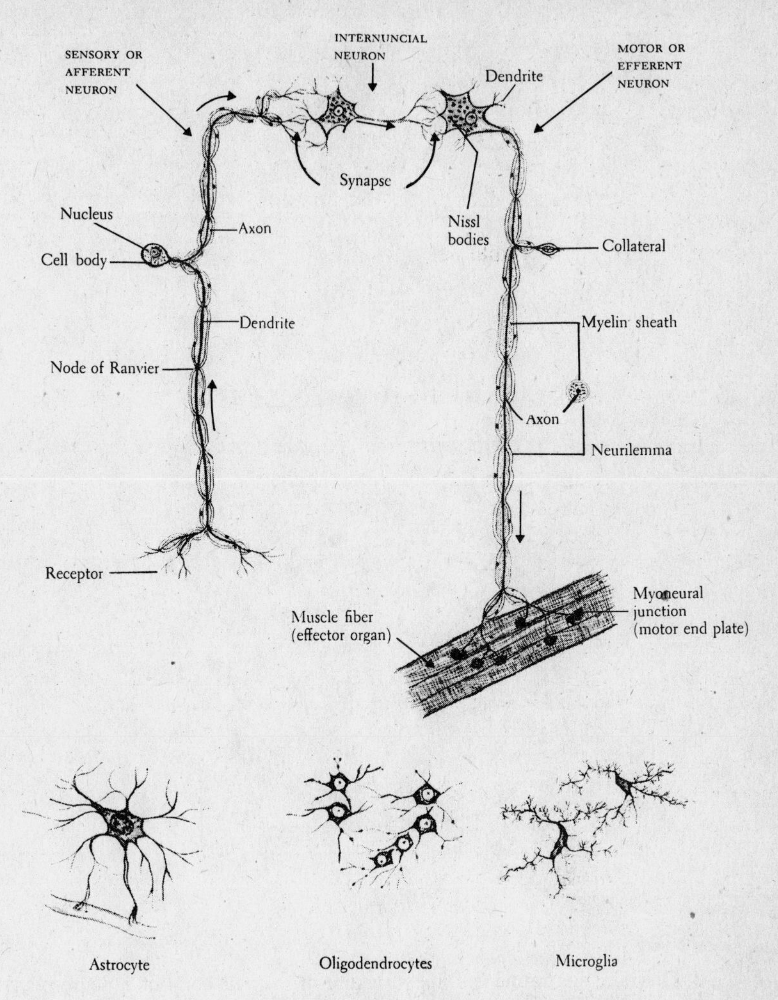

Fig. 3-1. Types of neurons and neuroglia cells.

HISTOLOGICAL STRUCTURE OF THE NERVOUS SYSTEM

Nervous tissue consists of nerve cells or *neurons* and supporting cells or *neuroglia* (Fig. 3-1).

Neurons. A neuron consists of a central portion or *cell body* (*perikaryon*), which bears one or more cytoplasmic projections called *cell processes* (axons and dendrites).

CELL BODY. The cell body (Fig. 3-2) consists of a nucleus and cytoplasm.

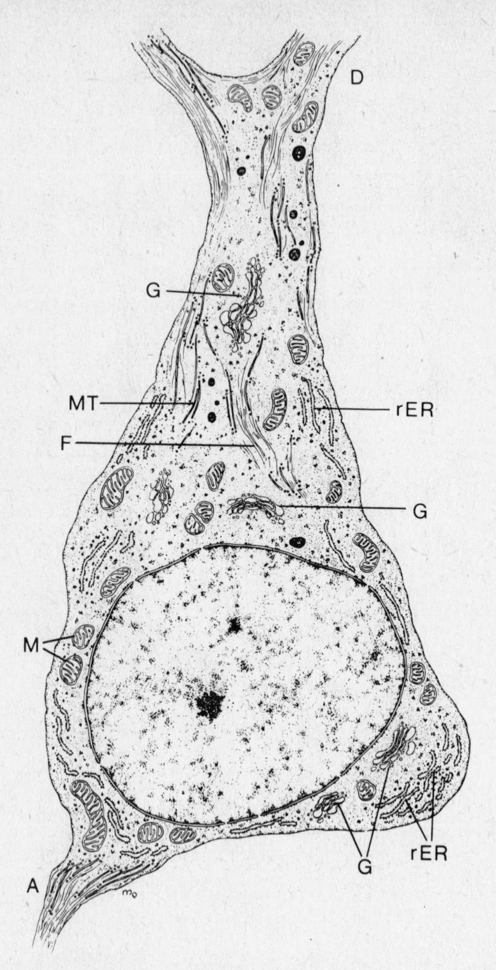

Fig. 3-2. Diagram illustrating the fine structure of the cell body of a neuron. (*A*) Axon hillock. (*D*) Dendrites. (*rER*) Rough-surfaced endoplasmic reticulum. (*G*) Golgi stacks. (*MT*) Microtubules. (*M*) Mitochondria. (*F*) Filaments. (Courtesy of C. P. Leblond.) (Reprinted with permission of J. B. Lippincott Co. from A. W. Ham and D. H. Cormack, *Histology,* 8th ed., 1979.)

Nerve cell bodies are extremely variable in form and size. Some are ovoid, others stellate; some are extremely small, others large.

Nucleus. The nucleus is a large, spherical structure, usually centrally located. Within the nuclear membrane, the chromatin granules are small and uniformly dispersed and hence not readily seen, thus giving the nucleus the appearance of being empty except for a usually prominent *nucleolus.* In the cells of females, a small body, the *Barr body,* composed of sex chromatin, usually lies adjacent to the nucleolus.

Cytoplasm. The cytoplasm, also called *neuroplasm,* surrounds the nucleus and contains a number of structures. Among these are *neurofibrils, chromophil substance, Golgi apparatus, mitochondria,* and *inclusions.* In immature cells, a *centrosome* may be present.

Neurofibrils are slender, delicate fibrils consisting of groups of parallel *neurofilaments.* Composed of proteins, they form an interlacing network within the cell body and extend into the cell processes.

Chromophil substance consists of material that stains with basic dyes and forms discrete bodies of various shapes called *Nissl bodies.* These bodies consist of slender tubules and flattened cisternae of rough, granular endoplasmic reticulum bearing ribosomes, which are the site of protein synthesis. Proteins pass into the cell processes, especially the axon, and replace those lost in cellular metabolism. Nissl bodies are scattered throughout the cytoplasm and may be present in the dendrites of larger neurons; however, they are absent in axons and the axon hillock. They vary in appearance in various physiologic states, and in pathological conditions they may disappear, a process called *chromatolysis.* If an axon is injured or severed, temporary chromatolysis may occur.

The *Golgi apparatus* or *complex* consists of a coarse network of fibrils that may form an arclike structure near the nucleus or may completely encircle the nucleus. It is considered to be tubular agranular endoplasmic reticulum, with tubules connected to the granular endoplasmic reticulum of the Nissl bodies.

Mitochondria are rodlike or filamentous structures scattered throughout the cytoplasm of both cell body and processes. They serve their usual function of energy release.

Inclusions include various granules and crystals. Dark brown or black pigment granules of *melanin* occur in certain cells of the brain, for example, the substantia nigra of the midbrain. Their significance is unknown. More commonly found in neurons are golden-brown pigment granules of *lipofuscin,* thought to be a product of normal metabolic activity. They tend to increase with age.

CELL PROCESSES. Neurons have two kinds of processes, dendrites and axons.

Dendrites. These short processes branch in treelike fashion. They are extensions of the parikaryon and contain neurofibrils, Nissl bodies, and mitochondria, but they lack a myelin sheath and neurilemma. Most possess short, spinelike projections called *gemmules.* Dendrites provide an extensive area where synaptic connections are made with the axons of other neurons. Impulses are conducted by dendrites *to* or *toward* the cell body.

In some neurons such as sensory neurons, whose cell bodies lie in cranial or spinal ganglia, typical dendrites are lacking. Impulses are conducted to or toward the cell body through a single process, which has the structure of an axon. Such processes are called *axonlike dendrites* and constitute the sensory fibers of cranial and spinal nerves.

Axons. In contrast to dendrites, a neuron possesses only one *axon* or *axis cylinder,* which arises from a specialized region of the cell body, the *axon hillock.* The cytoplasm of the axon, called *axoplasm* contains neurofibrils, neurotubules, axoplasmic vesicles, and mitochondria and is enclosed by a plasma membrane, the *axolemma.*

Axons vary in length. Some are very short (less than 1 mm); others are long (1 m or more in length). They vary in diameter from less than 1 μm to several micrometers. They are of uniform diameter and may bear side branches called *collaterals,* which come off at right angles. Axons may be enclosed in one or more sheaths, a *myelin sheath,* a *neurilemma,* or both. Strictly speaking, these sheaths are not parts of the axon, but they play an important role in its physiological activities. Axons end in terminal arborizations called *telodendria* that, both in the central nervous system and in ganglia, synapse with the cell bodies or dendrites of other neurons or terminate in effector organs.

Axonal Flow. It has been clearly demonstrated that, in neurons, there is a flow of cytoplasm distally from the cell body into the axon. Certain cytoplasmic structures (Golgi apparatus, ribosomes, endoplasmic reticulum) that are essential for protein synthesis or the secretion of transmitter agents are lacking in axons; consequently, these substances are synthesized by ribosomes in the cell body and migrate distally within the axoplasm of the axon. Labeled proteins have been determined to move along an axon at a rate of 1.5 mm/24 hr. Other substances may travel at a much faster rate (100 to 500 mm/24 hr). These are designated *slow-flowing* and *fast-flowing components.* Axonic cell organelles, as well as other substances (sugars, hormones, neurosecretory granules, phospholipids) involved in physiologic activities at the axon terminal, are transported along with the proteins.

NERVE FIBERS AND THEIR SHEATHS. Nerve fibers are axons or axonlike dendrites. They are widely distributed in the body but are concentrated in nerves, the spinal cord, and the brain. They usually possess a myelin sheath and neurilemma, but they may lack one or the other of these sheaths. *Unmyelinated* (*nonmedullated* or *gray*) fibers lack a myelin sheath but possess a neurilemma; *myelinated* (*medullated* or *white*) fibers have a myelin sheath but may or may not possess a neurilemma.

Myelin Sheath. The myelin sheath (Fig. 3-1) is a laminated membrane that closely invests an axon. It is composed of proteins and lipids. Lipids, being highly refractive, give nerves their whitish appearance. The myelin sheath is formed from Schwann cells, the cytoplasm of which grows concentrically about a nerve fiber, completely investing it by a number of layers (Fig. 3-3). In the process, the cytoplasm is squeezed into the outer nucleated region of the cell, leaving a sheath composed of layers (lamellae) of double cell membranes. At points along the fiber, the myelin sheath is

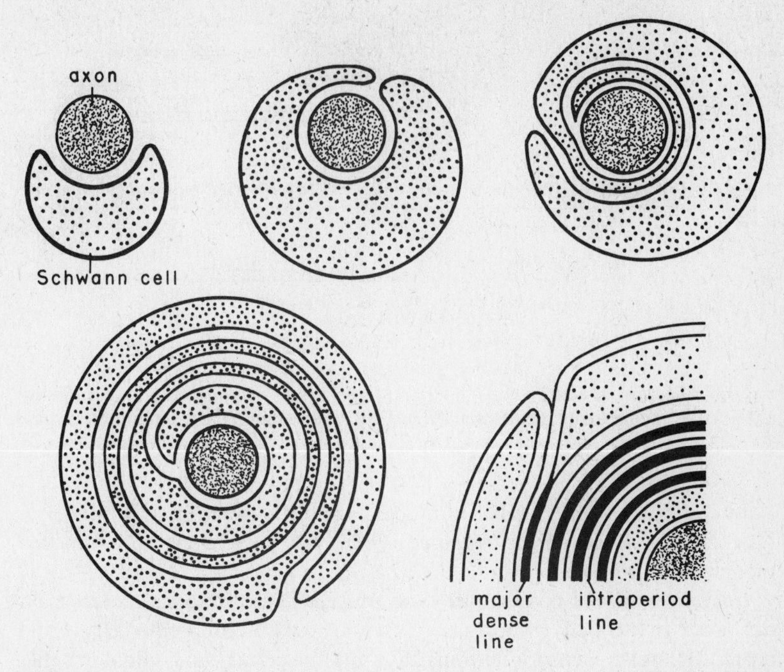

Fig. 3-3. Diagram showing formation of the myelin sheath. (Reprinted with permission of W. B. Saunders Co., Philadelphia, from C. R. Leeson and T. S. Leeson, *Histology,* 3d ed., 1976.)

lacking, giving the fiber a segmented appearance. These points are called *nodes of Ranvier.* Between the nodes, the sheath bears small diagonal incisures known as *clefts of Schmidt-Lanterman.* The myelin sheath serves as an insulating layer that prevents impulses from stimulating adjacent fibers. It also increases the rate of conduction of nerve impulses.

Neurilemma. In a myelinated fiber, the *neurilemma* or *sheath of Schwann* (Fig. 3-4) lies external to and closely invests the myelin sheath. It consists of *Schwann cells,* one of which occupies the space between two nodes of Ranvier. The internodal distance averages about 1 mm. Schwann cells are neuroglial cells and function in the formation of the myelin sheath as described in the preceding paragraph. They are also essential for the regeneration of nerve fibers. Schwann cells may also enclose unmyelinated fibers, sometimes several.

Nerve fibers in the brain and spinal cord have no neurilemma, it being replaced by glia cells (oligodendrocytes), which produce the myelin sheath as the Schwann cells do in peripheral nerves.

Classification of Neurons. Neurons can be classified in two ways:

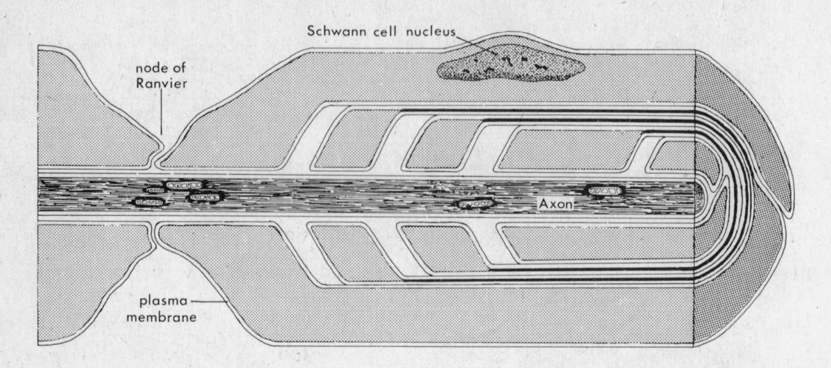

Fig. 3-4. Diagram showing a nerve fiber and its coverings. (Reprinted with permission of W. B. Saunders Co., Philadelphia, from C. R. Leeson and T. S. Leeson, *Histology*, 3d ed., 1976.)

(1) on the basis of the number of processes and (2) on the basis of function.

1. On the Basis of the Number of Processes. Neurons are unipolar, bipolar, or multipolar.

Unipolar neurons possess only one process. True unipolar neurons do not occur in the body (with rare exceptions, such as amacrine cells of the retina). However, sensory neurons (Fig. 3-1) of cranial and spinal ganglia, which do not have typical dendrites, possess a single process and are considered unipolar, although they are actually *pseudounipolar.* This structure results from a fusion in embryonic development of a single axon and a single dendrite to form a T-shaped process, which, a short distance from the cell body, divides into *central* and *peripheral* branches. These are described under afferent or sensory neurons in the following paragraphs.

Bipolar neurons have two processes: a single axon and a single dendrite. A cell body is usually fusiform in shape, with processes arising from opposite ends of the cell. These neurons are found in the retina, in certain ganglia (vestibular and cochlear), and in the olfactory epithelium.

Multipolar neurons have a single axon and a variable number of dendrites (for example, the motor neurons in the gray matter of the spinal cord).

2. On the Basis of Function. Neurons are afferent, efferent, or internuncial.

Afferent or *sensory neurons* carry sensory impulses from the periphery. They are mostly pseudounipolar, having globular cell bodies located in the dorsal root ganglia of spinal nerves or on the sensory roots of cranial nerves. Each possesses a single T-shaped process, which divides into a *central branch,* functionally an axon, that terminates in the brain or spinal cord, and a *peripheral branch* that terminates in sensory receptors in the body wall or viscera. The peripheral branch, though functioning as a dendrite, has the structure of, and is indistinguishable from, a typical axon.

It is an *axonlike dendrite* and in some cases can conduct impulses in the reverse direction (*antidromic conduction*).

Efferent neurons carry motor impulses from the central nervous system or from ganglia to effector organs or structures, which respond. Depending on the nature of the effect and the response produced, efferent neurons are either (*a*) *motor neurons,* whose axons end in voluntary muscles, which contract when stimulated; (*b*) *secretory neurons,* whose axons end in glands, which secrete when stimulated; (*c*) *accelerator neurons,* whose axons end in visceral or cardiac muscles, which either initiate or speed up activity; or (*d*) *inhibitory neurons,* whose axons end in visceral or cardiac muscle, which retard or stop muscular contraction.

Internuncial (intercalated, association, or *connector) neurons,* also called *interneurons,* lie within the central nervous system. They conduct impulses from one neuron to another and so are principally involved in integrative activities.

Neuroglia (Fig. 3-1). The term *neuroglia* is applied to cells comprising the interstitial tissue of the nervous system. These cells include *ependyma, neuroglia proper, satellite* or *capsular cells,* and *neurilemmal cells of Schwann.*

EPENDYMA. Ependymal cells line the ventricles of the brain and the central canal of the spinal cord.

NEUROGLIA PROPER (GLIA). Glia cells are present in the brain and the spinal cord, lying between the neurons. There are three types of glia cells:

1. *Astrocytes* have numerous processes, some with expanded ends that may be attached to blood vessels.
2. *Oligodendrocytes* are similar to astrocytes, but their processes are fewer and thinner and lack expanded endings.
3. *Microglia* are small, many-branched cells with processes bearing numerous tiny points or spines.

Astrocytes and oligodendrocytes, like neurons, are ectodermal in origin. Microglia are thought to be mesodermal in origin, making their way to the central nervous system by way of blood vessels.

Neuroglia cells serve to support neurons and probably play an important role in their normal metabolism. They are active in pathological processes. Some, especially the microglia, may become actively ameboid and phagocytic; they are involved in degenerative and regenerative processes following injury and are the principal cells involved in the formation of brain tumors. Oligodendrocytes are involved in the formation of myelin in the central nervous system.

SATELLITE OR CAPSULAR CELLS. These small cells surround the cell bodies of sensory neurons in peripheral ganglia.

NEURILEMMA. Cells of Schwann comprise the neurilemma surrounding the peripheral processes of cranial and spinal nerves. They function in the

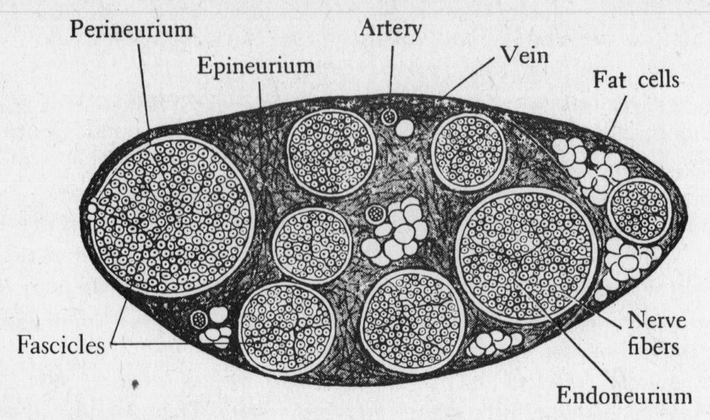

Fig. 3-5. Cross section of a nerve showing fascicles of nerve fibers and their surrounding membranes. (Reprinted with permission of The Macmillan Company from D. C. Kimber et al., *A Textbook of Anatomy and Physiology,* 13th ed., 1955.)

formation of the myelin sheath that surrounds the axon of neurons lying outside the central nervous system.

Nerves. *Nerves* (Fig. 3-5) are glistening white structures lying outside the central nervous system through which impulses are conducted. A nerve is made up of nerve fibers, principally axons and their coverings. The fibers may be medullated or nonmedullated. The fibers comprising a nerve are grouped together into bundles called *fasciculi* or *fascicles.* Each fasciculus is encased in a thin covering of connective tissue known as *perineurium,* from which thin strands of connective tissue extend inward among the nerve fibers, constituting the *endoneurium.* The fasciculi are in turn enclosed by a layer of connective tissue composed mainly of collagenous fibers, forming the *epineurium.* The epineurium comprises the membranous covering of the nerve; it contains blood vessels and, usually, some adipose tissue. The epineurium is also supplied with sensory nerve fibers, the *nervi nervorum.*

Nerves and nerve trunks vary greatly in size. Near the brain and spinal cord, they contain many fibers and are relatively large. Distally, they divide and subdivide until at the periphery they may be extremely small.

Within a nerve, nerve fibers may pass from one bundle to another, or they may split into branches that join the branches of other nerves, forming an anastomosing network called a *nerve plexus.* This property is especially pronounced in the nerves that supply the extremities.

Tracts. In the brain and the spinal cord, nerve fibers are grouped into functional pathways called *tracts.* In the brain, these tracts serve to connect one part with another or to conduct impulses to or from the spinal cord. In the cord, certain tracts (*ascending tracts*) carry afferent impulses *to* the brain; others (*descending tracts*) carry efferent impulses *from* the brain.

Nerve tracts in the brain contain three types of fibers:

1. *Projection fibers* connect the cerebral cortex with the brain stem or the spinal cord.

2. *Commissural fibers* connect one side of the brain with the other.

3. *Association fibers* connect different parts of the brain on the same side.

The *cell bodies* of the fibers forming these tracts lie in the gray matter of the brain and the spinal cord.

Gray Matter and White Matter. In the brain and spinal cord, two kinds of matter can be distinguished grossly by their color, namely, the gray matter and the white matter.

GRAY MATTER. This consists principally of cell bodies and their dendrites, nonmyelinated nerve fibers, and the supporting neuroglia elements. It is found in the cortex, the basal ganglia, and the nuclei of the brain; in the gray columns of the spinal cord; in peripheral ganglia; and in the retina of the eye.

Gray matter constitutes the nerve centers where impulses are received, connections made, and impulses discharged back to effector organs. Its primary functions are *integration, correlation,* and *coordination* of body activities. It provides the physical basis for the accomplishment of higher nervous functions, such as perception, thinking, judgment, and emotion.

WHITE MATTER. This is composed principally of myelinated nerve fibers and their supporting neuroglia cells. It is found in nerves, the commissures and tracts of the brain, and in the fiber tracts of the spinal cord. It contains few, if any, nerve cell bodies.

In general, the white matter serves to carry impulses from the peripheral portions of the body to and from the central nervous system or between various parts of the brain and the spinal cord. Its primary function is *conduction.*

Ganglia. A ganglion is an aggregation of nerve cell bodies and the proximal portions of their processes, together with their supporting glia cells.

Important ganglia are the *basal ganglia* of the brain, *sensory ganglia* on the roots of cranial and spinal nerves, and *autonomic ganglia.* Ganglia are usually found outside the central nervous system.

Nuclei. A nucleus is a localized mass of nerve cell bodies in the brain or the spinal cord whose axons form certain nerves, fasciculi, tracts, or commissures. Nuclei are of two types: *nuclei of origin* and *nuclei of termination.*

NUCLEI OF ORIGIN. These are accumulations of cell bodies whose axons form a nerve root or fiber tract. An example is the dorsal motor nucleus of the vagus nerve located in the medulla.

NUCLEI OF TERMINATION. These are accumulations of cell bodies in which the axons of a nerve or fiber tract terminate. Through synapses the cell bodies receive impulses from terminating axons of afferent neurons.

An example is the nucleus cunneatus in the medulla, which receives and relays sensory impulses of touch.

Peripheral Nerve Endings. All the fibers in peripheral nerves, whether motor, sensory, secretory, or inhibitory, terminate in some peripheral organ or structure. The fiber may terminate singly or it may branch repeatedly, forming *terminal arborizations.* It may end as a *free* or *naked nerve ending* or in a specialized structure.

Depending on their particular functions, peripheral nerve endings may be designated as either receptors or effectors.

RECEPTORS. *Receptors* or *sensory end organs* are nerve endings or specialized sensory cells that respond to stimuli. They comprise two types: exteroceptors and interoceptors.

Exteroceptors. Exteroceptors respond to stimuli originating from *outside* the body. These receptors are generally located on or near the surface of the body and provide information on changes taking place in the external environment. Exteroceptors include the end organs of touch, pressure, heat, cold, and pain, which are widely distributed over the surface of the body. They also include the end organs of the special senses of hearing, sight, taste, and smell. These are located in the cochlea of the ear, retina of the eye, mucosa of the oral cavity, and olfactory epithelium of the nasal cavity. They are described in detail in Chapter 5.

Interoceptors. Interoceptors respond to stimuli originating from *within* the body. They include two types: visceroceptors and proprioceptors.

Visceroceptors are located in various internal organs, such as those of the gastrointestinal tract, the lungs, and the carotid body and sinus. They respond to stimuli that give rise to sensations such as internal pain, hunger, thirst, and nausea. Most impulses arising from their stimulation are below the level of consciousness, but they are important in the regulation of normal body functions.

Proprioceptors include (1) end organs of equilibrium (*cristae* and *maculae*) located in the inner ear; (2) end organs located in joints, tendons, and muscles, which include *neuromuscular spindles, neurotendinous spindles,* and *Pacinian corpuscles;* and (3) *baroreceptors,* stretch receptors located in the carotid sinus, aortic arch, the walls of blood vessels entering the heart, and the atria and ventricles of the heart. These receptors provide information about the position of parts of the body and changing conditions in muscle tone and tension that is essential in the control of voluntary movements. They are also of vital importance in the subconscious regulation and coordination of muscular movements and in the control of circulatory activities.

The term *receptor,* as related to the nervous system has several meanings. It has been applied to (1) a sensory nerve ending; (2) a specialized receptor cell, such as an olfactory or a gustatory cell or a pain receptor; (3) an entire sense organ, such as the eye or ear; (4) an entire sensory neuron, as that in a reflex arc. In this volume, only the first two meanings are applied.

EFFECTORS. Effectors are the terminal portions of efferent neurons that transmit impulses to an effector organ, such as a muscle or a gland. They may be naked nerve endings or motor end plates (myoneural junctions).

Naked Nerve Endings. These terminate in smooth or cardiac muscles or in glands. Their nerve fibers may have slight terminal enlargements. The fibers come into close contact with the cells, but apparently they do not actually penetrate the protoplasm.

Motor End Plate. This structure is found at the junction (*myoneural, neuromuscular*) between an axon of a motor neuron and the fiber of a striated muscle. It consists of a mass of muscle sarcoplasm lying just beneath the sarcolemma in which the naked axis cylinder (axon) ends. A number of muscle nuclei may be grouped about the plate. The portion of the plate in contact with the contractile substance of the fiber is called the *sole.*

The term *effector* has several meanings. It has been applied to (1) the terminal portion of the axis cylinder of an efferent neuron; (2) an entire efferent neuron, as that in a reflex arc; (3) an organ in which a response is elicited, as a muscle or a gland. In this volume, only the first meaning applies.

Degeneration and Regeneration in the Nervous System. After neurons become differentiated in embryonic development, they lose their ability to multiply. Accordingly, after birth, neurons that are destroyed cannot be replaced. This applies to neurons in their entirety for, under certain conditions, processes of neurons may be regenerated if the cell body remains intact.

Neurons in the central nervous system may die as the result of injury, infections, irradiation, poisoning, or aging. When this occurs, loss of function may ensue. In some instances, however, other neurons have the capacity for taking over the function that was performed by the destroyed neurons.

WALLERIAN DEGENERATION. Outside the central nervous system, when a nerve is cut or injured, fibers in the distal portion of the nerve undergo degeneration. This is especially true of the axons or axonlike dendrites. *Degeneration* begins at the point of injury and progresses distally. The axon becomes tortuous and fragmented, then disappears. The myelin sheath breaks down, and the fat accumulates in the form of globules. For a period of several weeks, it can be stained by the Marchi method. The neurilemma persists, but in an altered state. Neurilemmal cells proliferate and form a tubelike structure. Because of a change in its staining properties, the degenerated myelin can be traced easily; hence the course of nerve fibers undergoing Wallerian degeneration can be readily followed. Degeneration of this type is usually completed in two or three weeks, after which the degenerating myelin is resorbed.

Regeneration occurs through the growing out of a new axis cylinder from the central portion of the injured fiber. Several fibers, as many as 50, may develop from the injured stump. If one of the many fibers enters the neurilemmal tube, regeneration proceeds at a rate of 1 to 4 mm/day. If the two ends of a severed nerve are separated by a gap exceeding 3 mm, the fibers may form a swollen, interlacing network called a *neuroma,* and regeneration will fail to occur. Bringing the two ends of a severed nerve close to each other facilitates the regeneration process. Since new fibers follow the course of the old neurilemma, the presence of neurilemma is essential for the regeneration of nerve fibers. Furthermore, regeneration does not occur in the central nervous system, where the neurons lack a neurilemma.

RETROGRADE DEGENERATION. This term is applied to changes that occur in the cell body and in the proximal portions of an axon that has been cut or injured. When retrograde degeneration takes place, the Nissl bodies disintegrate, and the chromophil substance becomes scattered in the form of granules through the cytoplasm. Changes may also occur in the nucleus, and the water content of the cell may increase, leading to swelling and vacuolization. This process is called *chromatolysis.* Usually the cell dies, but it may recover.

THE NEURON THEORY: THE SYNAPSE

Formulated by Wilheim von Waldeyer in 1891, the *neuron theory* postulates that the nervous system is made up of neurons and that a neuron, with its processes (dendrites and axon), is the cytological and trophic unit of the nervous system. According to this theory, all parts of the body are connected by complex series of neurons that carry impulses to and from the various parts. Impulses pass through the dendrites of a single neuron to its cell body, then from the cell body out through the axon that makes connections with the dendrites or the cell body of another neuron or neurons, these in turn conducting the impulse to still other neurons or effector organs. The area of contiguity of one neuron with another is called the *synapse.*

Synapses. A synapse is the junction between an axon terminal of one neuron (*presynaptic neuron*) and the cell body or dendrite of another neuron (*postsynaptic neuron*) (Fig. 3-6). Each axon terminal ends in a slight enlargement, a *synaptic knob* (*end foot, end bulb, bouton terminal*). Separating the two cells is a minute space, the *synaptic cleft.* The area underlying the cleft is the *synaptic membrane.*

A single postsynaptic neuron may provide a surface for hundreds or thousands of synaptic junctions (Fig. 3-6), thus receiving impulses from many presynaptic neurons. This represents *convergence* of neural input. However, a single presynaptic neuron may synapse with several postsynaptic neurons. This represents *divergence* of neural input.

TYPES OF SYNAPSES. An axon may make contact with any part of a

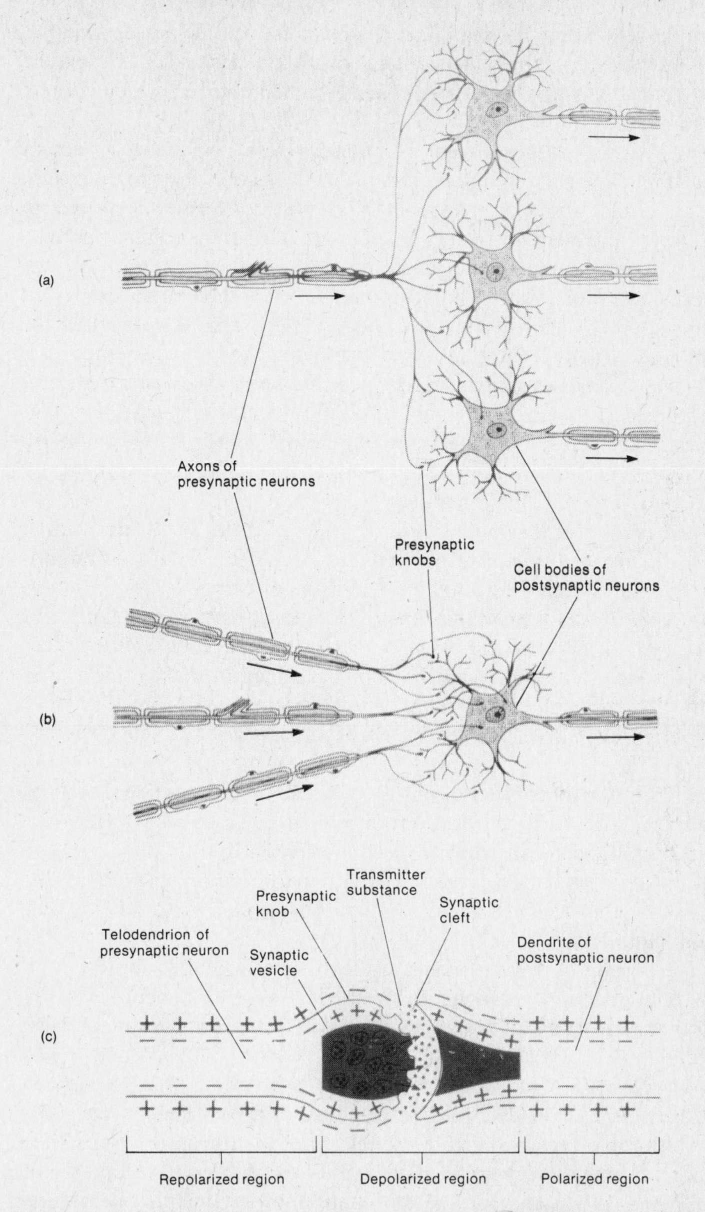

Fig. 3-6. Impulse conduction at synapses. (*a*) Divergence of synapses. (*b*) Convergence of synapses. (*c*) Impulse conduction from a presynaptic knob to a postsynaptic dendrite across a synapse. (Reprinted with permission of Harper & Row, Publishers, Inc., from G. J. Tortora and N. P. Anagnostakos, *Principles of Anatomy and Physiology,* 3d ed., © 1981.)

postsynaptic cell except the region of the axon hillock. Types of synapses include (1) *axodendritic,* in which contact is with a dendrite; (2) *axosomatic,* in which contact is with the cell body; and (3) *axoaxonic,* in which contact is with an axon.

SYNAPTIC CONDUCTION. A nerve impulse, which consists of an action potential transmitted along a nerve fiber, does not cross the synaptic cleft. Instead, it brings about a release of a chemical *transmitter* stored in *synaptic vesicles* present in the synaptic knob. This transmitter substance passes into the synaptic cleft, where it acts on receptor sites on the postsynaptic cell membrane. The postsynaptic cell is then either excited or inhibited. If *excited,* the interior of the cell is depolarized, permitting the inflow of sodium ions. When threshold level is reached, a nerve impulse is initiated, which passes through an axon to another neuron or an effector organ. If the postsynaptic cell is *inhibited,* the transmitter substance fails to depolarize the cell membrane below the synaptic cleft. It opens the cell membrane to an outflow of potassium ions, and as a result, no action potential (nerve impulse) is generated.

TRANSMITTERS. Acetylcholine has been found to be a transmitter substance in myoneural junctions, in the synapses of certain autonomic ganglia, and at certain postganglionic terminals. There is evidence that *norepinephrine* is the transmitter agent released at certain postganglionic sympathetic nerve endings and in the brain. In most neuronal transmission, the specific transmitter substance has not been identified. Possible neurotransmitter agents include *serotonin, gamma-aminobutyric acid (GABA), dopamine, glutamic acid, vasoactive intestinal peptide (VIP),* and *somatostatin.*

As soon as the transmitter agent is released, it is essential that it be inactivated as soon as its effect is produced. In the case of acetylcholine, this is accomplished by an enzyme, *acetylcholinesterase.*

CHARACTERISTICS OF SYNAPSES. Phenomena common to synapses are polarity, synaptic delay, fatigability, and susceptibility to anoxia and metabolic inhibitors.

Polarity. A state of polarity exists at the synapse whereby impulses pass in only one direction, namely, from the axon of one neuron to the dendrites or cell body of another; impulses *never* pass in the reverse direction.

Synaptic Delay or Resistance. Following stimulation of a presynaptic neuron, there is a slight delay in the response of a postsynaptic cell. This is the result of the time involved in the release of the transmitter substance, its diffusion through the synaptic cleft, and its attachment to receptor sites.

Fatigability. When the rate of stimulation of presynaptic neurons is increased, a point is reached at which a neuron will fail to transmit an impulse at the synapse. It is assumed that under these conditions, a neuron is unable to synthesize the transmitter rapidly enough or in sufficient quantities to activate the postsynaptic membrane. As a consequence, the postsynaptic cell will fail to respond.

Susceptibility to Anoxia and Metabolic Inhibitors. It is well known that oxygen is involved in synaptic conduction. In its absence or depletion, metabolic processes within neurons are impaired, especially the synthesis of ATP, upon which cellular activity depends, and the secretion of the transmitter substance essential for synaptic conduction. The effect of metabolic inhibitors, for example, iodoacetic acid (IAA), which interferes with glycolysis, is that of decreasing oxygen consumption, which impairs synaptic conduction. Various drugs can alter the actions of transmitter agents at synapses.

Mechanism of Impulse Conduction. A *nerve impulse,* also called an *action potential,* is a self-propagated wave of "electrical negativity" transmitted along the cell membrane of a neuron. The changes that occur in the initiation of an impulse and its passage along a nerve fiber are as follows.

The resting nerve cell is a *polarized structure* bearing positive ions on the outer surface of the cell membrane and negative ions on the inner surface. A *resting* or *membrane potential* of minus 70 millivolts (mV) exists due to the excess of sodium ions in the extracellular fluid and an excess of potassium ions within the cell. This difference is maintained through the action of a *sodium pump,* an active transport mechanism that pumps sodium ions out of the cell and allows potassium ions to diffuse into the cell.

When a stimulus is applied to a neuron, the permeability of the cell membrane is altered. Positive sodium ions enter the cell, and with their reduction in the extracellular fluid, a transient state of *depolarization* is established. As they continue to enter the cell, an excess of positive ions develops within the cell and an excess of negative ions outside the cell. This condition reverses the resting potential, which now becomes positive, and creates an *action potential,* which acts as a stimulus for the next adjacent portion of the nerve fiber, which then becomes depolarized. The action potential, which is rapidly propagated along a nerve fiber, thus constitutes the *nerve impulse.*

It should be noted that the change in membrane potential (from −70 to +30 mV and return to normal) occurs with extreme rapidity (in as short a time as 0.5 to 2 msec). The passage of the action potential along a nerve fiber consequently appears as an almost instantaneous process.

PHYSIOLOGY OF NERVE CONDUCTION

To understand the function of the nervous system, the following topics involving nerves and impulse conduction must be considered. These include threshold stimulus, the all-or-none principle, refractory periods, direction of impulse movement, nerve fatigue, and the velocity of and blocking of nerve impulses.

Threshold Stimulus. If a weak stimulus is applied to a nerve fiber, a slight degree of depolarization may occur, but not enough to generate an action potential. Only when a stimulus is strong enough to depolarize the membrane completely will an action potential be generated and a nerve impulse initiated. A stimulus of this strength is called a *threshold stimulus.* One below this strength is a *subthreshold stimulus.*

A single subthreshold stimulus will not initiate an impulse, but if a second subthreshold stimulus is applied within a millisecond, an action potential may be generated. This effect is known as *summation.*

All-or-None Principle. When a nerve fiber is stimulated, it conducts to its fullest extent, regardless of the strength of the stimulus. The action potential initiated by a strong stimulus is identical to that arising from a weak stimulus of threshold strength. The action potential, once initiated, is *maximal.* However, all impulses are not of the same magnitude, for altered conditions within a fiber, such as those brought about by subjecting the fiber to a narcotic, may alter its capacity for conductivity.

Although an individual fiber acts on the all-or-none principle, a nerve that is made up of several fibers does not. This accounts for *graded responses,* some weak and others strong. Depending on the strength of the stimulus, a weak stimulus may stimulate only a few fibers, each conducting to its fullest extent, whereas a strong stimulus may stimulate many fibers.

Refractory Period. When a threshold stimulus is applied to a nerve fiber and an action potential is generated, a second identical or even stronger stimulus applied immediately afterward within 0.5 msec will fail to produce a second action potential. This period in which the fiber is unresponsive is called the *absolute refractory period.* This occurs because the fiber must resume its resting potential before a second action potential can be triggered. However, following an absolute refractory period in which the fiber will not respond to a normal stimulus, there is a period in which a much stronger stimulus may produce a second action potential. This period, which may last from 10 to 30 msec, is called a *relative refractory period.*

Direction of Impulse Movement. A nerve fiber may conduct impulses toward or away from the cell body. Normally, the conduction is *toward* the cell body in dendrites, *away* from the cell body in axons (*orthodromic conduction*). At a synapse, however, impulses can pass in only one direction, namely, from the axon of one neuron to the dendrite(s) or the cell body of another.

In some cases, a nerve fiber may conduct impulses in a reverse direction, that is, peripherally in a sensory fiber or centrally in a motor fiber. Such conduction is called *antidromic conduction* and is limited to a single cell. Impulses do not pass in a reverse direction in a synapse.

Nerve Fatigue. From a practical standpoint, a nerve fiber does not become fatigued. Thousands of impulses can be carried over a fiber without impairing its capacity to conduct additional impulses. As long as the energy required for maintenance of the sodium-potassium pump mechanism is available to establish a resting potential, the ability of a fiber to propagate an action potential will persist. However, such abnormal conditions as oxygen deprivation or the effects of toxins or anesthetics may diminish the excitability of nerve fibers and impair their ability to conduct impulses.

Velocity of Nerve Impulses. The velocity of a nerve impulse depends on a number of factors. Among these are temperature, size of the fiber, and presence or absence of a myelin sheath. The speed of conduction is greater in warm-blooded than in cold-blooded animals and greater in fibers of large diameter than in fibers of small diameter. Velocity is also higher in myelinated fibers than in unmyelinated fibers. This is due to the fact that in myelinated fibers, action potentials are developed only at the nodes of Ranvier, and the impulse more or less jumps from node to node (*saltatory conduction*) rather than passing continuously along a fiber as in unmyelinated fibers.

Characteristic rates of conduction are approximately 120 m (394 ft)/sec in large myelinated fibers and approximately 1 m (3.28 ft)/sec in nonmyelinated fibers.

Blocking of Nerve Impulses. The process by which the passage of nerve impulses is stopped or obstructed is referred to as *blocking* or *deadening* a nerve. It can be accomplished (1) by cooling, (2) by application of pressure, (3) by application of electric current, and (4) with chemical substances.

1. *Cooling.* Nerve impulses can be obstructed by cooling to 0°C.
2. *Pressure.* Loss of sensation in a limb, as when the limb "goes to sleep," is due to pressure on the nerve. Removal of the pressure acts as a stimulus and gives rise to the "prickly" feelings that are referred to the distal end of the extremity. Crushing of a nerve is another form of pressure. Crushing of a motor nerve results in temporary paralysis of the muscles.
3. *Electric current.* In certain circumstances, an electric current is applied for the relief of pain from neuralgia or muscle cramps.
4. *Chemical substances.* Chemical substances that are used to block nerve impulses are classifed as anesthetics, sedatives, and hypnotics. Also included are analgesics, substances that relieve pain (analgesia being the absence of the normal sense of pain). Intoxication, a form of poisoning, also impairs the passage of nerve impulses.

ANESTHESIA. *Anesthesia* is a state of partial or complete loss of sensation with or without the loss of consciousness. In *local anesthesia,* sensation is lost in a limited area with the application of such substances as cocaine, procaine, and novocaine to specific nerves or nerve trunks. In *regional anesthesia,* an extensive portion of the body, or a region, is involved. When an anesthetic is injected into the membranes of the spinal cord (as in spinal block), sensation is lost in all structures innervated by nerves below the level of the injection. Injections in or around a ganglion block the nerve impulses to all structures supplied by the ganglion. In *general anesthesia,* the loss of sensation is complete, and consciousness is lost;

also, control of voluntary muscles disappears. It is accomplished by the inhalation of such anesthetics as ether, chloroform, or nitrous oxide.

The exact mode of action of anesthetics is not known. Several theories have been proposed for their action. Many anesthetics are soluble in lipids, hence it is thought that their effects result from their action on the lipids in cell membranes. Others are thought to act on cell membranes to prevent depolarization, upon which excitability depends. Specific receptor sites for certain anesthetics, especially opiates, have been identified. Some are thought to act through reduction of oxygen supply to brain cells; others through the inhibition of cyclic adenosine monophosphate formation.

NARCOTICS. A narcotic is a drug that numbs the senses, relieves pain, and, in large doses, induces drowsiness and coma. Commonly used narcotics include opium and its derivatives (morphine, heroin, codeine). Narcotics are addictive drugs.

SEDATIVES. *Sedatives* are substances that reduce nerve irritability and thus have a quieting effect. They reduce motor activity without inducing sleep. Some are general in their effects; others, such as cardiac, respiratory, gastric, and intestinal sedatives, are specific. Examples are certain bromides, chloral hydrate, pilocarpine, and belladonna.

HYPNOTICS. *Hypnotics* induce sleep and dull the senses. Barbiturates (derivatives of barbituric acid or their salts) are the most commonly used hypnotic and sedative drugs. Prolonged use of barbiturates can result in habituation and physical dependence.

TRANQUILIZERS. A *tranquilizer* is a drug that reduces mental tension and anxiety without interfering with normal functions of the cerebral cortex such as sensory perception, muscular coordination, thought, memory, and other mental processes. They are central nervous system depressants. Drugs in common use include chlorpromazine, meprobamate, promazine, and reserpine.

ANTIDEPRESSANTS. These drugs, also called *psychostimulants* or *energizers,* stimulate the central nervous system. They are used in cases of depression, melancholy, and psychomotor retardation. Some are monoamine oxidase (MAO) inhibitors, which retard the metabolism of amines essential in the functioning autonomic nervous system. Widely used non-MAO inhibitors include the amphetamines and tricyclic antidepressants. Cocaine is a pronounced stimulant.

Stimuli. A *stimulus* is an agent or an environmental change that initiates an activity or brings about a response in an organism or a part. Any condition that initiates a nerve impulse constitutes a stimulus. An *exteroceptive stimulus* is one received from outside the body; an *interoceptive stimulus* is one that originates within the body. Stimuli may be of several types: physical (mechanical), chemical, radiant, thermal, or electrical.

PHYSICAL (MECHANICAL) STIMULI. Contact with objects, pressure within an organ, the stretching of a tendon or muscle, changes in blood pressure, and movement of fluid within the semicircular ducts of the ear are examples of physical stimuli.

CHEMICAL STIMULI. Substances such as acids or alkalies in contact with the skin or mucous membranes, volatile gases within the nasal cavity, food within the mouth, or carbon dioxide and hydrogen ions within the blood are examples of chemical stimuli.

RADIANT STIMULI. Light rays, to which the rods and cones of the retina respond, are radiant stimuli.

THERMAL STIMULI. Changes in temperature of the air or substances that come into contact with the body are examples of thermal stimuli.

ELECTRICAL STIMULI. An electric current, that is, the flow of electrons from a battery, an induction coil, or a generator, serves as an effective stimulus and is widely utilized in experimental work. *Electronic stimulators,* now available, enable the experimenter to control precisely the frequency, amplitude, and duration of stimuli. *Frequency* refers to the number of stimuli applied per unit of time, usually seconds; *amplitude* refers to the intensity of the stimuli, measured in millivolts; *duration* is the length of time the stimulus is applied, measured in milliseconds.

Characteristics of a Stimulus. To act effectively as a stimulus, an environmental change must possess the following characteristics:

1. It must *have a certain strength or intensity.*
2. It must *act for a certain minimum length of time.*
3. It must *occur at a sufficiently rapid rate.*

THE REFLEX ARC—REFLEX ACTIONS

The reflex arc is the *functional unit of the nervous system.* It is a neuromuscular or neuroglandular mechanism through which most of the activities of the body are initiated.

A reflex arc consists of a series of neurons over which impulses are conducted from a receptor or sense organ to the central nervous system (brain or spinal cord) and then to an effector organ (a muscle or a gland). When a receptor is stimulated and a response, such as muscular contraction or glandular secretion, occurs involuntarily, the action is called a *reflex action.*

If a person touches a hot object, the hand is quickly and involuntarily withdrawn; if a bright light is directed into the eyes, the pupils contract and the eyelids close. A loud sound will cause a person to jump and to direct attention to the source of the sound. All these actions are reflex actions.

In a relatively simple reflex action such as the withdrawal of the hand, the following structures, which comprise a *reflex arc,* are involved (Fig. 3-7):

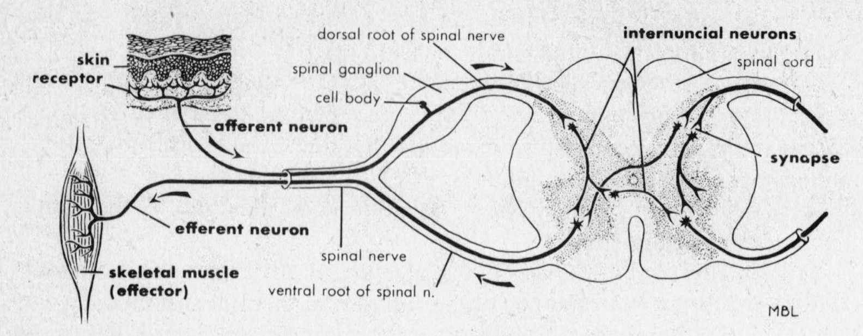

Fig. 3-7. Diagram of a reflex arc. (Reprinted with permission of Lea & Febiger, Philadelphia, from J. E. Crouch, *Functional Human Anatomy,* 1965.)

1. A *receptor* consisting of the peripheral ending of an afferent neuron.
2. An *afferent* or *sensory neuron,* which conducts impulses to the spinal cord. Its cell body lies in a dorsal root ganglion, and its axon enters the cord through the posterior root of a spinal nerve. Within the spinal cord, this neuron synapses with an internuncial or association neuron.
3. An *internuncial, association,* or *connecting neuron.* This neuron lies entirely within the spinal cord. Its axon synapses with a motor neuron.
4. An *efferent* or *motor neuron* whose cell body lies in the gray matter of the spinal cord. Its axon passes out through the ventral root of a spinal nerve and terminates in an effector organ.
5. An *effector organ,* in this case a skeletal muscle.

In the hand-withdrawal reflex, three neurons were involved in the reflex arc. In the simplest reflex, the knee jerk, only two are involved, since an internuncial neuron is lacking. Most reflexes are much more complex and may involve a number of intermediate or connector neurons located in the spinal cord or brain. A region where connections are made between afferent and efferent neurons is called a *reflex center.*

Reflex arcs involving glands or smooth or cardiac muscle utilize neurons of the autonomic nervous system. Such reflex actions are more complex because at least two neurons are involved in the efferent pathway: one whose cell body lies within the central nervous system, the other whose cell body lies in an autonomic ganglion. (See "Autonomic Reflex Arc," pages 149–151.)

Failure of Reflexes to Occur. If an adequate stimulus is applied, an expected reflex action may fail to occur for any of the following reasons: (1) an anatomic or physiologic defect in the receptor, (2) the severing or inactivation of afferent nerve fibers, (3) a defect in the central nervous system preventing connections from being made with efferent neurons, (4) the severing or inactivation of efferent nerve fibers, (5) failure of the transmitter agent at the neuroeffector junction, or (6) an anatomic or physiologic defect in the effector organ. To these may be added the possibility of *cortical inhibition.* The testing of reflexes is an important tool used in medicine to determine the integrity of various parts of the nervous system.

Complex Reflex Actions. Reflex actions may be of relatively simple nature (such as the blinking of an eyelid), they may be more complex (coughing or vomiting), or they may be extremely complex (the multiple responses involved in jumping from the path of an approaching automobile in response to the sounding of its horn).

CHARACTERISTICS OF REFLEX ACTIONS. The following characteristics apply to reflex actions in general.

1. *Reflex actions are involuntary.* The individual may or may not be conscious of the reflex act; in most cases, awareness of the act is recognition by the brain that the reflex action has occurred.
2. *Reflex actions are purposeful and adaptive.* In the majority of instances, they are actions that are essential for the protection and general well-being of the body. Most of the activities concerned with locomotion, posture, and the movements of individual parts are reflexly controlled; the activities of the internal organs are primarily under reflex control.
3. *Reflexes are specific and predictable.* When a given stimulus brings about a specific reaction, it can be predicted with a high degree of reliability that repetition of the stimulus will evoke the same reaction.
4. *Reflexes have a measurable reaction time.* In simple reflex actions, the response takes place much more quickly than in more complex reactions. Nevertheless, they do not occur instantaneously, since a measurable time always elapses between the application of the stimulus and the occurrence of the response. This time is called the reflex time or *reaction time.* Reaction time is variable, depending on a number of factors; the principal of these are (*a*) strength of the stimulus (in general, the stronger the stimulus, the shorter the reaction time); (*b*) complexity of the reaction (the more complex the activity, the longer the reaction time because of the greater number of synapses involved); (*c*) the fundamental nature of the central nervous system (individual differences among organisms result in more rapid responses in some individuals than in others); and (*d*) the physiologic condition of the central nervous system (for example, as a result of fatigue or overindulgence in alcohol, the time required for reflex actions is significantly increased).

SOME SPECIAL PHENOMENA RELATING TO REFLEX ACTIONS. Even simple reflex actions are not without complicating phenomena. Such phenomena are inhibition, irradiation, facilitation, and summation, as well as the abnormal actions seen in convulsive reflexes.

Inhibition of Reflexes. A stimulus that reflexly induces the contraction of certain muscles may at the same time inhibit the contraction of other muscles. For example, in the patellar reflex or knee jerk, the extensor muscles contract and extend the leg. But for this to happen, the flexor muscles, which act antagonistically, must relax; that is, they must be *inhibited* from contracting. This is accomplished in the following manner: Sensory impulses are initiated by tapping the patellar tendon. These pass through sensory neurons whose axons terminate in the gray matter of the spinal cord. At synapses, *excitatory* transmitter substances are released that stimulate (1) motor neurons that carry impulses to the extensor muscles (quadriceps femoris), which contract, and (2) interneurons that release *inhibitory* transmitter substances at their synapses with motor neurons, which pass to the opposing flexor muscles, the hamstrings. Since no action potentials are generated in these fibers, no contraction occurs, and the muscles remain in a state of relaxation. This action, which involves the contraction of one or more muscles and the simultaneous inhibition of their antagonists, is called *reciprocal innervation.* Coordinated muscle activity would not be possible without a mechanism of this nature operating.

In certain complex activities, a stimulus may cause the contraction of muscles on one side of the body and inhibit their contraction on the other side. For example, when one steps on a sharp object with the left foot, the flexor muscles of the left thigh contract, and this is accompanied by inhibition of the opposing extensors; at the same time, however, the extensors of the right thigh contract, and the opposing flexors of this side are inhibited.

At a much higher level of the nervous system, reflexes may be inhibited voluntarily, as in an effort to restrain coughing or crying. The prick of a sharp object normally elicits a cry of pain and movement away from the offending stimulus, but if the stimulus occurs under conditions of strong social pressure (as in a church or a public library), the normal response will probably be completely suppressed or greatly reduced. Pressure on the upper lip will usually inhibit a sneeze, and pressure on the eyeball sometimes puts an end to hiccoughing.

Irradiation. When a weak stimulus is applied, the impulses tend to leave the cord in the same segment in which they entered, producing a reaction on the same side of the body, usually near the point of stimulation. This is a *unilateral reflex.* A stronger stimulus may give rise to impulses that cross to the opposite side of the spinal cord, leaving the cord in efferent neurons of the same segment. This is a *contralateral reflex.* A still stronger stimulus may give rise to impulses that pass both up and down the spinal cord to issue at points several segments above or below the point where the afferent impulses enter. This spreading of impulses throughout the central nervous system is known as *irradiation,* a phenomenon that accounts for the increased responses resulting from increased stimulation.

Facilitation of Reflexes. Certain reflex actions can be amplified or facilitated by bringing into play two or more stimulus-response mechanisms. In the knee-jerk (patellar) reflex, for example, if the leg is dangling loosely when the patellar tendon is struck, the leg will extend, but if the subject clinches a fist before the stimulus is applied, the amplitude of the response will be greatly increased.

Summation of Reflexes. A reflex action may be initiated by the stimulation of two different receptors. For instance, a bright light directed into the eyes may initiate a sneeze. Cooling of the skin may produce the same effect. When either of these stimuli is inadequate to produce the response, both occurring together will usually bring about the reaction. Such an effect is called *spatial summation.* In some cases, a weak or subliminal stimulus is incapable of inducing a reflex response, but if this stimulus is repeated several times, there is an addition or summation of effects, which eventually brings about the response. This is called *temporal summation.*

Convulsive Reflexes. The law of reciprocal innervation is susceptible to several conditions that interfere with its operation. These conditions include strychnine poisoning and certain diseases. Strychnine is a marked stimulant of the central nervous system; when it has been administered in overdoses, a slight stimulus tends to produce exaggerated reactions involving unrelated muscle groups. Coordinated excitation and inhibition are lacking. The result is a paroxysm of involuntary reflex contractions and relaxations called a *convulsion, spasm,* or *seizure.* It may involve the entire body or be localized. Some of the abnormal conditions that may give rise to convulsions are eclampsia, meningitis, uremia, dietary deficiencies, epilepsy, hysteria, heat cramps, and food poisoning.

Classification of Reflexes. Reflex actions may be classified on the basis of heredity, number of neurons involved, number of segments of the spinal cord involved, types of muscle response, and clinical significance.

On the basis of heredity, reflexes are *instinctive* or *inherited* (natural, inborn, unconditioned) or *conditioned* (acquired, learned).

INSTINCTIVE OR INHERITED REFLEXES. These include the fundamental reflex activities that are concerned with food-getting, self-protection, sound production, mastication, swallowing, defecation, micturition, and reproduction. Regulation and control of the visceral organs are maintained through reflex actions.

CONDITIONED (ACQUIRED) REFLEXES. Responses that occur as the result of association and training may or may not be directly related to the stimulus. For example, in a classic experiment, if food is placed before a hungry dog, the animal's salivary glands are reflexly stimulated and begin to secrete saliva. Early in training, the sound of a bell will not bring about this secretion. However, if a bell is sounded on numerous occasions when the food is presented to the dog, the sound begins to be associated with the presence of the food, and eventually the sounding of the bell even in the absence of food will induce the secretion of saliva.

Many learned actions involve the establishment of conditioned reflexes. Indeed, most habits and a large part of an individual's behavior are the result of conditioned reflexes.

On the basis of the number of neurons involved, reflexes are monosynaptic or multisynaptic. *Monosynaptic reflexes* involve only two neurons, a sensory and a motor, as the knee jerk. *Multisynaptic reflexes* involve three or more neurons.

On the basis of the number of segments of the spinal cord involved, reflexes are segmental, intersegmental, or suprasegmental. *Segmental reflexes,* as the knee jerk, involve neurons that enter and leave the spinal cord in the same segment; *intersegmental reflexes* involve more than one segment of the cord; *suprasegmental reflexes* involve centers in the brain as well as those in the spinal cord.

On the basis of types of muscle response, reflexes may be flexion or stretch. A *flexion reflex* involves the action of flexor muscles of a limb, as withdrawal of a limb upon contact with a sharp object; a *stretch reflex* is the quick contraction of a muscle in response to a stretch, for example, the knee jerk or ankle jerk. Note that a stretch reflex may occur in a flexor muscle.

On the basis of clinical significance, reflexes used in the diagnosis of abnormal conditions are superficial, deep, visceral, or pathological. *Superficial reflexes* result from the stimulation of the surface of the body. Examples are the plantar and corneal reflexes. *Deep reflexes* include muscle and tendon reflexes, which result from stimulation of deeply placed receptors. *Visceral reflexes* are those involving the internal organs, such as the pupillary, carotid, and micturition reflexes. *Pathological reflexes* result from an abnormal condition, such as the Babinski reflex: Stroking the sole of the foot results in dorsiflexion or extension of the toes. While this response is regarded as normal in infants, in adults it is indicative of a lesion involving the corticospinal tracts.

Specific Reflex Actions. The failure of a normal reflex to take place is usually associated with a disturbance in the reflex pathway. If afferent and efferent pathways are found to be intact, the lesion must exist in a reflex center in the brain or spinal cord or in the receptor or effector organ involved. A knowledge of specific reflexes, such as those listed in Table 3-1, is therefore of value in diagnosing disorders of the nervous system or responding organs.

Reflex Centers. Reflex centers are areas in the spinal cord or the brain where connections are made between the afferent neurons and efferent neurons of a reflex arc. Examples of reflex actions that involve various reflex centers are as follows:

1. *Reflexes involving centers in the spinal cord.* The knee-jerk and flexor reflex illustrate this type. Others include defecation, micturition, erection, and ejaculation reflexes.
2. *Reflexes involving centers in the medulla oblongata.* Several important centers through which body activities are regulated are located in the medulla; among them are the vital cardiac, vasomotor, and respiratory

TABLE 3-1 SPECIFIC REFLEXES*

Reflex	Stimulus / Response
Knee-Jerk (Patellar) Reflex	**S**—Tapping patellar tendon **R**—Extension of leg
Flexor Reflex	**S**—Pinching foot or pricking foot with a pin **R**—Flexion of thigh, withdrawal of foot
Achilles Reflex	**S**—Tapping tendon of Achilles **R**—Extension of foot
Chin-Jerk Reflex	**S**—Tapping or suddenly depressing lower jaw when it is half open **R**—Sudden closing of jaw
Corneal Reflex	**S**—Touching cornea **R**—Closing of eyelid
Pupillary Reflex	**S**—Light striking retina **R**—Constriction of pupil
Ciliospinal Reflex	**S**—Pinching nape of neck **R**—Dilatation of pupil
Cremasteric Reflex	**S**—Stroking skin on inner side of thigh **R**—Elevation of testis
Plantar Reflex	**S**—Stroking sole of foot **R**—Plantar flexion of toes
Abdominal Reflex	**S**—Stroking skin of abdomen **R**—Contraction of abdominal muscles

* **S** = stimulus; **R** = response

centers, as well as those for swallowing, vomiting, coughing, sneezing, and winking.

3. *Reflexes involving centers in the cerebellum and midbrain.* Although specific centers have not been identified in the cerebellum, all reflex activities associated with locomotion and maintenance of posture depend on impulses that pass through the cerebellum and, in some cases, the midbrain. Terminating in these areas are afferent fibers of the vestibular branch of the vestibulocochlear nerve which play an important role in postural reflexes.

4. *Reflexes involving centers in the diencephalon.* The hypothalamus contains reflex centers concerned with the regulation of body temperature. These centers regulate heat production reflexly by increasing the metabolic rate or the activity of striated muscles (shivering). Heat loss is reduced by constriction of peripheral vessels and erection of hairs ("gooseflesh"); it is increased by vasodilatation and secretion by the sweat glands. Bodily activities associated with manifestations of various emotions such as fear, anger, rage, and pleasure, are mediated through reflex centers in the hypothalamus.

5. *Reflexes involving centers in the cerebral hemispheres.* The cerebral cortex presents a wide range of possible connections between receptors and effector organs. Some of the reflex actions are relatively simple, for example, the closing of the eyelids in response to a bright light. Others are complex, for example, jumping and screaming upon viewing a frightening scene. In the latter, visual impulses, on reaching the visual centers, are relayed through association neurons to motor neurons, which may bring about

widespread responses throughout the body. All conditioned reflexes involve connections made in the cerebral cortex.

Summary of Reflex Actions. Some general principles concerning reflex actions should be borne in mind in the study of the nervous system.

1. Reflexes may be simple or complex. The simplest reflexes involve centers in the spinal cord only; the more complex have centers in the brain stem or cerebellum; the most complex, including conditioned reflexes, have their centers in the cerebral hemispheres.
2. Reflexes may involve any afferent neuron, any efferent neuron, or any other part of the central nervous system. They may bring about responses in striated, smooth, or cardiac muscle or in glands.
3. A single stimulus may produce a very limited response, or it may bring into play many efferent neurons, producing widespread responses by the organism. Impulses may enter the central nervous system from several different receptors and be channeled through a common efferent pathway. This is called *convergence.* It may result in a limited response. Or impulses arising from stimulation of a single receptor may be channeled through many afferent pathways, resulting in numerous and widespread responses. This is called *divergence.*

4: THE CENTRAL, PERIPHERAL, AND AUTONOMIC DIVISIONS OF THE NERVOUS SYSTEM

The nervous system (Fig. 4-1) functions as the master coordinating or integrating system of the body, maintaining the unity and harmony of the organism as a whole and adjusting its internal state to the external environment. As described in the preceding chapter, the nervous system has two principal anatomic divisions: the central and the peripheral.

The *central nervous system,* comprising the brain and the spinal cord, serves as a sort of switchboard, receiving impulses from receptor organs and making connections by which the impulses are dispatched to effector organs, the responding mechanisms. It is the control center through which all body activities except those under chemical control are regulated.

The *peripheral nervous system,* comprising the cranial and spinal nerves and the sympathetic division of the autonomic nervous system, serves to connect peripheral organs with the central nervous system. Afferent nerve fibers conduct impulses centrally; efferent fibers conduct them peripherally.

The *autonomic nervous system* is a functional rather than an anatomic division. It includes nerves and ganglia that innervate all visceral organs and structures that function involuntarily.

EMBRYONIC DEVELOPMENT

An understanding of the early development of the nervous system is helpful in the study of neural structure and function.

Development of the Central Nervous System (Fig. 4-2). The nervous system arises in the embryo from the surface layer, the *ectoderm.* Along the longitudinal axis in the middorsal region a thickening develops, forming a *neural plate.* As a result of unequal growth and folding, this plate gives rise to two longitudinal ridges, the *neural folds,* which enclose the *neural groove.* Along the margin of each fold lies a group of cells that form the *neural crest.* The neural folds meet along the midline and fuse, forming a *neural tube,* which becomes detached from the ectoderm and comes to rest just beneath the surface layer of the embryo. The neural tube develops into the brain and spinal cord. The *neural canal* (the cavity within the tube) develops into the ventricles of the brain and the central canal of the spinal cord.

Growth in the anterior end of the neural tube is more rapid than in the remaining portion. Three regions are soon distinguishable by slight constrictions; these are the *forebrain,* the *midbrain,* and the *hindbrain,* from

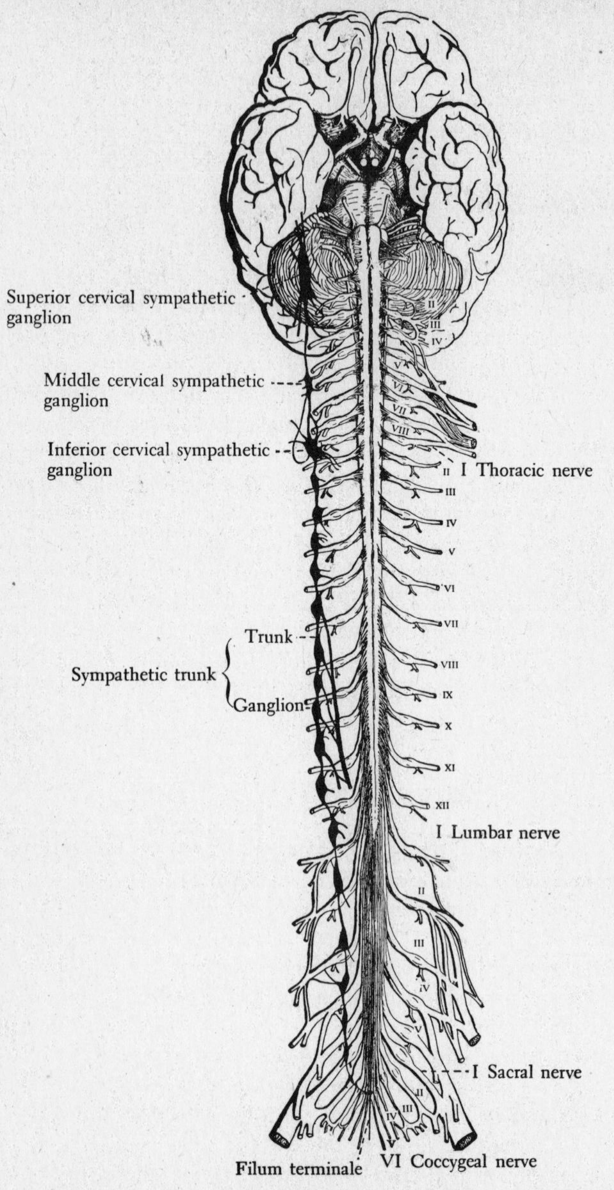

Fig. 4-1. Anterior view of the central nervous system with the sympathetic trunk and the proximal portions of the cranial and spinal nerves. Brain has been straightened dorsalward with reference to spinal cord. (After Allen Thomson from Rauber, modified.) (Reprinted with permission of Blakiston Division, McGraw-Hill Book Company, from *Morris' Human Anatomy,* 11th ed., ed. by J. P. Schaeffer, 1953.)

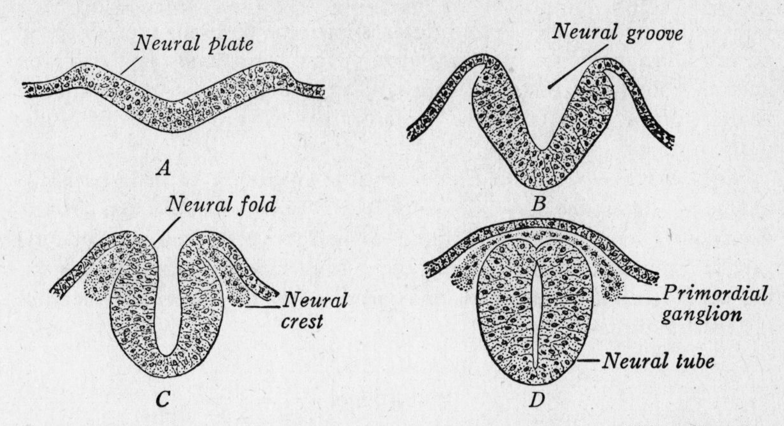

Fig. 4-2. Origin of the nervous system. (Reprinted with permission of W. B. Saunders Co., Philadelphia, from L. B. Arey, *Developmental Anatomy,* rev. 7th ed., 1974.)

which the various parts of the brain develop. The derivatives of each part are shown in the following table:

Primary Divisions	*Subdivisions*	*Derivatives*	*Cavities*
Prosencephalon (Forebrain)	Telencephalon	Rhinencephalon Corpora striata Cerebral cortex	Lateral ventricles Rostral portion of third ventricle
	Diencephalon	Epithalamus Thalamus Hypothalamus Optic chiasma Hypophysis Tuber cinereum Mammillary bodies	Most of the third ventricle
Mesencephalon (Midbrain)	Mesencephalon	Corpora quadrigemina Tegmentum Crura cerebri	Cerebral aqueduct
Rhombencephalon (Hindbrain)	Metencephalon	Cerebellum Pons	Fourth ventricle
	Myelencephalon	Medulla oblongata	

Development of the Peripheral Nervous System. During the formation of the neural groove, a longitudinal ridge of cells forms on each side at the junction of the ectoderm and the tube; these are the *neural crests* previously mentioned. After closure of the neural tube, the neural crests come to lie between the tube and the myotomes, and from these arise the cranial and spinal ganglia as well as the autonomic ganglia. Ganglion cells in the spinal ganglia develop processes, some of which grow centrally and penetrate the walls of the neural tube. These processes form the posterior roots of spinal nerves. Other processes of ganglion cells grow peripherally and join with anterior root fibers that grow outward from cell bodies located in the ventral portion of the tube. The union of these two groups of fibers results in the formation of a spinal nerve.

Sympathetic ganglia are formed by the migration of cells from the neural crest along the dorsal nerve roots and the nerve trunks to a position dorsal and lateral to the aorta, where they become organized into vertebral ganglia arranged as a pair of sympathetic trunks. By a similar mechanism, the other ganglia (prevertebral and terminal) of the autonomic nervous system are formed.

THE BRAIN

The brain (Fig. 4-3) is the part of the central nervous system that lies within the cranial cavity. For convenience it is divided into three regions: the brain stem, the cerebellum, and the cerebrum. The *brain stem,* the most inferior region, comprises the medulla oblongata, pons, midbrain, and diencephalon. The *cerebellum* lies posterior and superior to the brain stem. The *cerebrum* is a greatly expanded portion lying superior to and almost completely covering the brain stem and cerebellum.

The size and weight of the brain vary with race, age, sex, and size of the individual. A normal adult brain varies in weight from 1100 to 1700 g; average weight in a young adult male is 1360 g; in a female, 1250 g. The minimal weight compatible with intelligence is 950 to 1000 g. Above the minimal weight, there is only a rough correlation between degree of intelligence and weight of the brain. In general, the weight of the brain of persons of eminence tends to be slightly above the average.

General Functions of the Brain. In general, the brain serves the following functions:

1. *Regulatory center.* Through the activity of the brain, various bodily activities are integrated, regulated, and controlled. Sensory impulses are received, and motor or inhibitory impulses are discharged to muscles or glands, whereby adjustments are made to the changing internal or external environmental conditions. Additional regulation is accomplished through the secretion of hormones.
2. *Seat of consciousness.* The brain is the center of consciousness, which is the state of awareness of time, place, person, and, in greater or lesser degree, the activities of the body.
3. *Seat of sensations.* The brain interprets sensory impulses received from the various sense organs (e.g., eyes, ears, nose, taste buds, skin, proprioceptors), giving rise to sensations (e.g., sight, hearing, smell, taste, touch, pain, and movement).

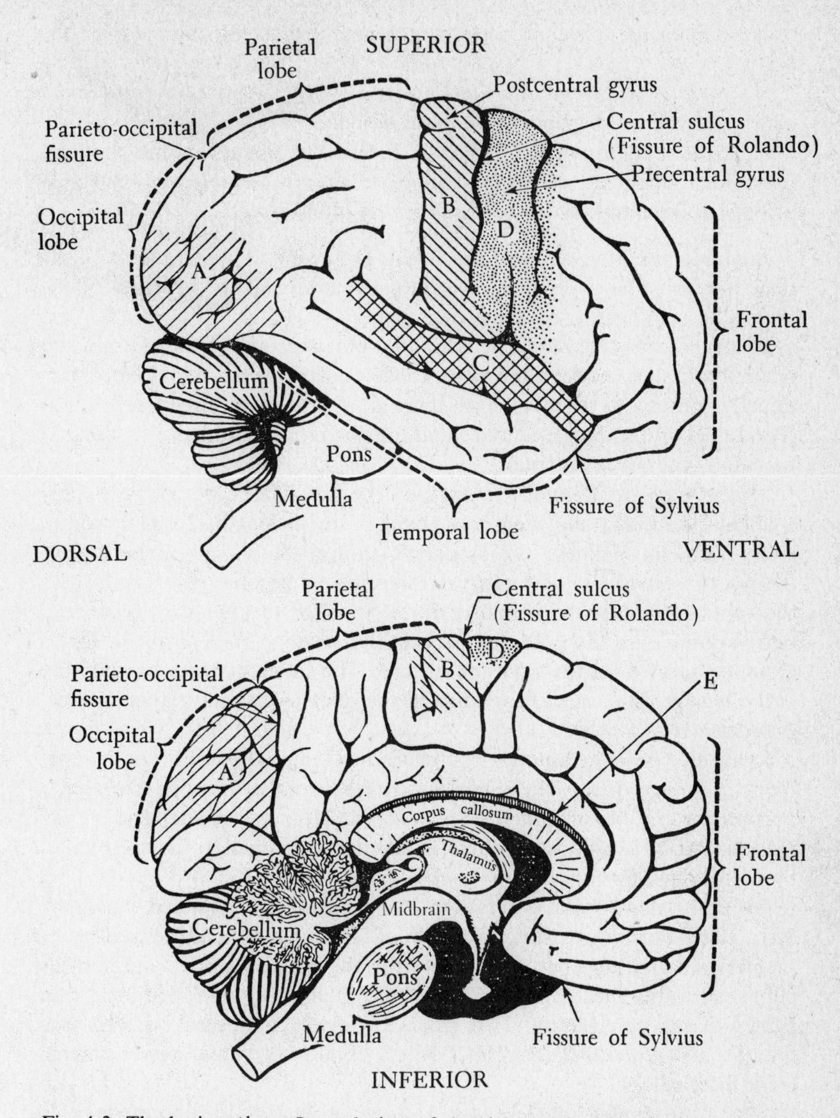

Fig. 4-3. The brain. *Above:* Lateral view of the right cerebral hemisphere, showing schematically the important fissures, lobes, and projection centers of the cortex, the medulla and pons, and the cerebellum. The dotted area (D) represents the motor area. Cross-hatched areas represent sensory areas for vision (A), somethesia (B), and audition (C). Unidentified parts are integration areas. *Below:* Mesial view of the left half of the brain, showing structures of the brain stem, cerebellum, and cerebral cortex. The thalamus, corpus callosum (composed of nerve fibers connecting the two hemispheres), and part of the projection area for olfaction and gustation (E) are visible. (From D. H. Fryer et al., *General Psychology,* Barnes & Noble, Inc., 1954.)

4. *Source of voluntary acts.* All voluntary acts are initiated by the brain.

5. *Seat of emotions.* The feelings, drives, or urges that profoundly affect behavior are dependent on brain activity.

6. *Seat of higher mental processes.* Perception, understanding, thought, reasoning, judgment, memory, and learning are possible only through activity of the brain. These are the basis of intelligence.

Anatomy and Physiology of the Brain. Following are descriptions of the structure and functions of the parts of the brain. (The coverings of the brain, the meninges, are discussed on pages 119–121.)

MEDULLA OBLONGATA. The medulla oblongata, the lowermost portion of the brain, lies between the pons above and the spinal cord below. It is directly continuous with the cord; there is no line of demarcation between it and the cord, since nerve tracts and fibers continue through the foramen magnum without interruption.

Structure of the Medulla Oblongata. The medulla oblongata is roughly triangular in shape and measures about 3 cm in length, 2 cm in width, and 1.5 cm in thickness. On its dorsal surface, there is a depressed area, the *fourth ventricle,* which narrows inferiorly to become continuous with the central canal of the spinal cord. A *posterior median fissure* forms a narrow groove on the posterior surface; the *anterior median fissure* forms a similar groove on the anterior surface. The latter of these extends the entire length of the medulla, ending at the pons in a small triangular area, the *foramen cecum.*

Lying alongside the anterior fissure are two longitudinal folds or enlargements, the *pyramids.* These contain nerve fibers connecting the upper portions of the brain with the spinal cord. Near the caudal end of the medulla about two-thirds of these fibers cross the median fissure; that is, they decussate, forming the *pyramidal decussation.*

On each lateral surface dorsal to the pyramid is a rounded mass, the *olive.* Each olive gives rise to a bundle of fibers that crosses the midline of the medulla and, together with fibers from the nucleus gracilis and nucleus cuneatus, forms the major portion of the inferior cerebral peduncle that passes to the cerebellum. Two shallow grooves may also be seen, the *anterior* and *posterior lateral sulci.* Roots of various cranial nerves emerge from these sulci.

The arrangement of the *gray* and *white matter* in the medulla is considerably different from that in the spinal cord. In the cord, the gray matter occupies the central portion; in the medulla oblongata, the presence of the fourth ventricle in the central dorsal portion causes the gray matter to be so distributed that it occupies the floor and sides of the ventricle. The gray matter contains a number of *nuclear masses* separated by masses of fibers that form the white matter. These nuclear masses are the *nuclei of origin* of efferent fibers and *nuclei of termination* of afferent fibers.

Among the important nuclei are the *nucleus gracilis* and the *nucleus cuneatus,* as well as various nuclei of the cranial nerves.

Superiorly, the medulla oblongata is connected dorsolaterally to the cerebellum by the *inferior cerebellar peduncles* or *restiform bodies.* It is separated from the pons by a furrow from which the abducens, facial, and acoustic nerves emerge.

Functions of the Medulla Oblongata. The medulla oblongata performs the following functions:

1. It contains the following nuclei: (*a*) the nuclei of termination of certain ascending tracts of the spinal cord, (*b*) the nuclei of termination of certain cranial nerves (V, VII, IX, X), and (*c*) the nuclei of origin of certain cranial nerves (IX, X, XI, XII).
2. It is the connecting pathway for ascending and descending fibers between the spinal cord and other portions of the brain.
3. It contains important reflex centers regulating certain vital activities, namely, the *cardiac center* (which regulates heartbeat), the *respiratory center* (which regulates the rate of respiration), and the *vasoconstrictor center* (which regulates the diameter of the blood vessels). There are other centers that mediate swallowing, vomiting, sneezing, coughing, and other reflexes.

PONS (PONS VAROLII). The pons lies directly above the medulla and anterior to the cerebellum. It is somewhat ovoid in shape and presents on its ventral surface a transverse band of fibers arched like a bridge (hence the name *pons*). These fibers converge on either side to form the middle *cerebellar peduncle* (*brachium pontis*), which connects the pons with the cerebellum. From the anterior surface of these peduncles emerge the roots of the trigeminal nerve (V); from the groove separating the pons from the medulla emerge the roots of cranial nerves VI (abducent), VII (facial), and VIII (vestibulocochlear).

Structure of the Pons. The pons has two regions: the basilar or ventral portion and the tegmental or dorsal portion.

The *basilar portion* consists of transverse fibers, longitudinal fibers, and pontine nuclei. The *transverse fibers,* in a thick layer on the ventral surface of the pons, together with some of the deep fibers, form the major portion of the brachium pontis. The *longitudinal fibers,* arranged in bundles or fasciculi and entering the pons from the cerebral peduncles, comprise three tracts—the *corticospinal tract* (fibers that continue through the pons into the pyramids of the medulla and on to the spinal cord), the *corticobulbar tract* (fibers that pass through the pons and end in motor nuclei of cranial nerves in the medulla), and the *corticopontile tract* (fibers that end in the pontine nuclei). The *pontine nuclei* comprise small masses of motor neurons located between the transverse fibers; axons pass to the opposite side and enter the cerebellum by way of the brachium pontis.

The *tegmental portion,* consisting of longitudinal and transverse fibers and gray nuclei, forms the upper portion of the floor of the fourth ventricle. Its longitudinal fiber tracts are in general continuous with those of the medulla oblongata and are arranged in a similar position. Nuclei located in this portion are *motor nuclei* of the abducens, facial, and trigeminal nerves and *sensory nuclei* of the trigeminal nerve and the vestibular and cochlear branches of the acoustic nerves. Also present are nuclei and fibers of the reticular formation.

Functions of the Pons. The pons varolii has the following functions:

1. It contains important fiber tracts connecting the medulla oblongata with the higher centers of the brain.
2. It contains motor and sensory nuclei of three cranial nerves: V (trigeminal), VI (abducent), and VII (facial). It also contains terminal sensory nuclei of the vestibular and cochlear branches of the VIII nerve (vestibulocochlear).
3. It contains portions of the reticular activating system.

MIDBRAIN (MESENCEPHALON). The midbrain or mesencephalon lies directly above the pons and anterior and slightly superior to the cerebellum. It connects the pons and the cerebellum to the diencephalon and cerebrum, which lie immediately above it. Passing through the central portion is a narrow canal, the *cerebral aqueduct,* which connects the third ventricle of the diencephalon with the fourth ventricle of the medulla oblongata. Surrounding the cerebral aqueduct is a *central gray substance* of a gelatinous nature.

Internal Structure of the Midbrain. The midbrain consists of two portions, a *ventral portion,* which includes two cerebral peduncles and their tegmental structures, and a *dorsal portion* or *tectum,* which comprises a four-part body, the *corpora quadrigemina.*

The *ventral portion of the midbrain* includes two broad bands of fibers, the *cerebral peduncles.* Inferiorly, the peduncles converge near the midline and enter the superior surface of the pons; superiorly, they enter the lower surface of the cerebral hemispheres. The depressed triangular area between them is called the *interpeduncular fossa.* From the medial surface of each peduncle, the oculomotor (III) nerve emerges, whereas from the posterolateral surface, in a groove between the peduncles and the pons, emerges the trochlear (IV) nerve. Seen in cross section, each peduncle consists of a dorsal portion (*tegmentum*) and a ventral portion (*base* or *crusta*), the two being separated by a deeply pigmented layer of gray matter, the *substantia nigra.*

The *tegmentum,* which contains a part of the reticular formation, is continuous below with the tegmental portion of the pons. It consists of white matter made up of longitudinal and transverse fibers and gray matter made up of nuclei and fiber tracts. Some of the important nuclei found here are the following.

1. *Oculomotor and trochlear nuclei,* motor nuclei of cranial nerves III and IV.

2. *Accessory* or *pupilloconstrictor nucleus,* which gives rise to efferent fibers that pass through the oculomotor nerve to the intrinsic muscles (ciliary muscle and sphincter pupillae) of the eye.

3. *Mesencephalic nucleus,* which contains the cell bodies of sensory fibers present in the maxillary branch of the trigeminal (V) nerve.

4. *Red nucleus* (*nucleus ruber*), the cells of which contain a red pigment. Afferent and efferent fibers connect this nucleus with the higher centers of the brain, the cerebellum, the pons, and the medulla. This nucleus serves as an important relay station for impulses to and from these regions. It contains the cells of origin of efferent fibers in the rubrospinal tract.

Immediately below the cerebral aqueduct is a mass of fibers, the *decussation* of the superior cerebellar peduncles. After crossing, the fibers pass to the red nucleus or ventral thalamus. Two other bands of fibers, the *lateral* and *medial lemnisci,* which carry sensory impulses, lie in the tegmentum.

The *dorsal portion of the midbrain* (*tectum*) is traversed by two grooves, a longitudinal groove and a transverse groove, which divide it into four parts or *colliculi,* two of them superior and two inferior, the whole forming the *corpora quadrigemina.* Extending laterally from each colliculus is a band of fibers or brachium. The *superior brachium* extends from the superior colliculus to the *lateral geniculate body* of the thalamus. The *inferior brachium* extends from the inferior colliculus to the *medial geniculate body* of the thalamus and the auditory area of the temporal lobe.

Functions of the Midbrain. The midbrain has the following functions:

1. It serves as a connecting region between the higher and lower centers of the brain.

2. It contains the motor nuclei of the oculomotor (III) and trochlear (IV) nerves, and other nuclei, among them the red nucleus.

3. It plays an important role in equilibrium and postural reflexes; the center for maintaining equilibrium is located near the red nucleus.

4. The colliculi serve as reflex centers, the superior mediating visual, auditory, and tactile impulses, the inferior mediating auditory impulses.

5. It plays a role in the functions of the reticular activating system to be discussed later.

DIENCEPHALON. The diencephalon lies between the midbrain and the cerebral hemispheres. It includes the *thalamus, epithalamus, subthalamus,* and *hypothalamus.* The general relationships between these regions can be observed in a diagram of the midsagittal section. In the center of the diencephalon lies a cavity in the form of a vertical cleft, the *third ventricle.*

Inferiorly, it communicates with the fourth ventricle by means of the cerebral aqueduct; laterally, it communicates with the lateral ventricles of the cerebral hemispheres by means of two small openings, the *interventricular foramina.*

Thalamus. The thalamus consists of two large ovoid masses, each about 4 cm in length, lateral to the third ventricle. Each consists largely of gray matter covered by a thin layer of white matter, the *stratum zonale.* The posterior end presents a medial enlargement, the *pulvinar,* which continues laterally as the *lateral geniculate body.* Directly beneath the pulvinar is an oval mass, the *medial geniculate body.* Connecting the two masses is a flattened band of gray matter, the *intermediate mass* (middle commissure); this mass is not present in all individuals.

The two geniculate bodies are small ovoid bodies lying in the posterolateral portion of the thalamus. The *medial body* receives acoustic fibers from the inferior colliculus of the midbrain and relays their impulses to the temporal lobe of the cortex; the *lateral body* receives fibers from the optic tract, about three-fourths of such fibers ending here. Impulses of the optic fibers are relayed to the visual areas in the occipital lobe of the cortex.

The thalamus contains a number of nuclei, grouped according to their position as follows: midline, anterior, medial, lateral, and posterior. In these nuclei, secondary, afferent fibers carrying sensory impulses from receptors located in all parts of the body synapse with neurons whose axons pass to the cortex or other parts of the brain. The thalamus thus serves as an important *relay station.* Neural connections exist between the cortex and the thalamus as well as between the thalamus and the cortex, providing reverberating circuits between these two regions. Impulses to the cortex may be modified or amplified for the proper response; impulses from the cortex usually exert an inhibiting effect on thalamic activity.

Epithalamus. This portion of the diencephalon includes the trigonum habenulae, the pineal body, and the posterior commissure.

The *trigonum habenulae* is a triangular depressed area located directly anterior to the superior colliculus at the base of the pineal stalk. It contains the habenular nuclei, the olfactory correlation centers.

The *pineal body* or *epiphysis* is a conical reddish-gray body about 8 to 10 mm in length attached by a narrow stalk to the posterior portion of the roof of the third ventricle. The ventricle extends a short distance into the stalk. The function of the pineal body is uncertain. As a vestigial homologue of a light-sensitive organ or "third eye" present in certain primitive reptiles (*Sphenodon*), it has been regarded as a rudimentary sense organ. There is some evidence that the pineal body or gland is an endocrine organ producing *melatonin* and related indols, which exert an inhibiting effect on the gonads. In this role, the pineal gland would serve as a neuroendocrine transducer mediating the effects of light on the gonads. Further details of its endocrine nature are given in the discussion of endocrine glands, pages 246–247.

The *posterior commissure,* a round band of white fibers connecting the two halves of the diencephalon, lies just below the pineal stalk and above the cerebral aqueduct.

Subthalamus (Ventral Thalamus). This includes the ventrolateral portions of the diencephalon lying anterior and lateral to the midbrain. The red nucleus and the substantia nigra of the midbrain may project into it. The subthalamus contains a band of fibers, the *tegmental field of Forel* and the *nucleus of Forel.* Other smaller nuclei (nucleus subthalamus and zona incerta) are present.

Hypothalamus. The hypothalamus comprises the major portion of the ventral region of the diencephalon and forms most of the floor of the third ventricle. It includes the *optic chiasma, mammillary bodies, tuber cinereum, infundibulum,* and *neurohypophysis.*

The *optic chiasma* is an X-shaped band of fibers located anteriorly. Nerve fibers from the optic nerves enter the chiasma and undergo partial decussation before continuing on to the lateral and medial geniculate bodies of the thalamus. The *mammillary bodies* are two rounded bodies lying in the floor of the third ventricle between the cerebral peduncles and immediately above the pons. They contain nuclei involved in relaying olfactory impulses. The *tuber cinereum* is a mass of gray matter lying between the optic chiasma and the mammillary bodies. It is bounded laterally by the optic tracts.

The *infundibulum* is a hollow, conical process extending inferiorly from the tuber cinereum. Its enlarged upper end is the *median eminence;* its lower end is expanded and forms the *neurohypophysis,* which constitutes the principal portion of the posterior lobe of the hypophysis. It becomes attached to a glandular structure, the *adenohypophysis,* and together they comprise the *hypophysis cerebri* or *pituitary gland,* an endocrine structure.

The hypothalamus contains several nuclei. Of special importance are the paired *supraoptic* and *paraventricular nuclei.* Neurons in these nuclei are the source of the neurohormones *oxytocin* and *vasopressin,* which pass through axons in the infundibular stem to the posterior lobe of the hypophysis, where they are released into the circulation.

Functions of the Diencephalon. The diencephalon has the following functions:

It gives rise to the optic nerve (II), fibers of which pass through the optic chiasma to the retina of the eye.

The thalamus serves as a *relay station.* Information received from sensory receptors (except olfactory receptors) throughout the body is processed and relayed to various cortical sensory areas. It is a center for crude, uncritical sensations. Heat, cold, pain, touch, light, and sound are recognized, but the ability to localize the source of the stimulus and to understand its significance is lacking.

The hypothalamus, although weighing only 4 g and comprising only $\frac{3}{1000}$

of the weight of the brain, plays a vital role in many bodily functions. Among these are the following.

1. General control and integration of organs innervated by the autonomic nervous system. This involves all visceral activities including the contraction of smooth and cardiac muscle and the secretion of glands.
2. Control of specific bodily activities. Hypothalamic control centers include a *temperature center,* which serves as a *thermostat* by which a constant body temperature is maintained; an *appetite center* or *appestat,* which is stimulated by hunger, thus initiating feeding; a *satiety center,* which is stimulated by food intake and acts to restrict excessive feeding; a *thirst center,* which responds to the fluid content of the blood and acts to maintain optimum water and salt balance; and a *sex center,* upon which normal sex drive and the functioning of the sex organs depend.
3. Control of feelings and reactions associated with emotional states such as pleasure, displeasure, pain, fright, and rage. This is probably accomplished through the production of *endorphins,* endogenous morphinelike substances. Endorphins, which include the *enkephalins,* are small peptides found in the brain and thought to arise from a precursor, *beta-lipotropin.* They mimic the action of opiates. Although their function in neurophysiology has not been fully determined, there is evidence that they play a role in the relief of pain and the generation of moods, and they may possibly have the ability to alter mental states and behavior. They also may serve as neurotransmitters.
4. Control of psychosomatic disorders. Persistent stressful situations may, through the automatic nervous system, bring about disorders of bodily functions such as gastrointestinal disorders, cardiovascular disturbances, chronic respiratory illness, arthritis, and malignancies.
5. Integration of the nervous and endocrine systems. The hypothalamus produces (*a*) *releasing factors* or *hormones,* which regulate the production and release of hormones from the anterior lobe of the pituitary gland, and (*b*) *neurohormones* (oxytocin and vasopressin), which pass to the posterior lobe of the pituitary gland, where they are released into circulation.
6. Control of sleep. The hypothalamus and the reticular activating system act to control sleep and waking patterns. The hypothalamus contains activating regions that bring about transition from sleep to wakening and suppressor regions that induce somnolence.

Reticular Formation. This is a diffuse, poorly defined collection of nerve cells, nuclei, and fibers extending from the medulla upward through the pons and midbrain into the thalamus and hypothalamus and downward into the cervical region of the spinal cord. It occupies the central portion of the brain stem. This neuron network receives impulses from nearly all receptors in the body, both enteroceptive and exteroceptive, through direct fibers and also from collaterals of sensory fibers terminating in the cerebral cortex. In addition, it receives impulses from various parts of the brain, including the cerebellum, basal ganglia, thalamus, and portions of the cerebral cortex. Efferent impulses pass from the reticular formation upward to all parts of the cerebral cortex and downward into the spinal cord. The formation is profoundly influenced by carbon dioxide and such chemicals as hormones, anesthetics, tranquilizers, and narcotics.

The *reticular formation,* also called the *reticular activating system* (*RAS*), functions somewhat like a central switchboard operator who has the task of assigning priorities, that is, what messages are to be listened to, which are to be amplified, which ignored or minimized, and which held to be acted on later. Electrical stimulation of the RAS increases electroencephalographic (EEG) activity—hence it serves as an *arousal mechanism;* destruction results in depressed activity or coma during which the brain presents an EEG characteristic of the *sleeping state.* The RAS thus serves as a center of *consciousness.* When it is inactivated, as in anesthesia, it is unable to send impulses to the higher cortical centers of the brain by which an awareness of incoming sensory impulses is manifested. As a result, the patient is unconscious. There is evidence that the cortex has the ability to actively discriminate between sensory impulses even during periods of sleep or unconsciousness.

The reticular activating system is responsible for the condition of *habituation* observed in most organisms in which the monotonous repetition of the same stimulus leads to a decrease in response. A loud sound, when first heard, usually results in a person's becoming alert and directing attention to the stimulus. However, continuous repetition of the sound causes a person to become progressively less responsive and, after a time, the sound may be completely ignored. However, if the sound suddenly changes by being amplified or much reduced, the original alertness to the sound will again be manifested. This condition of habituation is not the result of receptor fatigue or adaptation but a function of the reticular formation.

Through efferent impulses, the reticular activating system plays an important role in the functioning of muscles and glands. In some way it may activate certain neural circuits, while other circuits are deactivated. Impulses dispatched down the spinal cord to muscles may increase or decrease the degree of response. The degree of contraction of resting muscles essential in the maintenance of general muscle tone and normal posture is largely controlled by the reticular formation.

Cerebellum. The *cerebellum* lies posterior to the pons and medulla oblongata and inferior to the occipital lobes of the cerebral hemispheres. It is a large, bilobular structure with its surface thrown into small folds or *folia* separated by narrow but deep *sulci.* Deeper transverse *fissures* divide the cerebellum into *lobules.*

The cerebellum consists of two lateral *hemispheres* and a median unpaired *vermis.* It is connected to other parts of the brain as follows: to the *medulla* by the inferior cerebellar peduncle (restiform body), to the *pons* by the middle cerebellar peduncle (brachium pontis), and to the *midbrain* by the superior cerebellar peduncle (brachium conjunctiva).

STRUCTURE OF THE CEREBELLUM. The cerebellum is composed of gray matter, which consists of an outer layer or *cortex* and several nuclear masses, and an irregular central portion of white matter, the *medulla.* Seen in sagittal section, the white matter has a treelike form, for which reason it is named *arbor vitae.*

The cortex of the cerebellum has two layers of cells: the outer gray *molecular layer* and an inner, rust-colored *nuclear* or *granular layer.* Between these layers lies a single row of *Purkinje cells.* These peculiar-shaped neurons are characteristic of the cerebellum. The cell body is large and flask-shaped; from its neck, one or two dendrites emerge and pass into the molecular layer, where they divide and subdivide into numerous branches, giving the whole cell a treelike appearance. A single axon emerges from the bottom of the "flask" and passes through the nuclear layer into the white substance of the medulla. The nuclei of the cerebellum include a large *dentate nucleus* and the smaller *globose, emboliform,* and *fastigial nuclei,* all of which are paired.

The white matter is composed of two types of fibers: *projection fibers* (axons and dendrites entering and leaving the cerebellum through the cerebral peduncles) and *fibrae propriae* (composed of *commissural fibers,* which connect the two hemispheres, and *association fibers,* which connect the folia).

FUNCTIONS OF THE CEREBELLUM. In general, the cerebellum is concerned with the regulation and coordination of complex voluntary muscular movements without, however, initiating them. (1) It plays a role in the maintenance of muscle tonus. (2) It is involved in reflexes concerned with the maintenance of normal posture and equilibrium, and it serves to mediate vestibular and postural (proprioceptive) impulses. (3) It is essential for the normal timing and integration of voluntary muscular movements, especially those involved in skilled activities. (4) It reinforces muscle contractions, especially those in muscles of the extremities.

The cerebellum functions in this manner: Voluntary motor impulses originating in the cerebral cortex or impulses resulting from inborn or conditioned reflexes initiate the contraction of specific muscles. An apparently simple action such as the raising of the hand to the face to avoid a blow necessitates the synchronized action of 50 or more muscles acting on 30 separate bones of the arm and hand. If the head is turned, many additional muscles would be involved. Proprioceptive impulses arising from stimulation of muscle spindles in contracting muscles or proprioceptors in joints pass to the cerebellum. Within the cerebellum, pons, and midbrain are "stored programs" of neural connections that automatically coordinate the contraction of prime movers and synergists and the relaxation of antagonistic muscles. This results in precise and purposeful movements.

Removal of or injury to the cerebellum leads to disorders in motor

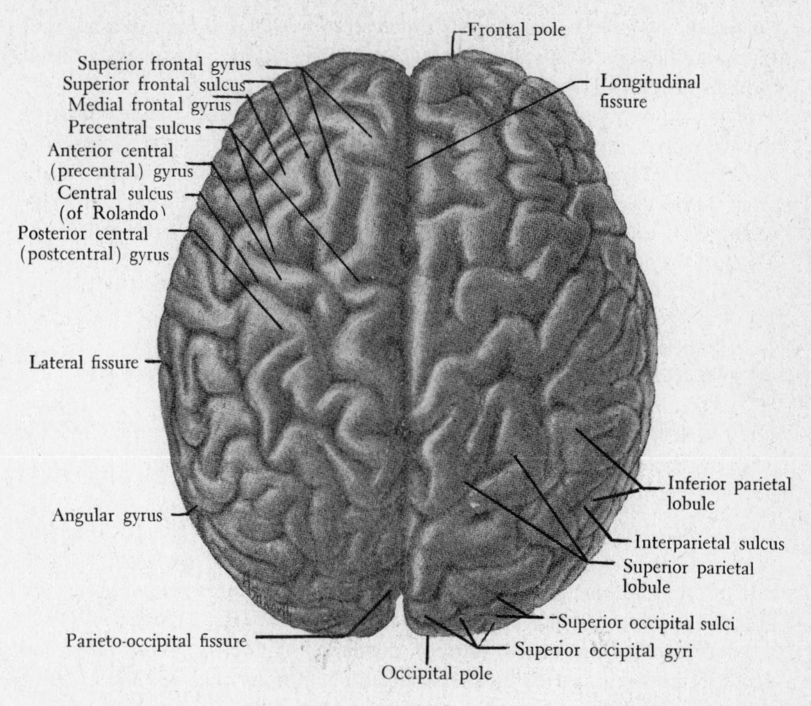

Fig. 4-4. Surface of the cerebrum. (Reprinted with permission of Blakiston Division, McGraw-Hill Book Company, from *Morris' Human Anatomy,* 11th ed., ed. by J. P. Schaeffer, 1953.)

performance. These may be disturbances in posture or postural tone or disturbances in voluntary movement. Maintenance of balance is difficult, and *hypertonia* (excessive muscular tone) or *hypotonia* (diminished muscular tone) may result. The rate, direction, and strength of voluntary muscle movements are affected, and muscle *tremor* may develop. The condition resulting from failure of muscle coordination is called *ataxia.*

Cerebrum (Cerebral Hemispheres) (Fig. 4-4). The cerebral hemispheres, the largest part of the brain, form a bilobular mass covering the other portions of the brain. The hemispheres are separated by a deep cleft, the *longitudinal cerebral fissure.* Each contains a cavity, the *lateral ventricle,* that communicates with the third ventricle by an opening, the *interventricular foramen.* The two hemispheres are connected across the midline by a broad band of white fibers, the *corpus callosum,* and two smaller bands, the *anterior* and *posterior commissures.* Each hemisphere consists of an outer layer of gray matter, the cerebral *cortex,* which encloses the inner *white matter.*

The two hemispheres, although similar in general structure, function differently. The left hemisphere is involved in speech and logical, analytical thinking, the right hemisphere with spatial relations and artistic and holistic thought processes.

SURFACE ANATOMY OF THE CEREBRUM. The surface of the cerebrum is thrown into a large number of folds called *convolutions* or *gyri.* These are separated by furrows, the deeper ones being known as *fissures,* the shallower ones, *sulci.* The gyri and sulci have fairly definite locations, but they vary to some extent in different individuals and even in the two hemispheres of a single brain. The principal fissures or sulci are as follows:

Fissure or sulcus	Location
Lateral fissure (fissure of Sylvius) Central sulcus (fissure of Rolando)	on the lateral surface
Parieto-occipital fissure	on the superior-dorsal surface
Calcarine fissure Cingulate fissure	on the medial surface
Collateral fissure Sulcus circularis	on the inferior surface

The *lobes* into which each hemisphere is divided by the fissures and sulci can all be seen superficially, with the exception of the insula, which lies internal to the other lobes. The lobes are (1) the *frontal lobe,* the anterior portion; (2) the *parietal lobe,* the superior-lateral portion; (3) the *temporal lobe,* the inferior-lateral portion; (4) the *occipital lobe,* the posterior portion; and (5) the *insula* (island of Reil, or central lobe), inferior to the frontal lobe and internal to the temporal lobe.

The *gyri* are the result of unequal growth processes that greatly increase the surface area of the cerebral cortex. The principal gyri are as follows:

Gyri as Seen from the Lateral Surface

In the frontal lobe:	Anterior central gyrus, which contains the motor area Frontal gyri (superior, middle, and inferior)
In the temporal lobe:	Superior, middle, and inferior temporal gyri
In the parietal lobe:	Posterior central gyrus, which contains sensory area Superior parietal lobule Inferior parietal lobule { Supramarginal gyrus, Angular gyrus }

Gyri as Seen on the Medial and Basal Surfaces

In the frontal and parietal lobes:	Superior frontal gyrus Gyrus cinguli (superior to the corpus callosum) Gyrus fornicatus Gyrus rectus
In the temporal lobe:	Inferior temporal gyrus Fusiform gyrus (Parahippocampal gyrus) Hippocampal gyrus
In the occipital lobe:	Cuneus Lingual gyrus

GRAY MATTER OF THE CEREBRAL HEMISPHERES. The parts of the cerebrum consisting principally of gray matter are the cortex, the basal ganglia, and the limbic system.

The *cerebral cortex* is the surface layer of the cerebral hemispheres. It covers the gyri and lines the sulci and fissures to form an area estimated at about 200,000 mm^2. The cortex varies in thickness from 1.5 to 4.0 mm, being thinner at the bottom of the sulci and in the occipital region. It contains neurons (cell bodies and processes), neuroglia, and blood vessels. It is estimated that the cortex contains the cell bodies of from 10 to 15 billion neurons.

MICROSCOPIC STRUCTURE OF THE CORTEX. The cortex consists of six layers as follows: (1) the outermost *molecular* or *plexiform layer* containing horizontal cells and their fibers; (2) the *outer granular layer* containing granule cells of small size; (3) the *pyramidal cell layer* containing principally large pyramidal cells and some granule cells; (4) the *inner granular layer* containing stellate and granule cells; (5) the *ganglionic or inner pyramidal layer* containing principally pyramidal cells, which in the motor cortex are extremely large and called *Betz cells;* and (6) a *polymorphic layer* containing cells of many shapes.

SURFACE ANATOMY AND FUNCTIONS OF THE CORTEX. The regions of the cortex (Fig. 4-2) differ structurally and functionally. They differ in thickness and composition of the layers, thickness of the cortex itself, number and position of the white fibers, and number of afferent and efferent fibers. On the basis of histologic, experimental, and clinical data, the areas can be classified as sensory, motor, and association areas.

The *sensory areas,* by way of the thalamus, receive impulses that originate in sense organs or receptors. It is in these areas that conscious sensations are evoked and perception and localization of the stimulus are accomplished. The more important of these areas are (1) the *visual area,* located in the occipital lobe in the vicinity of the calcarine fissure and concerned with visual sensations; (2) the *auditory area,* located in the temporal lobe and concerned with auditory sensations; and (3) the *somesthetic* or *sensory area,* located in the postcentral gyrus of the parietal lobe and concerned with sensations of touch, pressure, position, and temperature.

The *motor area* exercises control over skeletal muscles. It is located in the precentral gyrus immediately anterior to the fissure of Rolando. In this area are found the giant *pyramidal cells of Betz,* large motor neurons whose axons constitute fibers of the corticospinal and corticobulbar tracts. The axons of these neurons (called *upper motor neurons*) continue down the spinal cord and synapse with neurons located in the anterior horns of the gray matter of the spinal cord. Axons of the latter (called *lower motor neurons*) innervate the skeletal muscles. Impulses from the motor area control the action of specific muscles. Just anterior to the motor area is a *premotor area,* the neurons of which control whole series of movements, such as those involved in writing. Anterior to this area is the frontal eye-field area, which controls the direction of movement of the eyes.

Experiments show that the motor area of one hemisphere controls the contraction of muscles on the opposite side of the body. This is explained by the fact that most of the axons of the pyramidal cells in their passage downward through the brain and spinal cord cross to the opposite side. This occurs principally in the medulla in the *decussation of the pyramids.* Consequently, injury to the motor area in one hemisphere (as from skull fracture or cerebral hemorrhage) results in paralysis of muscles on the opposite side of the body.

The *association areas* comprise most of the cerebral cortex that is not included in the sensory and motor areas. They are connected with one another, with motor and sensory areas, and with similar areas in the opposite hemisphere. They also receive numerous fibers from the thalamus. Damage or injury to certain association areas result in defects of speech and in failure to understand word meanings. A specific area known as *Broca's area* lies in the left inferior frontal gyrus just above the lateral fissure; it was described by Broca as the center for articulate speech. This *motor speech center* sends impulses to the premotor and motor areas to bring about coordinated activities of muscles of the tongue, lips, pharynx, larynx, and chest that enable a person to express thoughts in words. Destruction of or injury to this area results in *expressive aphasia* or inability to speak. *Receptive aphasia* is the difficulty in understanding spoken or written words; *amnesic aphasia* is the inability to remember the correct words to use.

Association areas are thought to be involved in the higher mental processes, which include thinking, reasoning, memory, and judgment. The frontal lobes, although they play no role in the physical activities of the body, function in intellectual activities, especially in the learning of new concepts. They are active in the formulation of abstract ideas, the fixation of attention, and the organization of mental activity. They play a role in the development of moral and social values, the development of personality traits, and memory.

Memory, the ability to recall past experiences, is of two types, *short-term memory* and *long-term memory.* It has been theorized that memory depends on electrical circuits (engrams) encoded in neural networks. However, it has not been possible to identify any specific part of the brain, whether a neuron or a group of neurons, as a repository of stored information. Removal of the frontal lobe does, however, impair memory of recent events. Damage to the temporal lobe also interferes with memory. The mechanism of recall is at present unknown. Loss of memory, which may be total or partial, is called *amnesia.* It may result from head injury as from a concussion, cerebral anoxia, electroconvulsive shock, or the effects of drugs.

The *basal ganglia* (Fig. 4-5) include several masses of gray matter located in the basal portion of each cerebral hemisphere. They are the corpus

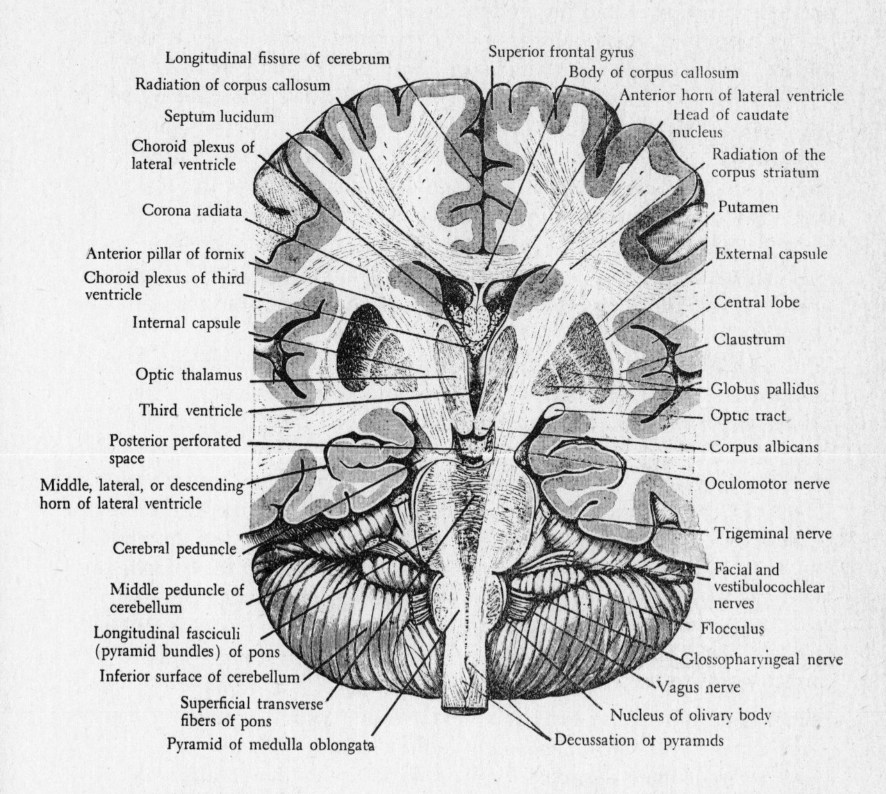

Fig. 4-5. Basal ganglia. Transverse section of the brain in the direction of the medulla oblongata and the cerebral peduncles. The course of the pyramidal tract from the decussation of the pyramids upward, through the pyramid of the medulla oblongata, the pons varolii, and the crusta of the cerebral peduncle or crus cerebri into the internal capsule, where it enters the peduncle of the corona radiata. In the medullary center or white matter of the cerebrum, we see the radiation of the corpus callosum emerging with the fibers of the corona radiata as they diverge from the internal capsule and with the fibers of the radiation of the corpus striatum. Note the two parts of the corpus striatum (the caudate nucleus and the lentiform nucleus, which is divided into the putamen and the globus pallidus); the white fibers of the internal capsule, which separates the putamen from the caudate nucleus; the fibers of the external capsule, which separates the putamen from the triangular claustrum; the large, ovoid thalamus (optic thalamus), which is separated from its fellow on the opposite side by the third ventricle; the septum pellucidum, which separates the anterior, posterior, and inferior horns of the lateral ventricle; the choroid plexus, which is located in the roof of the third ventricle; and the choroid plexus of the lateral ventricle. (Reprinted with permission of The Macmillan Company from C. V. Toldt and A. D. Rosa, *Atlas of Human Anatomy for Students and Physicians,* 1948.)

striatum, the amygdaloid nuclei, and the claustrum.

The *corpus striatum* consists of two portions: a dorsal part (*caudate nucleus*) and a ventral part (*lentiform nucleus*); the latter is divided into the *putamen* and the *globus pallidus*. The caudate nucleus and the putamen are separated by a band of white fibers, the *internal capsule,* which gives the structure a striated appearance; hence the name *corpus striatum*. The internal capsule also separates the lentiform nucleus from the thalamus, which is located medially to it. Lateral to the lentiform nucleus is another band of fibers, the *external capsule,* which separates it from the *claustrum,* a triangular mass lying beneath the cortex of the insula. The *amygdaloid nuclei* consists of a mass of nuclei located in the roof of the inferior horn of the lateral ventricle.

Closely associated with the basal ganglia, both anatomically and functionally, are the *subthalalamic body* of the diencephalon and the *red nucleus* and *substantia nigra* of the midbrain.

It has not been possible to attribute specific functions to specific regions of the basal ganglia, but through clinical observation it has been noted that lesions of these structures are associated with a number of disorders involving motor activity of the body. Most of these disorders are characterized by tremor and rigidity of voluntary muscles. Among these conditions are (1) *paralysis agitans* (*parkinsonism, shaking palsy*), characterized by tremor, muscle rigidity, slowness, and weakness in voluntary movements and loss of associated movements, as swinging the arms when walking; (2) *chorea* (*St. Vitus' dance*), in which movements are rapid, jerky, and purposeless; and (3) *athetosis,* in which involuntary movements, especially those in the hands, are slow, sinuous, and writhing.

All three of these disorders result from failure of the basal ganglia and related structures in the thalamus and midbrain to inhibit impulses discharged from the cerebral cortex. As a consequence, the cortex spontaneously discharges impulses that result in abnormal movements. In lower animals, the basal ganglia can initiate voluntary movement. This possibility exists in humans, but it has not been demonstrated.

Rhinencephalon or Olfactory Brain. The rhinencephalon is the portion of the cerebrum involved in the reception and integration of olfactory impulses. It includes the following structures:

1. *Olfactory bulb,* an oval mass lying beneath the frontal lobe and directly above the nasal cavity. It receives the olfactory nerves, which pass upward through the cribriform plate from the olfactory region of the nasal cavity.
2. *Olfactory tract,* a band of fibers passing posteriorly from the bulb. On entering the substance of the cerebrum it divides into two bands: the *lateral* and *medial olfactory striae*.
3. *Olfactory trigone,* a small triangular area lying between the diverging striae. Posterior to it is the *olfactory tubercle*. This forms a part of the

anterior perforated substance, which extends from the olfactory striae to the optic tract.

4. *Piriform area,* an area including the *hippocampal gyrus,* the *uncus,* and the *lateral olfactory sulcus.*

5. *Hippocampal formation,* a structure that includes the *subcallosal gyrus,* the *supracallosal gyrus,* the *longitudinal striae* of the corpus callosum, the *diagonal band of Broca,* the *dentate gyrus,* and the *hippocampus.* These parts lie along the medial wall of each cerebral hemisphere. The hippocampus is an elevated area lying along the floor of the inferior horn of each lateral ventricle.

6. *Paraterminal body,* a triangular area of the cortex anterior to the *lamina terminalis,* which forms the anterior wall of the third ventricle.

7. *Fornix,* a paired, curved structure lying over the thalamus and under the corpus callosum. It consists of *two crura* whose anterior ends, the *fimbriae,* lie adjacent to the hippocampus. As the crura curve upward over the thalamus, they unite in the median plane to form the *body of the fornix.* Anteriorly, the crura diverge and curve downward to form the *anterior pillars.* Posteriorly, the crura are connected by the *hippocampal commissure;* anteriorly, by the *anterior commissure.*

The *septum pellucidum* is a thin sheet of nervous tissue consisting of two layers or laminae. It is triangular in shape and is attached to the corpus callosum above and the fornix below. Each lamina forms the medial wall of the lateral ventricle of each hemisphere. Between the two laminae is a narrow cleftlike cavity, the so-called fifth ventricle, which is not a true ventricle.

FUNCTION OF THE RHINENCEPHALON. The rhinencephalon or "nose brain" is highly developed in lower animals and plays an important role in the basic activities of food-getting, nutrition, and reproduction. The olfactory sense warns of danger, guides an animal to food, and is of importance in recognition and mating activities. In humans, the sense of olfaction is of less importance, but through connections with the hypothalamus, its role in the expression of emotions, especially rage, has been established.

Limbic System. The portion of the brain designated the *limbic cortex* surrounds the brain stem and thus lies intermediate between the rhinencephalon and the major portion of the cortex. It includes the *limbic lobe* (*cingulate* and *hippocampal gyri*) and *hippocampus,* which are cortical structures. Subcortical structures include the *amygdala* (one of the basal ganglia) and the *septum.*

The limbic system plays an important role in the emotions. An *emotion* involves two aspects: an internal subjective "feeling" and a response to or manifestation of that feeling. Stimulation of or injury to these areas may produce or eliminate behavior patterns such as those exhibited in fear, rage, aggression, docility, pain, displeasure, pleasure, and sexual activity.

Autonomic responses elicited or altered include changes in heart rate, blood pressure, respiration, and glandular secretion, especially the secretion of hormones. These responses are mediated principally through the hypothalamus and the hypophysis.

White Matter of the Cerebral Hemispheres. The white matter of the cerebral hemispheres consists of three kinds of myelinated nerve fibers: *projection, association,* and *commissural fibers.*

Projection fibers connect the cortex with the lower parts of the brain and the spinal cord. They are of two types: (1) afferent or ascending fibers that arise from cell bodies located principally in the thalamus and (2) efferent or descending fibers that arise from cell bodies located in the motor area of the cerebral cortex and pass through the corona radiata and the internal capsule to the brain stem, where they form the *corticospinal, corticobulbar,* and *corticopontile tracts.*

Association fibers consist of axons that connect the gyri of the same hemisphere. Some are short fibers connecting adjacent gyri; others are long fibers connecting widely separated gyri.

Commissural fibers are transverse fibers leading from the gyri of one hemisphere to those of the other hemisphere. They are grouped into three bands: the corpus callosum and the anterior and posterior commissures. The *corpus callosum* is a thick transverse band of fibers that in a midsagittal section of the brain appears as a large, arched band lying in the central portion of the brain. It lies over the region of the thalamus and the septum pellucidum. Its anterior end is called the *genu,* its posterior end the *splenium.* The *anterior commissure* is a small transverse band of fibers lying anterior to the fornix in the anterior wall of the third ventricle. The *posterior commissure* is a band of fibers crossing the midline near the base of the stalk of the pineal body.

Blood and Energy Supply of the Brain. The brain receives its blood from two *internal carotid arteries* and two *vertebral arteries.* The latter unite to form the *basilar artery,* which joins with the others to form the *circulus arteriosus,* which encircles the hypophysis at the base of the brain. Branches from this arterial circle and the basilar artery enter the substance of the brain and divide into smaller arteries and capillaries, which pass to all parts of the brain. Arteries on the surface of the brain anastomose freely, thus assuring an equal supply of blood to all parts of the brain and an adequate supply in case one or more of the arteries become obstructed. However, within the brain tissue, the end vessels *do not* anastomose. This makes the brain especially susceptible to damage in case of occlusion of an artery as occurs in cerebral thrombosis. *Glucose* is the primary source of energy for brain cells, and a continuous supply is essential, since practically no glycogen is stored within the brain. If the brain is deprived of oxygen for 4 to 5 min or glucose for 10 to 15 min, irreversible brain damage may occur. Low blood sugar level, as in hypoglycemia, may

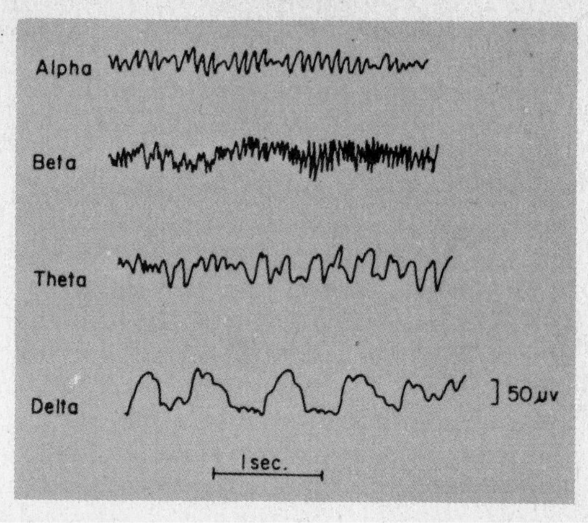

Fig. 4-6. Different types of normal brain waves. (Reprinted with permission of W. B. Saunders Co., Philadelphia, from A. C. Guyton, *Basic Human Physiology,* 2d ed., 1977.)

produce such mental symptoms as dizziness, apprehensiveness, and fainting. In severe cases, coma and death may result.

BLOOD-BRAIN BARRIER. The exchange of substances between the blood and the tissues of the brain and spinal cord differs from that in other parts of the body. Certain dyes injected into the bloodstream will diffuse into and stain all the tissues of the body except the brain and spinal cord. A *barrier* thus exists which prevents certain substances, especially macromolecules, from entering nervous tissue. This barrier may be due to differences in the structure of the capillary walls, the action of glial cells surrounding them, or physiologic differences in the transport mechanisms. The blood-brain barrier acts to maintain a constancy in the extracellular fluid surrounding the brain cells and to prevent unneeded or harmful substances from entering the brain. A similar barrier exists between the blood and the cerebrospinal fluid.

ELECTROENCEPHALOGRAMS. The electrical activity of the brain may be recorded by placing electrodes on the surface of the skull or within the brain tissue. With an apparatus called an *electroencephalograph,* a record can be made of changes in electrical potential. These changes constitute *brain waves* (Fig. 4-6) and reflect the activity of the cerebral cortex. The waves, amplified and recorded, constitute an *electroencephalogram* (*EEG*). Electroencephalograms reveal that the electrical activity of the brain varies with the degree of alertness, depending on whether the subject is relaxed, excited, drowsy, asleep, or in deep sleep. In normal individuals, four types of waves or rhythms are produced:

1. *Alpha waves.* Rhythmic waves occurring at a frequency of 8 to 10 cycles/sec. They occur in most normal individuals in a quiet resting state. They are absent during sleep.
2. *Beta waves.* Waves occurring at frequencies of 15 to 50 cycles/sec. They are recorded during periods of brain activity.
3. *Theta waves.* Waves occurring at frequencies of 4 to 7 cycles/sec. They occur during periods of emotional stress in adults; they may occur normally in children.
4. *Delta waves.* Waves occurring at frequencies of 1 to 3 cycles/sec. These waves are characteristic of deep sleep but may occur normally in an awake infant. When they occur in an awake adult, they indicate a serious brain disorder.

SLEEP. Sleep is a naturally recurring state of rest characterized by reduced physical and nervous activity, relative unconsciousness, and reduced responsiveness to external stimuli. Although sleep is an essential physiologic activity, its functional significance is not known.

There are two kinds of sleep: *slow-wave sleep* and *paradoxical sleep.* As a person becomes drowsy and falls asleep, the EEG pattern changes. The alpha waves, characteristic of the awake state, become less pronounced, and larger, slower waves appear. During *slow-wave* or *synchronized sleep,* the record may suddenly change, the waves becoming fast and of low voltage. The EEG pattern is that of a person awake and alert, despite the fact that the person is in a deep sleep. Tone in postural muscles is completely inhibited, stretch reflexes disappear, and eye movements underneath closed eyelids occur. This type of sleep, called *paradoxical, desynchronized,* or *rapid-eye-movement (REM) sleep,* usually lasts 10 to 15 min and occurs several times during a night. About 80 percent of sleep is slow-wave sleep, 20 percent paradoxical or REM sleep. It is during REM sleep that dreaming usually occurs. The restfulness of sleep is associated with the amount of REM sleep.

The hypothalamus and reticular activating system play important roles in the control of sleep. When a person is asleep, the reticular activating system is inactive. However, almost any kind of sensory stimulus, such as a loud sound or bright light or a pain impulse, can activate the system and bring about arousal. In experimental animals, injury to or stimulation of the lateral hypothalamus or the midbrain may produce *somnolence,* prolonged sleep, or even coma. Lesions of these areas resulting from tumors, cerebral hemorrhage, thrombosis, or infections such as sleeping sickness may result in a sustained sleeping state. As a consequence, it is thought that the brain contains an active mechanism that induces sleep. However, the hypothalamus also contains an *awakening region.* This region, located anteriorly, abolishes sleep and induces activity. The control of sleep apparently involves a sleep-awakening mechanism centered in the hypothalamus but responding to impulses from the reticular activating

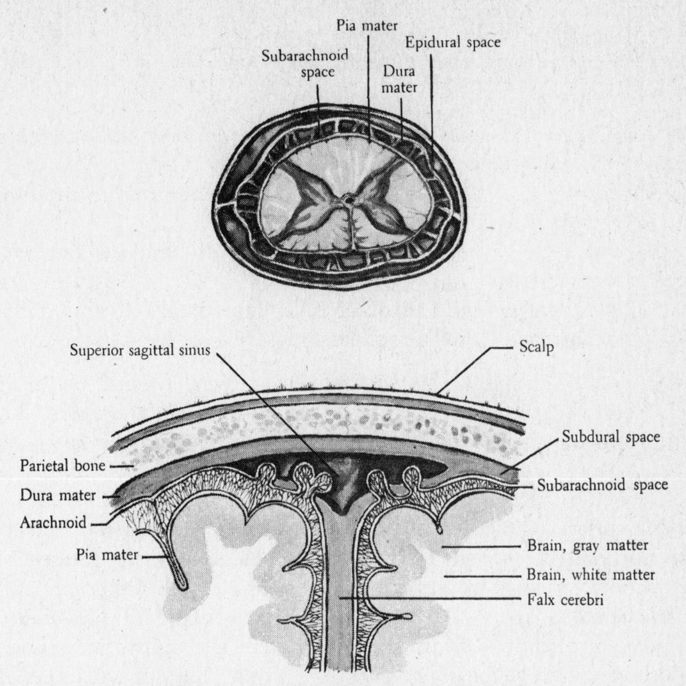

Fig. 4-7. Meninges of brain and spinal chord.

system. The cortex also plays a role in sustained wakefulness through impulses that act on the reticular activating system, triggering action that sustains a state of alertness.

THE MENINGES

The brain and the spinal cord are each enclosed by three membranes called *meninges;* these are the *dura mater,* the *arachnoid,* and the *pia mater.* The meninges of the brain differ in some respects from those of the spinal cord.

Cranial Meninges (Fig. 4-7). The membranes enclosing the brain have the following characteristics.

DURA MATER. This is a tough, fibrous membrane forming the outermost covering of the brain. It is dense and inelastic. The dura mater consists of two layers, an outer endosteal and an inner meningeal layer. The *endosteal layer* lines the bones of the cranial cavity and functions as the periosteum of these bones. At the foramina of the skull it forms an investing layer surrounding the cranial nerves and is continuous with the external periosteum on the cranial bones. The inner *meningeal layer* of the dura mater is smooth and lined with mesothelium.

Over the surface of the brain, folds or processes of the dura mater dip in between the various parts, forming partitions. The more important of these folds are:

1. *Falx cerebri,* a sickle-shaped fold that lies in the longitudinal fissure separating the two cerebral hemispheres.
2. *Tentorium cerebelli,* a fold that lies between the cerebellum and the occipital lobes of the cerebral hemispheres.
3. *Falx cerebelli,* a triangular fold that lies in the midline between the cerebellar hemispheres.
4. *Diaphragma sellae,* a small, circular, horizontal fold that encircles the *infundibulum* and forms a rooflike structure over the hypophysis.

ARACHNOID. This is a thin, delicate layer lying internal to the dura mater, from which it is separated by a very narrow *subdural space.* It does not dip into the fissures or sulci, with the exception of the longitudinal fissure.

Between the arachnoid and the pia mater is the *subarachnoid space.* Over the surface of the hemispheres the cavity is very narrow, but at the lower portions of the sulci triangular-shaped spaces are formed. The subarachnoid space is traversed by numerous delicate *trabeculae,* which give the space a spongelike appearance. At the base of the brain the arachnoid is separated from the pia mater, giving rise to spaces of considerable depth called *subarachnoid cisternae.* The subarachnoid space is filled with cerebrospinal fluid. It communicates with the fourth ventricle.

The arachnoid (and, in particular, the arachnoid cisternae) forms a protective structure that serves as a fluid-filled, weblike cushion separating the brain from the cranial bones, with which it would otherwise be contiguous.

At certain places, especially in the floor of the superior sagittal sinus, the arachnoid develops evaginations that, with the overlying dura mater, project into the cavity of the sinus. These are *arachnoid villi* through which the cerebrospinal fluid reenters the bloodstream. The arachnoid villi may project through the dura mater of the cerebral hemispheres into the inner surface of the cranial bones, giving the bones a pitted appearance. Here they are called *arachnoid granulations.*

PIA MATER. This innermost meninx is a highly vascular membrane that closely invests the brain, dipping into all the fissures, furrows, and sulci. It consists of fine, areolar tissue and mesothelial cells, and contained within it are numerous blood vessels supplying the brain. At certain places the pia mater is invaginated into the ventricles to form the *choroid plexuses* (described on page 122).

Spinal Meninges (Fig. 4-7). The three membranes that enclose the brain are continuous at the *foramen magnum* with the three membranes of the same names that enclose the spinal cord. However, they differ in some respects, as described here.

The *dura mater* in the vertebral canal is separated from the periosteum of the vertebrae by the *epidural space,* which contains loose connective and adipose tissue and many veins. On each side the dura is firmly connected to the cord by *dentate ligaments* of the pia mater. The dura mater forms tubular projections that invest the roots of the spinal nerves. These continue outward a short distance through the intervertebral foramina. The dura is separated from the arachnoid by a narrow space, the *subdural space.*

The *arachnoid* of the brain and the spinal cord are similar in structure. The *subarachnoid space* is wider in the cord than in the brain, especially in the lower part of the vertebral canal.

The *pia mater* of the spinal cord is denser than that of the brain and contains fewer blood vessels. Along the side of the cord, triangular processes extend from the pia mater and connect with the dura mater. These are extensions of a band that extends along each side of the spinal cord, separating the roots of the spinal nerves. This band is called the *dentate ligament.*

Sometimes the pia mater and the arachnoid are regarded as one membrane, in which case it is called the *pia-arachnoid* or *leptomeninges.*

Ventricles, Choroid Plexuses, and Cerebrospinal Fluid. In the development of the central nervous system (see page 95), the cavities of the primitive neural tube persist. In the brain these enlarge and form definite cavities or *ventricles* that communicate with each other and are continuous with the *central canal* of the spinal cord.

VENTRICLES OF THE BRAIN (Fig. 4-8). The *first* and *second* (or lateral) *ventricles* lie in the two cerebral hemispheres. Each consists of a body bearing three extensions: the *anterior, posterior,* and *inferior horns* (cornua). They are separated from one another by the septum pellucidum and communicate with the third ventricle by the *interventricular foramina.* The *third ventricle* is a narrow cleftlike cavity lying in the midsaggital plane between the thalami. It communicates with the fourth ventricle by the *cerebral aqueduct,* which passes through the midbrain. The *fourth ventricle* lies principally in the medulla. It is roughly quadrangular in shape, its upper and lower corners forming the *superior* and *inferior angles,* its lateral corners forming the *lateral recesses.* It is lined with epithelium. The *floor* of the fourth ventricle is formed by the pons and the medulla oblongata; its roof or dorsal wall is divided into three regions: the *superior* region, formed by the anterior medullary vela and superior peduncles; the *intermediate* region, formed by the cerebellum; and the *inferior* region, formed by the meninges. The *roof* of the fourth ventricle has three openings: a medial *foramen of Magendie* and two lateral *foramina of Luschka.* Through these openings the fourth ventricle communicates with the subarachnoid spaces of the meninges. At the superior angle, the fourth ventricle is continuous with the cerebral aqueduct; at the inferior angle, it is continuous with the central canal of the spinal cord.

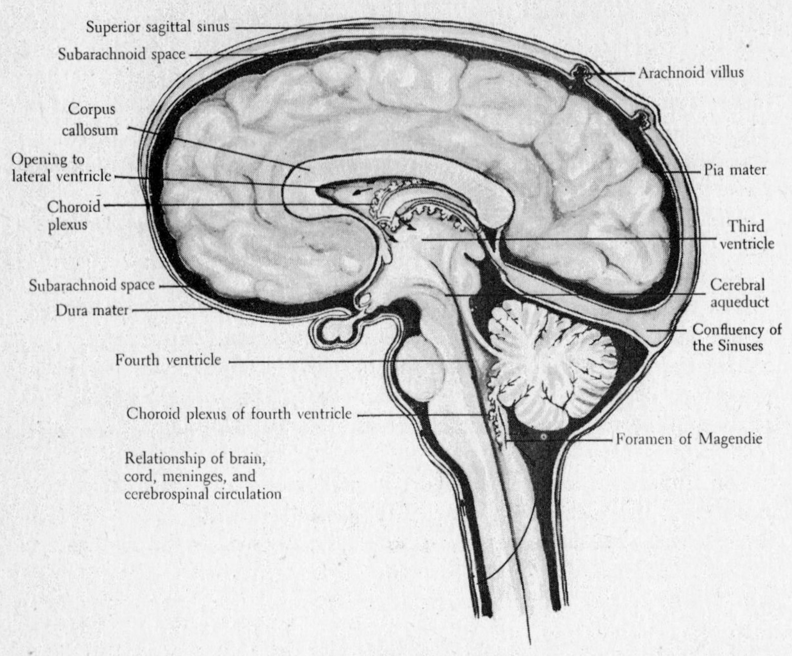

Fig. 4-8. Ventricles.

Choroid Plexuses. In certain areas, the brain wall retains its early embryonic character and persists as thin, nonnervous epithelium. At these places, the pia mater becomes modified, forming the *tela choriodea,* from which highly vascular structures, *choroid plexuses,* develop. Each choroid plexus is a much-folded structure that projects into a ventricle. It presents a relatively large surface to the cerebrospinal fluid that fills the ventricle. Choroid plexuses are found in the walls and roof of the lateral ventricles and in the roof of the third and fourth ventricles. They play an important role in the formation of the cerebrospinal fluid.

Cerebrospinal Fluid. The cerebrospinal fluid fills the ventricles of the brain, the central canal of the spinal cord, and the subarachnoid and subdural spaces. It is a clear, colorless fluid of watery consistency with a specific gravity of 1.006. It contains proteins (about 0.03 percent), glucose, urea, salts, and usually a few leukocytes (average 6 cells/mm^3). Sodium and chloride concentrations are higher and protein, potassium, and calcium concentrations lower in cerebrospinal fluid than in the blood plasma. A "blood–cerebrospinal fluid barrier" exists between the blood and the cerebrospinal fluid that prevents or slows down the exchange of substances of physiologic importance.

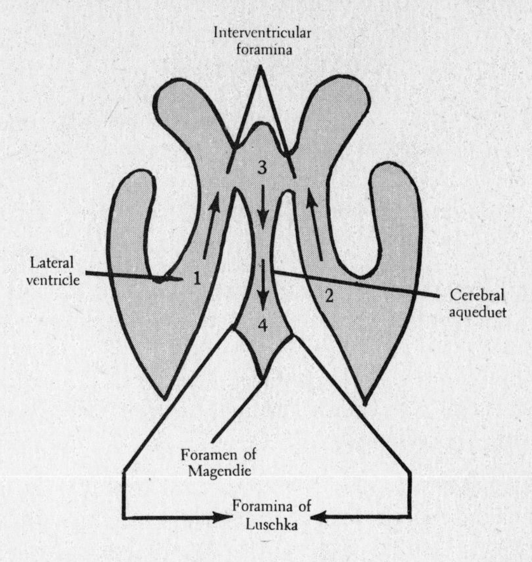

Fig. 4-9. Diagram of the ventricles showing the flow of cerebrospinal fluid.

Function of Cerebrospinal Fluid. The cerebrospinal fluid, which fills the subarachnoid space, functions as a fluid cushion that protects the brain and the spinal cord from mechanical shocks and also serves as a source of nutritive substances.

Formation and Circulation of Cerebrospinal Fluid (Fig. 4-9). The cerebrospinal fluid is formed principally by the choroid plexuses of the ventricles. It is believed to come about by the processes of filtration and dialysis, although active secretion by the epithelial cells may also take place. The fluid formed in the lateral ventricles passes (circulates) through the interventricular foramina to the third ventricle. There more is added, and it flows on through the cerebral aqueduct to the fourth ventricle, where further fluid is added. Fluid accumulating in the fourth ventricle escapes into the subarachnoid space through the two foramina of Luschka and the foramen of Magendie. It passes *downward* in the subarachnoid space on the dorsal surface of the spinal cord and *upward* in the space of the ventral surface. Eventually, the cerebrospinal fluid makes its way through the subarachnoid spaces of the brain to the arachnoid villi, which extend into the superior sagittal sinus, where it is reabsorbed into the bloodstream. It may also reenter the blood by way of the perilymphatic vessels that drain the brain.

Cerebrospinal fluid is being formed continually. Normally, it is reabsorbed as rapidly as it is formed. The total amount present at any one time averages from 100 to 150 ml. About 400 ml of fluid is formed every 24 hr.

THE SPINAL CORD

The spinal cord is the portion of the central nervous system that lies outside the cranial cavity. It is located within the vertebral canal of the spinal column.

Functions of the Spinal Cord. The spinal cord performs two general functions:

1. *It is a conducting pathway.* Afferent impulses from the periphery and efferent impulses from the brain are conducted through the fibers of the white matter of the cord.
2. *It serves as a reflex center.* In the gray matter of the cord, connections can be made between afferent and efferent neurons that provide the basis for reflex action.

Structure of the Spinal Cord. The spinal cord is an elongated cylindrical structure extending from the foramen magnum to the 2nd lumbar vertebra. It averages 42 to 45 cm in length and is slightly longer in males than in females. Its average weight is 30 g.

The cord consists of a series of 31 segments, each of which gives rise to a pair of spinal nerves. Each nerve is attached to the cord by two roots: a *dorsal root* that contains afferent or sensory fibers and a *ventral root* that contains efferent or motor fibers. The segments of the spinal cord are continuous one with the other, no external lines of demarcation being visible between them.

Superiorly, the spinal cord is continuous with the medulla oblongata; inferiorly, it tapers to a conical portion, the *conus medullaris,* whose extreme tip forms a delicate filament, the *filum terminale* extending to the coccyx.

The spinal cord bears two enlargements, an upper or cervical and a lower or lumbar. The *cervical enlargement* lies between the 3rd cervical and the 2nd thoracic vertebrae; from this region, nerves pass laterally to the upper extremities. The *lumbar enlargement* lies at the level of the lower thoracic vertebrae, being broadest at the 12th; from it extend the roots of the nerves leading to the lower extremities, but instead of passing laterally, the nerves follow an oblique course downward in the vertebral canal to reach their points of exit in the lower vertebrae and the sacrum. The nerves emerging from the lower portion of the spinal cord pass downward through the vertebral canal. Their roots collectively constitute the *cauda equina.*

In a cross section of the spinal cord (Fig. 4-10), the following can be noted: It consists of two symmetrical halves joined in the midregion by a transverse band of fibers, the *white* and *gray commissures,* in the center of which is a small opening, the *central canal.* The halves are separated by the *anterior median fissure* and the *posterior median septum.* On the

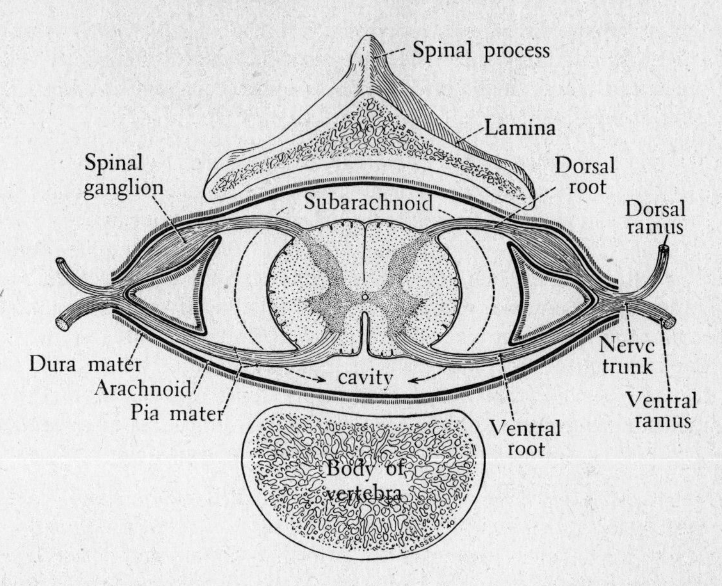

Fig. 4-10. Cross section of the spinal cord. (Reprinted with permission of W. B. Saunders Co., Philadelphia, from B. G. King and M. J. Showers, *Human Anatomy and Physiology,* 6th ed., 1969.)

posterior surface is a median depression, the *posterior median sulcus.*

Gray Matter and White Matter of the Spinal Cord. The substance of the spinal cord consists of gray matter and white matter enclosed within the meninges. The *gray matter,* composed of neuroglia, neurons, and fine interlacing nerve fibers, is centrally located. The *white matter* forms the outer substance and encloses the gray matter. It is composed of neuroglia and nerve cell processes or fibers, principally medullated axons. The white matter also contains some blood vessels and inward extensions of the pia mater.

PROPORTIONS OF WHITE AND GRAY MATTER. The proportion of white and gray matter varies at different levels of the spinal cord. In the regions of the cervical and lumbar enlargements, the amount of gray matter is relatively greater due to the presence of large numbers of neurons whose axons pass to the extremities. In the upper portions of the cord, the amount of white matter is relatively greater because it contains a much larger number of afferent fibers received from the various spinal nerves.

GRAY MATTER. The gray matter of the spinal cord consists of an H-shaped mass comprising two halves connected by a transverse band in the center of which is an opening, the *central canal.* The portion of the transverse band in front of the central canal is the *anterior white commissure;* that behind it is the *posterior gray commissure.* The central canal extends the entire length of the cord. Superiorly, it opens into the fourth ventricle of the brain; inferiorly, the canal terminates in the filum terminale. Each of the lateral halves of the gray matter is somewhat crescent-shaped and consists of three regions:

1. The *anterior* or *ventral column.* Also called the "ventral horn," this short rounded structure does not approach the surface of the cord; it is occupied principally by the cell bodies of large motor neurons.
2. The *posterior* or *dorsal column.* This longer and more slender portion is directed laterally and extends almost to the surface of the cord. Its cells are internuncial neurons, also called adjustor cells or secondary sensory neurons, which receive and transmit impulses from the primary sensory neurons of the dorsal roots of spinal nerves.
3. The *lateral column.* Seen only in the upper cervical, thoracic, and midsacral portions of the spinal cord, this lateral-protruding mass contains the cell bodies of preganglionic neurons of the autonomic nervous system.

WHITE MATTER. The white matter in each half of the spinal cord is divided by the gray matter into three general regions, or *funiculi:* the *dorsal funiculus,* the region between the dorsal median septum and posterolateral sulcus (the latter is a groove along which the dorsal roots of the spinal nerves connect with the cord); the *lateral funiculus,* the region between the lines of attachment of the dorsal and ventral roots of the spinal nerves; and the *ventral funiculus,* the region between the ventral root fibers and the anterior median fissure.

The nerve fibers of the white matter are of three types: (1) those that conduct impulses to the brain (sensory fibers) and from the brain (projection fibers), (2) those that connect various segments of the cord with one another (*intersegmental* or *association fibers*), and (3) those that cross the midline (*commissural fibers*).

The fibers to and from the brain are grouped into bundles called *fiber tracts.* In these tracts, the fibers are principally axons whose cell bodies are located in one of three regions: (1) the dorsal root ganglia of spinal nerves, (2) the gray matter of the brain, and (3) the gray matter of the spinal cord. The fibers of a single tract generally arise from a common region, carry the same type of impulses, and have a common termination. In a funiculus, fibers of one tract may intermingle with those of a similar tract. Such a mixed bundle is called a *fasciculus.* The names of tracts generally indicate their origin and destination and the direction in which the impulses travel. The lateral corticospinal tracts, for example, which carry impulses originating in the cerebral cortex, descend and terminate in the gray matter of the cord and are located in the lateral funiculus.

There are two types of tracts: (1) *ascending tracts,* containing afferent fibers that carry impulses toward the brain, and (2) *descending tracts,* containing efferent fibers that carry impulses from the brain to the lower levels of the cord. (See Fig. 4-11.)

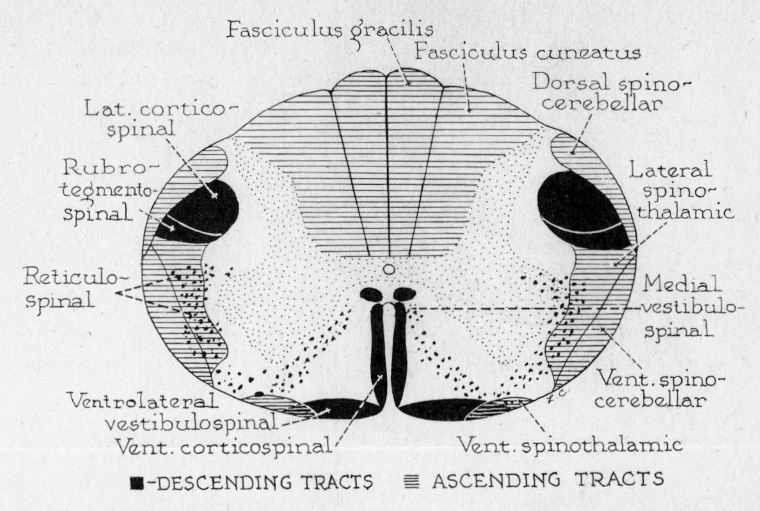

Fig. 4-11. Diagram of a cross section of the spinal cord, showing the position of the principal fiber tracts in the white matter. (Reprinted with permission of W. B. Saunders Co., Philadelphia, from B. G. King and M. J. Showers, *Human Anatomy and Physiology,* 6th ed., 1969.)

The principal fiber tracts are given in the following table:

Location	*Ascending Tracts*	*Descending Tracts*
Anterior funiculus	Ventral spinothalamic tract	Ventral corticospinal tract Vestibulospinal tract Tectospinal tract
Lateral funiculus	Dorsal spinocerebellar tract Ventral spinocerebellar tract Lateral spinothalamic tract Spinotectal tract	Lateral corticospinal tract Rubrospinal tract
Posterior funiculus	Fasciculus gracilis Fasciculus cuneatus	

ASCENDING TRACTS AND FASCICULI (Fig. 4-12). The locations and functions of these tracts are as follows:

Ventral Spinothalamic Tract. This tract consists of fibers whose cell bodies lie in the gray matter in the opposite side of the cord. Most of the fibers end in the thalamus, but some terminate in the medulla or upper portion of the cord. They convey impulses of tactile sensibility.

Dorsal and Ventral Spinocerebellar Tracts. These pathways consist of fibers whose cell bodies lie in the gray matter of the spinal cord. The fibers may or may not cross to the opposite side of the cord. They reach the cerebellum by way of the superior and inferior cerebellar peduncles, conveying to it proprioceptive impulses from muscles and tendons.

Lateral Spinothalamic Tract. This tract consists of fibers whose cell

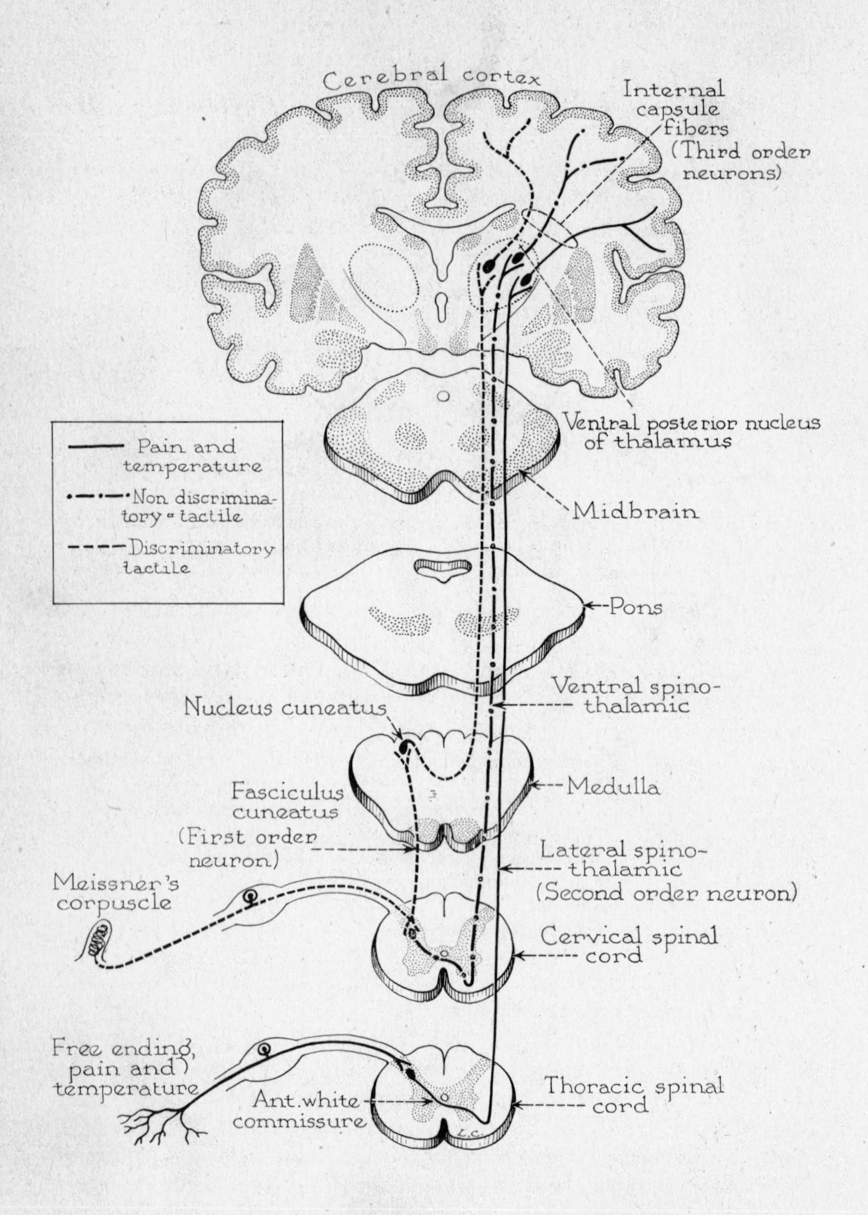

Fig. 4-12. Diagram of sensory mechanisms for pain, temperature, and touch. (Reprinted with permission of W. B. Saunders Co., Philadelphia, from B. G. King and M. J. Showers, *Human Anatomy and Physiology*, 6th ed., 1969.)

bodies lie in the gray matter of the opposite side of the cord. They convey impulses that give rise to sensations of pain and temperature. The ascending fibers end in the thalamus.

Spinotectal Tract. This tract consists of fibers that arise from cells in the posterior column of the gray matter. They cross to the opposite side of the cord and pass upward, terminating in the roof (tectum) of the midbrain. They convey impulses of pain, temperature, and touch.

Fasciculus Gracilis and Fasciculus Cuneatus. These pathways are composed of fibers whose cell bodies lie in the dorsal root ganglia of spinal nerves. They convey proprioceptive impulses from receptors located in muscles, tendons, and joints and tactile impulses from receptors in the skin. The fibers terminate in the nucleus gracilis and nucleus cuneatus located in the medulla oblongata. Here secondary neurons convey the impulses to the thalamus on the opposite side, where they are relayed to the sensory areas of the cerebral cortex.

DESCENDING TRACTS (Fig. 4-13). The locations and functions of these tracts are as follows:

Lateral and Ventral Corticospinal Tracts (Pyramidal Tracts). The fibers of these tracts arise from pyramidal cells of the motor areas of the cerebral cortex. On passing through the medulla, the fibers undergo partial decussation, the majority crossing to the opposite side in the decussation of the pyramids and continuing downward in the spinal cord as the *lateral corticospinal tract.* Fibers extend as far as the fourth sacral segment. In their course they give off collaterals that enter the gray matter and synapse with internuncial neurons or lower motor neurons. The *ventral corticospinal tract* consists of fibers that do not cross in the medulla but continue downward on the same side of the cord. The fibers terminate in the same fashion as those of the lateral corticospinal tract.

Vestibulospinal Tract. The fibers of this tract arise from cells of the lateral vestibular nucleus located in the medulla oblongata. They pass downward in the anterior funiculus, forming a narrow band near the surface of the cord. They carry impulses involved in muscle tonus and equilibrium.

Tectospinal Tract. The fibers of this tract originate from cell bodies located in the roof of the midbrain. The fibers cross to the opposite side and descend in the anterior funiculus. Collaterals and terminal endings enter the gray matter, where they synapse with motor neurons. They are involved in optic and auditory reflexes.

Rubrospinal Tract. The fibers of this tract originate from cells in the red nucleus of the midbrain. They cross to the opposite side and pass downward in the cord in the lateral funiculus. The tract is not well developed in man.

DETERMINATION OF PATHWAYS. The conduction pathways of the spinal cord have been determined by the following methods:

1. Stimulation of specific bundles or areas in sections and noting resulting actions or effects.

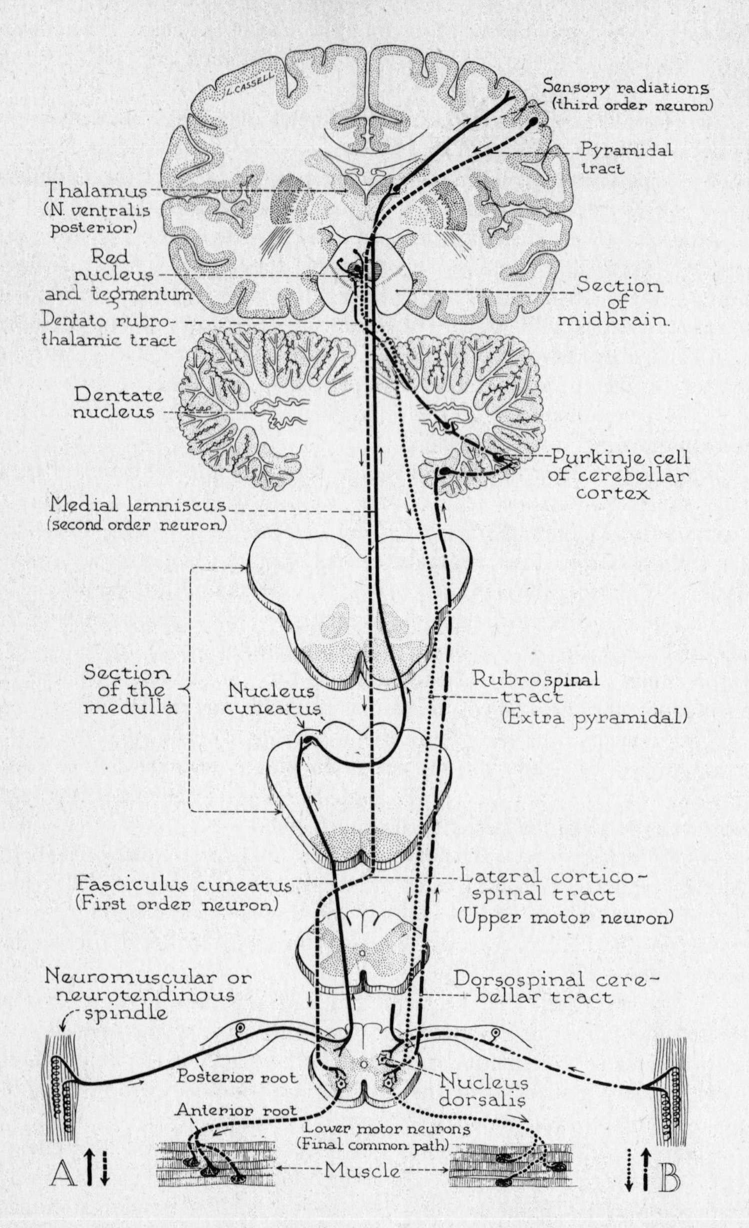

Fig. 4-13. Diagram showing pathways of muscle sense impulses. (*A*) Conscious muscle sense impulses. (*B*) Unconscious muscle sense impulses. (Reprinted with permission of W. B. Saunders Co., Philadelphia, from B. G. King and M. J. Showers, *Human Anatomy and Physiology,* 6th ed., 1969.)

2. Section of fiber tracts in experimental animals, noting resulting actions or effects.

3. Utilization of the principle of Wallerian degeneration, in which an axon severed from its cell body degenerates. Fibers in ascending tracts degenerate *above* the point of injury, those in descending tracts *below* the point of injury.

4. Observation of the embryologic development of axons.

5. Use of differential staining methods, which allow axons to be traced to their cells of origin.

6. Observation of the effects of lesions, such as those resulting from infections, tumors, or trauma. Lesions involving ascending tracts bring about loss of sensation; those involving descending tracts, loss of movement.

THE PERIPHERAL NERVOUS SYSTEM

The peripheral nervous system encompasses the structures that connect the various parts of the body to the brain and the spinal cord. It comprises the craniospinal nerves and the sympathetic division of the autonomic nervous system.

Craniospinal Nerves. As previously described, a nerve is composed of nerve-cell processes or fibers. Nerves may be classified as (1) *afferent* or *sensory* (carrying impulses from receptors to the brain or the spinal cord), (2) *efferent* or *motor* (carrying impulses from the brain or spinal cord to effector organs), and (3) *mixed* (containing both afferent and efferent fibers).

The *functional components* of nerves are the nerve fibers that transmit impulses. These fibers are either afferent or efferent.

AFFERENT FIBERS. These fibers conduct impulses *to* the brain or spinal cord. They are of the following types:

1. *General somatic afferent* fibers conduct impulses from end organs of touch, pressure, heat, cold, and pain and from proprioceptors.

2. *Special somatic afferent* fibers conduct impulses from the ear and the eye.

3. *General visceral afferent* fibers conduct impulses from sensory nerve endings in the viscera.

4. *Special visceral afferent* fibers conduct impulses from end organs of taste and smell.

EFFERENT FIBERS. These fibers conduct impulses *from* the brain or spinal cord. They are of the following types:

1. *General somatic efferent* fibers conduct impulses to all striated muscles with the exception of branchiomeric muscles.

2. *General visceral efferent* fibers conduct impulses to smooth muscles, cardiac muscles, and glands.

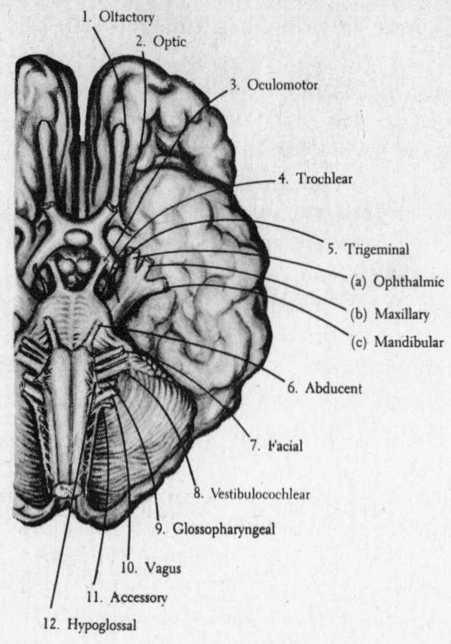

Fig. 4-14. Origins of cranial nerves.

3. *Special visceral efferent* fibers conduct impulses to branchiomeric (striated) muscles derived from the branchial arches of the embryo. These are muscles of the jaws, face, pharynx, and larynx. Although designated "visceral," the nerves supplying them do not belong to the autonomic system.

Cranial Nerves. Twelve pairs of cranial nerves arise directly from the brain (Fig. 4-14). They pass through foramina in the base of the cranium to reach their terminations. Most are mixed nerves, having both motor and sensory fibers, although a few contain only sensory fibers. The sensory or afferent fibers have their cell bodies in ganglia lying outside the brain; the cell bodies of the motor or efferent fibers lie in the nuclei of origin within the brain. The cranial nerves, identified by numbers as well as names, are the *olfactory* (I), *optic* (II), *oculomotor* (III), *trochlear* (IV), *trigeminal* (V), *abducent* (VI), *facial* (VII), *vestibulocochlear* (VIII), *glossopharyngeal* (IX), *vagus* (X), *accessory* (XI), and *hypoglossal* (XII). A small nerve, the *terminal nerve* (unnumbered), is the most anterior cranial nerve. It passes with the olfactory nerve to the septum of the nose. Its functions are unknown.

ESSENTIAL FACTS ABOUT THE CRANIAL NERVES. The following tables summarize the essential facts relating to the cranial nerves: type (sensory, motor, or mixed), superficial origin, deep origin, exit from cranium, functional components, structures innervated by afferent fibers and sensation so mediated, and structures innervated by efferent fibers and reaction initiated. (See Tables 4-1 and 4-2.)

SPECIAL PHENOMENA RELATING TO THE CRANIAL NERVES. The following special phenomena have been noted with relation to the cranial nerves:

1. All cranial nerves except the olfactory (I) and the optic (II) arise from the brain stem.

2. The olfactory tract and bulb and the optic nerve and tract are not true peripheral nerves (as are the other cranial nerves). Each develops as an evagination of the prosencephalon and is thus fundamentally comparable to a tract of the central nervous system. These nerves conduct only afferent (sensory) impulses. The olfactory nerve, instead of being a single nerve, consists of about 20 bundles of separate fibers.

3. Three of the cranial nerves (III, IV, and VI) are involved in the innervation of the musculature of the eyes. Formerly these nerves were regarded as strictly motor, but it is now well established that they conduct sensory proprioceptive impulses.

4. The trochlear nerve (IV) is the cranial nerve of smallest diameter, the trigeminal (V) the largest.

5. The vestibulocochlear nerve (VIII) is the only cranial nerve that does not pass completely through the cranium. It terminates at the ear within the temporal bone.

6. The trigeminal nerve (V) is the principal sensory nerve of the superficial and deep portions of the head and face. It is also the motor nerve for the muscles of mastication.

7. The vagus nerve (X) has the most extensive distribution. Only three small branches (meningeal, auricular, and pharyngeal) reach structures in the head; the main portion of the nerve passes through the neck into the thorax, where branches pass to the larynx, trachea, bronchi, lungs, heart, and pericardium. Branches of the right and left vagus form branching networks of fibers about certain structures; these fibers constitute the *pulmonary, cardiac,* and *esophageal plexuses.* On passing through the diaphragm, the branches enter the stomach, liver, spleen, pancreas, kidney, small intestine, and ascending and transverse colon. Important plexuses are the *gastric,* the *splenic,* and the *celiac.*

8. Four of the cranial nerves (III, VII, IX, and X) carry fibers of the autonomic nervous system and constitute a part of the craniosacral (parasympathetic) division of that system.

Spinal Nerves. Thirty-one pairs of spinal nerves arise from the spinal cord. There are eight pairs of *cervical nerves,* twelve pairs of *thoracic nerves,* five pairs of *lumbar nerves,* five pairs of *sacral nerves,* and one pair of *coccygeal nerves.* The first cervical nerve emerges between the occipital bone and the atlas; all others, except the sacral nerves, emerge from the

TABLE 4-1 THE CRANIAL NERVES—PART I

Nerve	*Type*	*Superficial Origin*	*Deep Origin of Efferent Fibers*	*Exit from Cranium*
I Olfactory	Sensory	Olfactory bulb	None	Cribriform plate of ethmoid
II Optic	Sensory	Optic chiasma	None	Optic foramen
III Oculomotor	Principally motor	Medial surface of cerebral peduncle of midbrain	Motor nucleus in floor of midbrain	Supraorbital fissure
IV Trochlear	Principally motor	Between cerebral peduncle and pons	Trochlear nucleus in midbrain	Supraorbital fissure
V Trigeminal	Mixed	From side of pons near superior border	Motor nucleus in pons	Three branches through supraorbital fissure, foramen rotundum, and foramen ovale
VI Abducent	Principally motor	From a groove between pons and medulla	Abducent nucleus in pons	Supraorbital fissure
VII Facial	Mixed	(Same as VI)	Motor nucleus in pons	Internal acoustic (auditory) meatus and stylohyoid foramen
VIII Vestibulocochlear	Principally sensory	(Same as VI)	None	Internal acoustic meatus
IX Glossopharyngeal	Mixed	Posterolateral sulcus of medulla	Nucleus ambiguus and inferior salivary nucleus in medulla	Jugular foramen
X Vagus	Mixed	Posterolateral sulcus of medulla	Nucleus ambiguus and dorsal motor nucleus in medulla oblongata	Jugular foramen
XI Accessory	Mixed	Posterolateral sulcus of medulla	Nucleus ambiguus of medulla	Jugular foramen

TABLE 4-1 (*Continued*)

Nerve	*Type*	*Superficial Origin*	*Deep Origin of Efferent Fibers*	*Exit from Cranium*
XII Hypoglossal	Mixed	Anterolateral sulcus	Hypoglossal nucleus in medulla	Hypoglossal foramen

intervertebral foramina between adjoining vertebrae. Sacral nerves emerge from the posterior sacral foramina.

STRUCTURE AND ATTACHMENT OF SPINAL NERVES. Each spinal nerve is attached to the spinal cord by a posterior root and an anterior root, each composed of many small rootlets. A short distance from the spinal cord, the roots unite to form the main trunk of the spinal nerve.

Roots of Spinal Nerves. On the *posterior* (*sensory afferent* or *dorsal*) *root* is an ovoid mass of nerve cells, the *spinal ganglion.* Nerve fibers in this root have their origin in cells of this ganglion. The neurons of spinal ganglia are pseudounipolar. From each cell body emerges a single process that bifurcates a short distance from the cell body. The *peripheral branch* passes laterally toward the periphery; although it conveys impulses *to* the cell body and therefore is functionally a dendrite, structurally it possesses a myelin sheath and neurilemma and thus resembles an axon. For this reason, it is called an axonlike dendrite. The *central branch* passes into the spinal cord. On approaching the spinal cord, the posterior root of each spinal nerve breaks up into a number of root filaments that pass through the dura mater and enter the cord.

Each *anterior (motor efferent* or *ventral) root* consists of several root filaments that emerge from the anterior surface of the cord. These filaments are made up of efferent fibers (axons) of neurons whose cell bodies lie in the gray matter of the spinal cord.

Branches of Spinal Nerves (Fig. 4-15). A typical spinal nerve, for example, an intercostal nerve, passes through the intervertebral foramen and almost immediately divides into four branches.

The *recurrent branch* is a small branch that turns medially and reenters the vertebral canal. It innervates the meninges and vessels of the spinal cord.

The *dorsal ramus* or *posterior primary division* turns posteriorly and innervates the skin, muscles, and fascia of the back.

The *ventral ramus* or *anterior primary division* passes laterally and continues ventrally in the body wall. It divides into a *lateral cutaneous branch,* which innervates the skin, and an *anterior branch,* which innervates the muscles of the body wall and the linings of the body cavities. The latter terminates near the midline in an *anterior cutaneous branch,* which innervates the skin on the ventral surface of the body.

TABLE 4-2 THE CRANIAL NERVES—PART II

Nerve	*Functional Components*	*Structures Innervated by Afferent Fibers and Sensation Mediated*	*Structures Innervated by Efferent Fibers and Reaction Initiated*
I Olfactory	Special visceral afferent	Olfactory epithelium (smell)	None
II Optic	Special somatic afferent	Retina of eye (sight)	None
III Oculomotor	General somatic efferent General visceral efferent General somatic afferent	Eye muscles Superior rectus Medial rectus Inferior rectus Inferior oblique Eyelid muscle Levator palpebrae (muscle sense)	Eye muscles (movement) Superior rectus Medial rectus Inferior rectus Inferior oblique Eyelid muscle (movement) Levator palpebrae Ciliary muscles (accommodation) Iris (constriction)
IV Trochlear	General somatic efferent General somatic afferent	Eye muscle Superior oblique (muscle sense)	Eye muscle (movement) Superior oblique
V Trigeminal			
1 Ophthalmic	General somatic afferent	Cornea, ciliary body, lacrimal gland, conjunctiva; mucous membranes of nasal cavity and sinuses; skin of eyelids, eyebrows, forehead, and nose (pain, cold, heat, touch)	None
2 Maxillary	General somatic afferent	Dura mater; gums and teeth of upper jaw, upper lip; nasal mucosa, orbit	None
3 Mandibular	General somatic afferent Special visceral efferent	Anterior two-thirds of tongue (taste); gums and teeth of lower jaw; skin and mucous membrane of cheek and lower lip (pain, temperature, touch); muscles of mastication (muscle sense)	Muscles of mastication (movement)

TABLE 4-2 (*Continued*)

Nerve	*Functional Components*	*Structures Innervated by Afferent Fibers and Sensation Mediated*	*Structures Innervated by Efferent Fibers and Reaction Initiated*
VI Abducent	General somatic efferent General somatic afferent	Lateral rectus muscle (muscle sense)	Lateral rectus muscle (movement)
VII Facial	General somatic efferent General somatic afferent General visceral efferent Special visceral efferent	Anterior two-thirds of tongue (taste); muscles of neck, face, jaw, scalp, auricle (muscle sense)	Muscles of neck, face, jaw, scalp, auricle (movement); submaxillary, sublingual, lacrimal, nasal, palatine glands (secretion)
VIII Vestibulo-cochlear	Special somatic afferent Special visceral efferent	Cochlea of ear (hearing); vestibule and semicircular canals (sense of equilibrium)	Muscles of middle ear; hair cells of cochlea
IX Glossopharyngeal	General visceral afferent Special visceral afferent General visceral efferent Special visceral efferent	Mucous membrane of pharynx, fauces, palatine tonsil, posterior third of tongue (taste, swallowing reflex); carotid sinus (cardiac reflex); skin of ear (cutaneous senses)	Striated muscles of pharynx (swallowing movements); parotid gland (secretion)
X Vagus	Same as IX	Mucosa of pharynx, larynx, trachea, bronchi (respiratory reflexes); lungs (Herring-Breuer reflex); aortic arch (cardiac reflex); abdominal viscera (hunger, pain)	Muscles of palate, pharynx, esophagus (swallowing); cardiac muscle (inhibition); smooth muscles of thoracic and abdominal viscera (contraction and inhibition); glands of stomach, intestine, pancreas, liver (stimulation, inhibition)

TABLE 4-2 (*Continued*)

Nerve	*Functional Components*	*Structures Innervated by Afferent Fibers and Sensation Mediated*	*Structures Innervated by Efferent Fibers and Reaction Initiated*
XI Accessory	General visceral efferent Special visceral efferent General visceral afferent	Sternomastoid and trapezius, uvula, levator veli palatini (muscle sense)	Muscles of neck and shoulder (movement); soft palate (movement)
XII Hypoglossal	General somatic efferent General somatic afferent	Muscles of tongue (muscle sense)	Muscles of tongue (movement)

The *rami communicantes* (sing., *ramus communicans*) are short, threadlike filaments that pass ventrally from the anterior primary division (ventral ramus) and connect it with the autonomic ganglia of the sympathetic trunk. These rami are of two types: the white ramus and the gray ramus. The *white ramus* is composed principally of myelinated axons of neurons

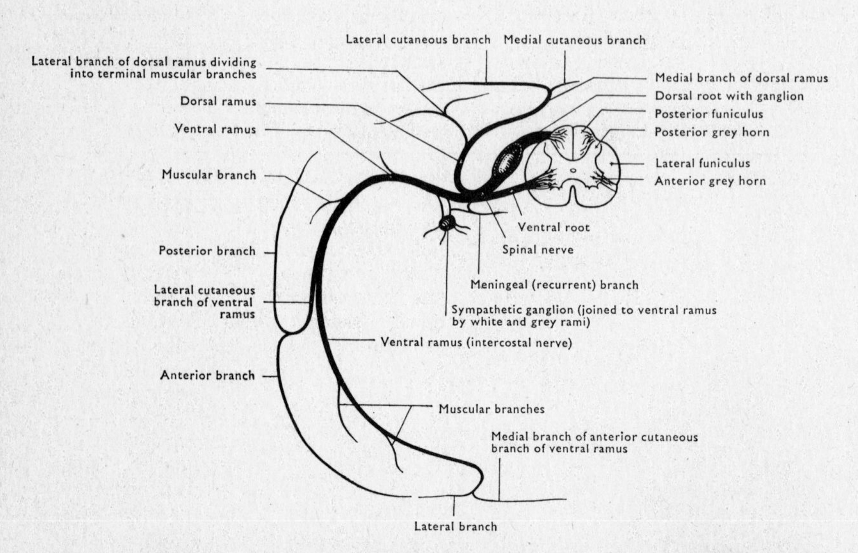

Fig. 4-15. Branches of a spinal nerve. (Reprinted with permission of Oxford University Press from Cunningham's *Textbook of Anatomy,* ed. by G. J. Romanes, 11th ed., 1972.)

whose cell bodies lie in the lateral horn of the gray matter of the cord (the neurons are called *preganglionic neurons*). These axons synapse with neurons in autonomic ganglia. The white ramus may also contain visceral afferent fibers whose cell bodies lie in spinal ganglia. The *gray ramus* is made up of unmyelinated fibers (axons) of neurons (*postganglionic neurons*) whose cell bodies lie in autonomic ganglia. On entering the anterior division of the spinal nerve, these fibers pass peripherally to smooth muscles and glands of the body wall or the extremities.

FUNCTIONAL COMPONENTS OF SPINAL NERVES. A typical spinal nerve contains the following functional components:

Afferent Fibers. These may be either *general somatic afferent fibers* (fibers that conduct sensory impulses from exteroceptors or proprioceptors to the spinal cord) or *general visceral afferent fibers* (sensory fibers that conduct sensory impulses from enteroceptors to the spinal cord). The cell bodies of both types of fibers lie in the spinal ganglia.

Efferent Fibers. These may be either *general somatic efferent fibers* (motor fibers that conduct impulses to striated muscles and whose cell bodies lie in the anterior column of the gray matter of the spinal cord) or *general visceral efferent fibers* (motor fibers that conduct motor and secretory impulses to smooth muscle, cardiac muscle, and glands). Impulses arise from preganglionic neurons (primary neurons) in the lateral horn of the gray matter of the cord and pass through the anterior root and white ramus to ganglia of the sympathetic trunk. At this point they synapse with postganglionic neurons (secondary neurons), whose axons may (1) pass to visceral organs through autonomic nerves, or (2) return to the anterior division of the spinal nerve through the gray ramus and pass to smooth muscles, especially those of blood vessels, and to glands located in the periphery of the body.

Concerning spinal nerves in general, the following may be noted:

1. The dorsal or posterior root contains only *afferent* fibers conducting impulses from the periphery to the spinal cord. Between the spinal ganglion and the cord, the dorsal root is composed of axons; between the ganglion and the trunk of the spinal nerve, it is composed of axonlike dendrites.
2. The ventral or anterior root contains only *efferent* fibers (axons of neurons whose cell bodies lie in the gray matter of the spinal cord).
3. The main nerve trunk contains *both afferent and efferent* fibers (axons and dendrites).

PLEXUSES OF SPINAL NERVES (Fig. 4-16). In the thoracic region, the spinal nerves are arranged segmentally, each pair innervating a more or less specific segment of the body. In the other regions of the spinal cord, however, the ventral rami of the spinal nerves anastomose with adjacent spinal nerves, forming plexuses. These are the *cervical plexus,* the *brachial plexus,* the *lumbosacral plexus,* and the *coccygeal plexus.*

Cervical Plexus. Lying in the neck alongside the first four cervical

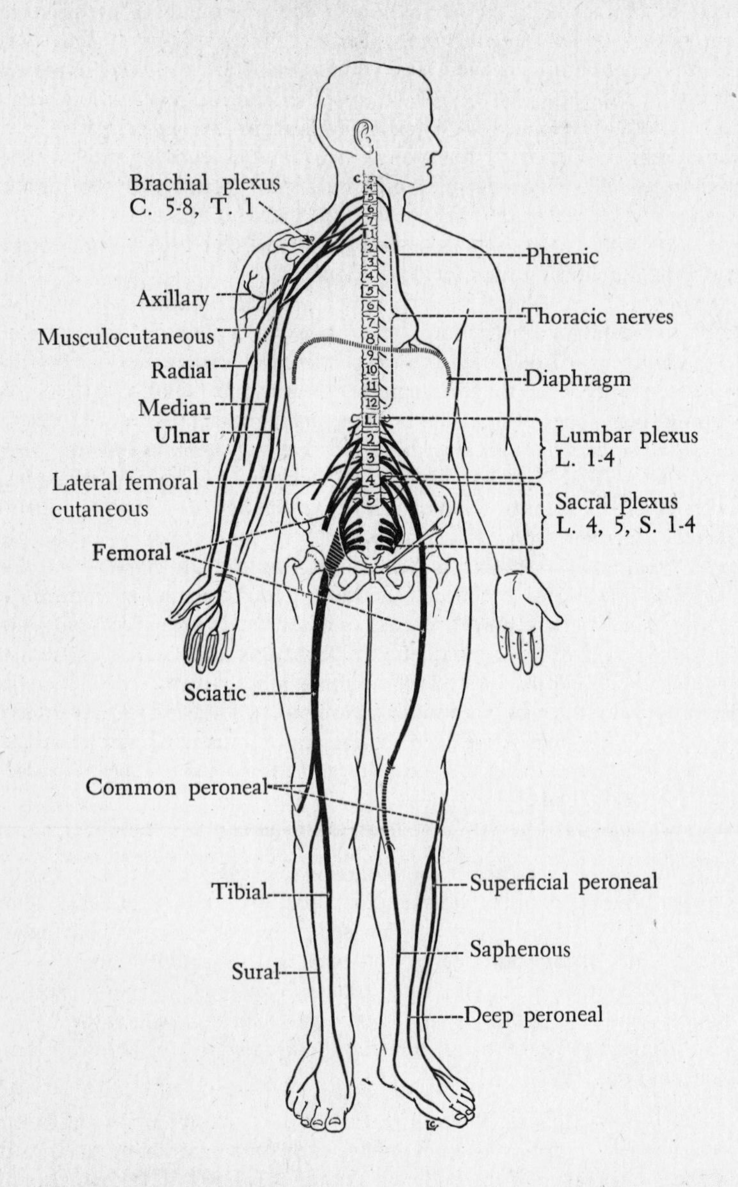

Fig. 4-16. Spinal nerves and plexuses. The *pudendal plexus* is a downward continuation of the sacral plexus. The *coccygeal plexus* includes the 5th sacral nerve, the coccygeal nerve, and a branch of the 4th sacral nerve. Cervical plexus is not shown. (Reprinted with permission of W. B. Saunders Co., Philadelphia, from B. G. King and M. J. Showers, *Human Anatomy and Physiology,* 6th ed., 1969.)

vertebrae, this plexus is formed by an anastomosis of the ventral rami of the first four cervical nerves. From it *superficial* and *deep branches* go to the skin of the head, neck, and shoulder and to certain muscles (trapezius, sternocleidomastoid, levator scapulae, and scalenus medius); one, the *phrenic nerve,* passes to the diaphragm. Communicating branches connect with cranial nerves X (vagus), XI (accessory), and XII (hypoglossal).

Brachial Plexus. This plexus is formed by the last four cervical nerves and the first thoracic nerve. The 5th and 6th cervical nerves join to form the *upper trunk;* the 7th forms the *middle trunk;* the 8th cervical and 1st thoracic join to form the *lower trunk.* The anterior divisions of the upper and middle trunks unite to form the *lateral cord,* whereas the anterior division of the lower trunk alone gives rise to the *medial cord.* Each of the posterior divisions from the three trunks combine to form the *posterior cord.* The branches of the cords constitute the nerve supply of the upper extremity. The principal branches of each cord are as follows:

Lateral cord: musculocutaneous nerve and lateral root of median nerve
Medial cord: ulnar nerve and medial root of median nerve
Posterior cord: axillary, radial, and subscapular nerves

The *median nerve* extends along the medial side of the arm and forearm to the hand, with fibers extending to the fingers. Efferent fibers go to the muscles on the anterior side of the forearm supplying in general the flexors and pronators. Afferent fibers supply the integument covering the central portion of the palm and a portion of the ulnar eminence. The *musculocutaneous nerve* supplies the coracobrachialis, biceps brachii, and brachialis muscles; cutaneous branches go to the skin. The *ulnar nerve* lies along the medial side of the arm, sending branches to the elbow joint, forearm, and hand. The *axillary nerve,* the largest branch of the brachial plexus, contains fibers of cervical nerves C5, C6, C7, and C8. It passes along the lateral side of the humerus to the region of the elbow, where it divides into *deep* and *superficial branches,* which extend to the digits. It innervates the triceps brachii, brachioradialis, extensor carpi radialis, and a part of the brachialis muscles. Cutaneous branches receive sensory fibers from the skin on the posterior surface of the arm and hand. The *subscapular nerves* (upper, middle or thoracodorsal, and lower) are small nerves supplying the subscapularis, latissimus dorsi, and teres major muscles.

Between the brachial and the lumbosacral plexuses, 12 pairs of *thoracic* or *intercostal nerves* are given off from the spinal cord. They pass laterally in the intercostal spaces, supplying fibers to the muscles and skin of the thoracic and abdominal walls. There being no plexus formation, each nerve follows an independent pathway. The first two intercostal nerves contribute fibers through the brachial plexus to the upper limbs. The last five supply the skin and muscles of the abdominal wall. Each of the thoracic nerves is connected to a corresponding ganglion of the sympathetic trunk by two rami: a *white* and a *gray communicating ramus.*

Lumbosacral Plexus. This plexus is formed by the anterior rami of the lumbar, sacral, and coccygeal nerves. For convenience this plexus is divided, in this description, into three parts: the *lumbar,* the *sacral,* and the *pudendal plexuses.*

The *lumbar plexus* is formed by the anterior rami of the first three lumbar nerves and a part of the fourth. Branches from this plexus supply the muscles and skin of the lower abdominal wall and a part of the lower extremity. The principal branches of the lumbar plexus and their distribution are: *iliohypogastric,* to the skin of the gluteal and hypogastric regions; *ilioinguinal,* to the muscles of the upper and medial portions of the thigh and to the external genitalia; *genitofemoral,* one branch to the scrotum (male) or the round ligament (female), another branch to skin of the upper anterior portion of the thigh; *lateral femoral cutaneous,* to the skin of the lateral portion of the thigh; *femoral,* the largest branch, formed by branches of lumbar nerves (L2, L3, and L4) and supplying the muscles and skin of the anterior and medial sides of the thigh and leg (principal branches are *intermediate* and *medial cutaneous nerves* and *saphenous nerve,* whose branches extend as far as the foot, where they are primarily sensory); *obturator,* which passes through the obturator foramen to the thigh, where it divides into *anterior* and *posterior branches* supplying the abductor muscles and the knee joint. In the thigh, branches of the saphenous nerve supply the quadriceps femoris muscle.

The *sacral plexus* is formed by branches of lumbar nerves (L4 and L5) and sacral nerves (S1, S2, and S3). The nerves of this plexus unite to form a single flattened band that passes through the *greater sciatic notch,* below which the band is known as the *sciatic nerve.*

The *sciatic nerve,* the largest single nerve in the body, measures about 2 cm in width. It passes along the posterior surface of the thigh, being crossed by the piriformis and biceps femoris muscles. In the region of the knee, it divides into two branches: the *tibial* and the *common peroneal.* These two branches actually retain their identity throughout the entire course of the sciatic nerve from its origin at the plexus. The tibial nerve passes along the posterior side of the leg, giving off articular, muscular, and cutaneous branches. The common peroneal nerve passes along the lateral surface of the leg, giving off similar branches.

Other branches of the sacral plexus are the *superior* and *inferior gluteal nerves,* which supply motor fibers to the gluteal muscles and the tensor fascia latae, and the *posterior femoral cutaneous nerve,* which supplies the skin and the posterior surface of the thigh and leg. Nerves of this plexus also include (1) a nerve to the quadratus femoris, (2) one to the obturator internus, and (3) another to the piriformis muscle. The *pudendal plexus* is a downward continuation of the sacral plexus.

The *pudendal nerve,* the main branch of the pudendal plexus, consists of fibers from sacral nerves S2, S3, and S4. It gives off the following branches: *inferior rectal (hemorrhoidal),* distributed to the lower end of the rectum and the muscles and skin surrounding the anus; *perineal,* supplying the scrotum and proximal portion of the penis (the labia majora in the female); and *dorsal nerve of the penis* (*clitoris* in females), supplying the dorsal and distal portions of the external genitalia.

Coccygeal Plexus. This plexus is sometimes considered a subdivision of the pudendal plexus. It consists of the 5th sacral nerve, the *coccygeal nerve,* and a branch of the 4th sacral nerve. It supplies the skin in the region of the coccyx.

THE AUTONOMIC NERVOUS SYSTEM

There is a lack of uniformity in the terminology applied to the part of the nervous system that innervates the viscera. It has been variously referred to as the *visceral motor system, involuntary nervous system,* and *vegetative nervous system.* The term *autonomic nervous system* has come into more or less general use because it signifies a *functional* unit, which may be conveniently studied, rather than an anatomic division.

General Description. The autonomic nervous system (Fig. 4-17) includes the structures concerned with innervation of smooth muscles, glands, and cardiac muscle. It consists of nerves, ganglia, and plexuses. In general, it innervates structures whose activities are essential to the life of the organism, activities that are involuntary and automatic. It supplies the digestive, respiratory, urinary, reproductive, circulatory, and endocrine systems—that is, all effector organs of the body with the exception of striated muscle. Examples of activities controlled by this system are accommodation for near vision, changes in size of the pupil, constriction and dilatation of blood vessels, rate and force of heartbeat, muscular activities of the digestive tract, emptying of the urinary bladder and gallbladder, erection of the penis, production of gooseflesh, and secretion by all glands under nervous control.

The autonomic nervous system is *entirely motor;* that is, it includes only efferent fibers that carry impulses to specific tissues (muscle or gland), either increasing or decreasing their activity. The nerve fibers are arranged in two opposed systems, both of which (with a few exceptions) supply a single structure. These two divisions are the *craniosacral* or *parasympathetic* and the *thoracolumbar* or *sympathetic.*

From the preceding paragraph it might be inferred that there are no afferent impulses from the visceral organs. Such is not the case; there do exist nerve fibers that conduct afferent impulses (e.g., pain impulses originating in the intestine). However, these fibers are the peripheral processes of sensory cells lying in the spinal ganglia; consequently, they belong to neurons that are not a part of the autonomic nervous system, although they may be involved in autonomic reflexes.

Visceral Efferent Neurons. In the autonomic nervous system, at least

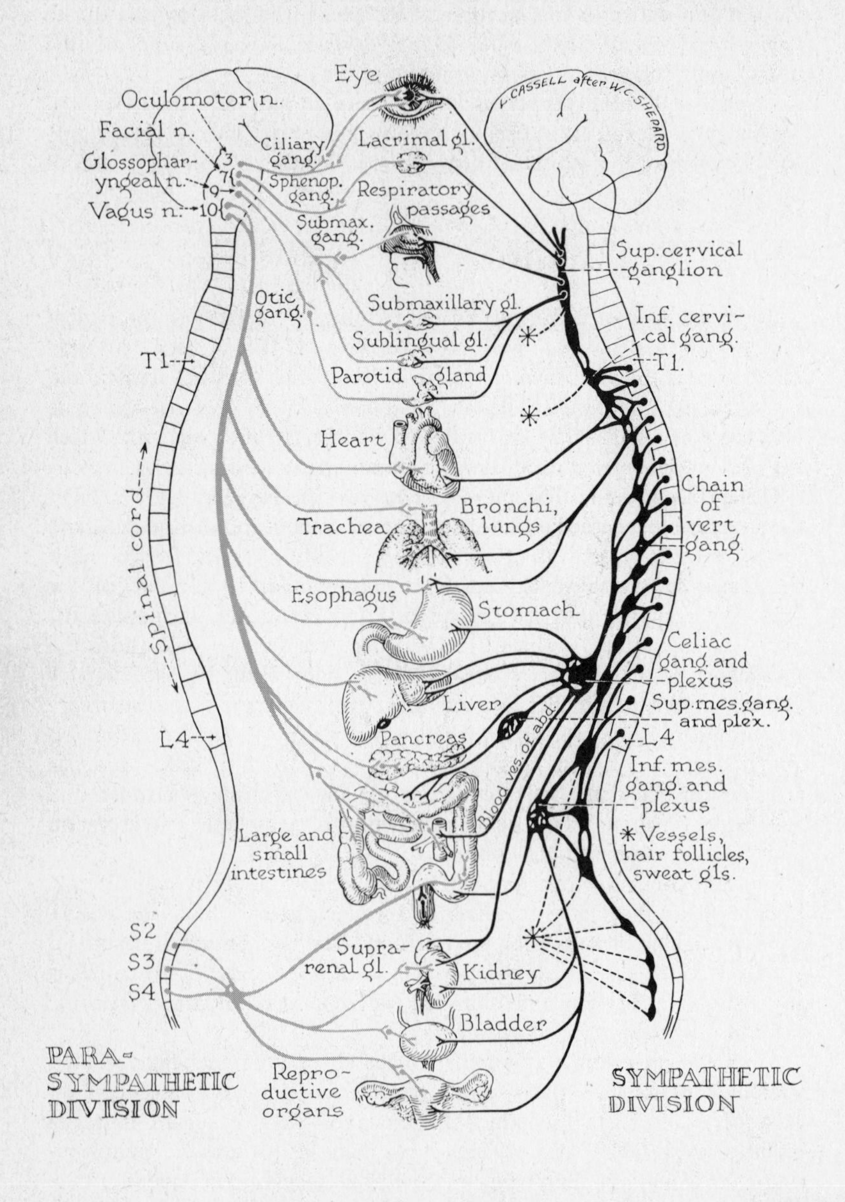

Fig. 4-17. Diagram of the autonomic nervous system. (Reprinted with permission of W. B. Saunders Co., Philadelphia, from B. G. King and M. J. Showers, *Human Anatomy and Physiology,* 6th ed., 1969.)

two neurons are always involved in the conduction of impulses from the cord or the brain to an effector organ. These neurons are known as *preganglionic* and *postganglionic neurons.*

PREGANGLIONIC NEURONS. The cell bodies of preganglionic neurons, also designated "first-order neurons," lie in the central nervous system (the brain and spinal cord). Their axons (called *preganglionic fibers*) pass through cranial or spinal nerves to ganglia, where they terminate. They are myelinated or white fibers.

POSTGANGLIONIC NEURONS. The cell bodies of postganglionic neurons, also designated "second-order neurons," lie in ganglia outside the central nervous system. Their unmyelinated axons (called *postganglionic fibers*) terminate in smooth muscles, the heart, and glandular tissue. The visceral structures they innervate are widely distributed, being found in nearly all parts of the body. Postganglionic fibers reach these structures in two ways:

1. *Through craniospinal nerves.* Visceral structures in somatic regions, such as cutaneous blood vessels and sweat glands, receive their fibers through the craniospinal nerves. These unmyelinated fibers have their origin in cells lying in autonomic ganglia.
2. *Through visceral autonomic nerves.* Visceral structures in the thorax, abdomen, and head receive autonomic fibers through the autonomic nerves. These nerves usually follow the artery that supplies blood to the organ innervated.

Ganglia of the Autonomic Nervous System. Three types of ganglia are present in the autonomic nervous system: *vertebral* or *lateral, prevertebral* or *collateral,* and *terminal* or *intramural ganglia.*

VERTEBRAL (LATERAL) GANGLIA. These ganglia lie in linear order on each side of the vertebral column close to the bodies of the vertebrae. They are united to form a chain or trunk, the *sympathetic trunk,* which extends from the base of the skull to the coccyx. There are 22 ganglia in each trunk, the most superior comprising the *superior, middle,* and *inferior cervical ganglia.* The last is frequently fused with the first thoracic ganglion to form the *stellate ganglion.*

PREVERTEBRAL (COLLATERAL) GANGLIA. These ganglia lie in the thorax, abdomen, and pelvis, near the aorta or branches of it. Important prevertebral ganglia are the *celiac, superior mesenteric,* and *inferior mesenteric.* They are found near the bases of the arteries after which they are named.

TERMINAL (INTRAMURAL) GANGLIA. These ganglia lie close to or within the structures their fibers innervate. In the *head* region, the ciliary, sphenopalatine, submaxillary, and otic ganglia contain the cell bodies of neurons that give rise to postganglionic fibers innervating the eyes, lacrimal glands, salivary glands, and mucous membrane of the mouth and pharynx. In the *thorax* and *abdomen,* terminal ganglia are present in the form of plexuses found close to or within the visceral organs. Examples of these are cardiac ganglia and *Auerbach's* and *Meissner's plexuses* in the intestine.

Divisions of the Autonomic Nervous System. Anatomically and functionally, the autonomic system can be divided into the *parasympathetic* and the *sympathetic* divisions. These are also known as the *craniosacral* and *thoracolumbar* divisions, respectively. In general, these two divisions have antagonistic effects on the organs they innervate. One acts to initiate or increase activity, the other to inhibit or slow it down. This rule does not apply to each division as a whole but to the individual organs or structures.

COMPARISON OF THE TWO DIVISIONS. The two divisions of the autonomic system differ in the following respects:

1. In general, visceral effector organs have a double innervation; that is, they are innervated by efferent fibers from both divisions. Impulses carried by the fibers of one division are antagonistic to those of the other. For example, impulses to the heart through the parasympathetic division slow down or inhibit the heartbeat, whereas impulses through the sympathetic division speed up the heartbeat. A few structures, such as the sweat glands, erector muscles of the hair follicles, and blood vessels of the digestive tract, receive fibers only from the sympathetic division.
2. Efferent fibers of the preganglionic neurons of the parasympathetic division are usually long and terminate in terminal ganglia located close to or within the organ innervated; postganglionic neurons are very short. Efferent fibers of preganglionic neurons of the sympathetic division end in vertebral or prevertebral ganglia.
3. At the termination of efferent neurons in effector organs, different chemical substances are produced. *Acetylcholine* is produced at parasympathetic endings, *norepinephrine* at sympathetic endings (with a few exceptions).
4. The two divisions differ in their responses to certain drugs. Some drugs affect the fibers of the parasympathetic division; others affect only those of the sympathetic division.

PARASYMPATHETIC (CRANIOSACRAL) DIVISION. In this division, the first-order neurons (preganglionic neurons) that send efferent fibers to the visceral organs lie in the brain stem or in the sacral region of the spinal cord. Their fibers (axons) pass through four cranial nerves (III, VII, IX, and X) or three sacral nerves (S2, S3, and S4) to the visceral structures they innervate. They end in terminal ganglia, where they synapse with second-order neurons (postganglionic neurons).

The parasympathetic division acts principally during periods of rest when restorative processes are being brought into play and energy supplies are being replenished. In general, parasympathetic effects are *specific;* that is, they tend to involve individual organs rather than the organism as a whole. Some of the parasympathetic effects are slowing of the heartbeat, storage of glycogen by the liver, increased tone and motility of the intestine, constriction of bronchioles (especially in the presence of noxious stimuli), and constriction of the pupils of the eyes.

SYMPATHETIC (THORACOLUMBAR) DIVISION. In this division, the first-order neurons (preganglionic) lie in the gray matter of the spinal cord in the thoracic and lumbar regions. Their fibers (axons) pass through the anterior roots and white rami of thoracic and lumbar spinal nerves to vertebral or prevertebral ganglia, where they synapse with second-order neurons (postganglionic neurons).

The sympathetic division acts rapidly to bring about generalized changes in the functioning of the organs that facilitate the quick mobilization and release of energy. Such changes are required under emergency conditions—that is, organs involved in activities associated with fright, flight, fighting, fear, or anger. Some of the sympathetic effects noted under such conditions are (1) increase in blood pressure accompanied by accelerated heartbeat, increased force of heartbeat, contraction of arterioles (especially in the skin and in visceral organs), and dilatation of coronary arteries and arteries going to skeletal muscles; (2) increase in blood sugar resulting from glycogen breakdown in the liver; (3) increased volume of blood resulting from splenic contractions; (4) dilatation of bronchi, increasing oxygen intake; and (5) dilatation of the pupils of the eyes. It will be seen that the foregoing changes enable the body to adapt itself to sudden emergencies that involve intense muscular activity or accompany emotional excitement in anticipation of such activity. In general, sympathetic effects are widespread, involving many organs or the body as a whole.

IMBALANCE BETWEEN THE TWO DIVISIONS. Normally, the parasympathetic and sympathetic divisions are in balance, the activity of one or the other coming into dominance according to the needs of the organism. In some individuals, there may be a marked shift toward one or the other, resulting in an imbalance and corresponding functional disorders. The influence of emotional states on this delicate balance of the autonomic nervous system is a matter of common observation. When the parasympathetic division tends to dominate, the condition is called *parasympathicotonia;* when the sympathetic division dominates, the condition is called *sympathicotonia.*

Functions of the Autonomic Nervous System. The autonomic nervous system regulates the activities of the visceral organs, which are essential to the existence of the organism and to its reproduction. These activities are involuntary; that is, they are, generally speaking, not susceptible to conscious control. Among the organs under autonomic control are the heart and blood vessels, respiratory organs, alimentary canal, kidneys and urinary bladder, reproductive organs, and endocrine glands.

Visceral organs normally function automatically. When conditions occur that alter the physiologic requirements of the organism, the visceral organs may need to function at an increased or decreased rate to meet the

TABLE 4-3 EFFECTS OF AUTONOMIC STIMULATION

Effector Organ	*Cranial-Sacral Outflow (Parasympathetic Division)*	*Thoracolumbar Outflow (Sympathetic Division)*
Muscles		
Iris	Contraction of circular fibers; constriction of pupil	Contraction of radial fibers; dilatation of pupil
Ciliary	Contraction in accommodation (lens becomes convex)	None
Of blood vessels (skin)	None	Vasoconstriction
Cardiac	Inhibition (heart rate slows)	Stimulation (heart rate accelerated)
Of bronchi	Constriction	Dilation
Of the digestive tract		
In the wall	Contraction (stimulates peristalsis)	Relaxation (inhibits peristalsis)
Sphincters	Relaxation	Contraction
Of the urinary bladder		
Bladder wall	Contraction (induces micturition)	Inhibition
Sphincters	Relaxation	Contraction (retention of urine)
Of the uterus (in pregnancy)	Inhibition	Contraction
Of the hair follicles	None	Contraction (erection of hairs)
Glands		
Salivary glands	Stimulates serum-secreting cells	Stimulates mucus-secreting cells
Digestive glands	Stimulates secretion	Inhibits secretion
Sweat glands	None	Stimulates secretion

situation. The autonomic system provides the mechanism by which such activities are regulated.

Some of the effects produced by the two divisions of this system are summarized in Table 4-3. It will be noted that, in general, the impulses from one division are antagonistic to those of the other.

MAINTENANCE OF BALANCE IN BODY FLUIDS. The functions of the autonomic nervous system are directed toward maintaining a constant, balanced internal environment (*homeostasis*). Changes take place continuously in many areas within the body, altering the body fluids in chemical composition, temperature, and distribution. The autonomic nervous system acts to regulate such activities. Through its regulation of the *heart* and of the caliber of *blood vessels,* the flow of blood to the various parts of the body is controlled. Through its regulation of *smooth muscle,* the size of tubes or openings and diameter of vessels are controlled. Through its regulation of the *digestive organs,* enzymes are produced and food is moved through the alimentary canal to facilitate digestion and absorption of energy-producing materials. Through its regulation of the *urinary organs,* waste products are eliminated from the body. Through its regulation of the *sweat glands* and other organs, a constant temperature is maintained. Through its regulation of such organs as the *pancreas,* the *adrenal glands,* and the *liver,* all of which work together, the blood sugar level is maintained. Through its regulation of the *endocrine glands* and their production of hormones, many functional activities are controlled. It can be seen, then, that (1) the body acts to maintain within narrow limits the composition of the body fluids that constitute the internal environment of the tissue cells (homeostasis) and (2) this depends on the proper functioning of the parasympathetic and sympathetic divisions of the autonomic nervous system.

CONTROL OVER THE AUTONOMIC NERVOUS SYSTEM. In spite of the picture of widespread regulation of many activities by the autonomic nervous system given in the preceding paragraph, it should be understood that this system does not operate separately and independently of the central nervous system. Control is exercised over it by centers in the brain, in particular the cerebral cortex, the hypothalamus, and the medulla oblongata. Although the autonomic nervous system carries out the functions in more direct relation to the affected areas, all body functions are regulated and coordinated through these higher controls. Autonomic activities are generally involuntary, but limited voluntary control is possible through *biofeedback mechanisms,* to be discussed later.

FUNCTIONAL LEVELS. Activities under the control of the autonomic nervous system occur at various functional levels. For example, *peristaltic movements* of the intestine take place even when segments of the intestine are removed from the body. Such movements are presumably the result of local (intramural) reflexes, although the afferent fibers involved have not been definitely identified. *Micturition* occurs reflexly from connections between afferent and efferent fibers in the same segment in the lower region of the spinal cord. Certain reflexes involving *digestion* are brought about by intersegmental connections in the upper portions of the spinal cord. *Respiratory* and *circulatory* reflexes have their centers in the lower portion of the brain stem (medulla oblongata). *Temperature* is controlled through reflex centers in the hypothalamus. *Cortical activities* may be accompanied by specific autonomic effects; examples are increased rate of heartbeat, rise in blood pressure, and pallor of the skin, as in fright; sweating of the palms, as in mental agitation; salivary and gastric secretion as a result of thought of food; and nausea and vomiting as a reaction to witnessing an unpleasant sight.

Autonomic Reflex Arc (Fig. 4-18). The activities under autonomic control are almost entirely reflex in nature. A stimulus initiates an impulse in an afferent neuron that is transmitted through the nervous system to

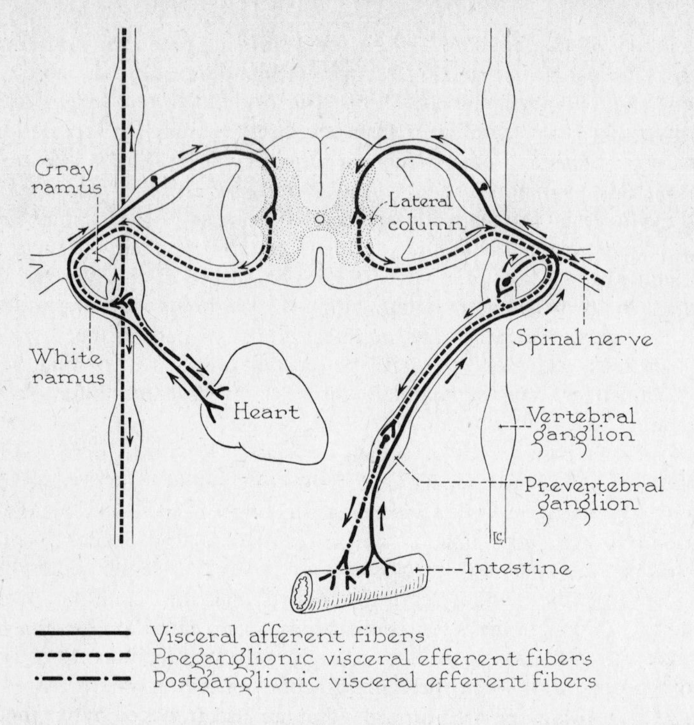

Fig. 4-18. Diagram of a reflex arc of the autonomic nervous system. (Reprinted with permission of W. B. Saunders Co., Philadelphia, from B. G. King and M. J. Showers, *Human Anatomy and Physiology,* 6th ed., 1969.)

the visceral effector organ or organs, bringing about a response. The parts involved in such a reflex are as follows:

1. A *receptor,* either somatic or visceral.
2. An *afferent neuron* (somatic or visceral) that conducts the impulse to the spinal cord or brain, where it synapses with an internuncial neuron.
3. An *internuncial neuron* within the spinal cord or brain. This neuron synapses with a visceral efferent neuron.
4. A *visceral efferent preganglionic neuron.* In the thoracic and abdominal regions, this neuron lies in the lateral horn of the gray matter of the cord. Its axon passes through the anterior root and white ramus of a spinal nerve to vertebral or prevertebral ganglia, where it synapses with postganglionic neurons. In the cranial and sacral regions, its axon passes to terminal ganglia, where it synapses with postganglionic neurons.
5. A *visceral efferent postganglionic neuron* (effector neuron), which transmits the impulse to an effector organ.
6. An *effector organ* (smooth muscle, cardiac muscle, or gland).

The autonomic reflex arc differs from that of the craniospinal system in that *two efferent neurons* are always involved and the cell body of the second or effector neuron lies outside the central nervous system.

Humoral or Chemical Transmission of Impulses. Nerve fibers innervating muscle or glandular tissue come into intimate relationship with only a few of the cells of the structure, yet nerve impulses to an effector organ bring about either excitation or inhibition of most or all the cells of that structure. It has been determined that such effects are induced by specific chemical substances elaborated at the nerve endings. These substances, called *neurohumors* or *neurotransmitters* (*transmitting agents*), are *acetylcholine, norepinephrine,* and *epinephrine.*

ACETYLCHOLINE. This substance is produced at the endings of the postganglionic fibers of the parasympathetic (craniosacral) division—for example, the vagus nerve endings in the heart. It is also produced at the terminations of preganglionic fibers in all autonomic ganglia and in the adrenal medulla, at sympathetic postganglionic terminations in sweat glands, and at motor end plates in skeletal muscle. Any fiber that liberates acetylcholine is said to be a *cholinergic fiber.* Acetylcholine is quickly destroyed or inactivated by an enzyme, *acetylcholinesterase,* produced within the tissues.

NOREPINEPHRINE AND EPINEPHRINE. These two substances are produced at the endings of postganglionic fibers of the sympathetic division, with the exception of those supplying sweat glands and the uterus. These substances, called *catecholamines,* are hormones that are also produced by the medulla of the adrenal gland. Nerve fibers that release these transmitter agents are called *adrenergic fibers.* These two substances, although similar in chemical structure, differ in amounts produced at various sites and in their effects on various effector organs. *Norepinephrine* is the principal transmitter agent at sympathetic nerve endings, while *epinephrine* is the principal substance released from the adrenal medulla. The apparent contradictory actions of these transmitters—for example, constriction of blood vessels in some cases and dilatation in others—are due to the fact that cells in the effector organs possess one of two types of adrenergic receptors (*alpha* and *beta*), which respond differently to the two substances. Following release, norepinephrine and epinephrine are inactivated in various ways. An enzyme, *monamine oxidase* (*MAO*), deaminizes both transmitters. However, it appears that the major portion of the transmitter, especially norepinephrine, diffuses back into the nerve endings and into the vesicles from which it originated to be stored there for future use. The specific effects of norepinephrine and epinephrine are discussed under the adrenal gland, page 242.

Autonomic Drugs. Drugs that act on the autonomic nervous system or their effector organs include the following:

PARASYMPATHOMIMETIC DRUGS (CHOLINERGICS). These act in the body at sites where acetylcholine is produced, their effects mimicking those resulting from parasympathetic stimulation. They stimulate autonomic ganglia and neuromuscular junctions. Drugs in this group include *pilocarpine, muscarine,* and *physostigmine.*

PARASYMPATHOLYTIC DRUGS (ANTICHOLINERGICS). These inhibit craniosacral effects by opposing the action of acetylcholine and cholinergic drugs and stimulating sympathetic activity. *Atropine,* for example, dilates the pupils of the eyes and the bronchioles, reduces secretions in the respiratory passageways, and inhibits movements of the alimentary canal. Other anticholinergic drugs include *scopolamine* and various antihistaminic drugs.

SYMPATHOMIMETIC DRUGS (ADRENERGICS). These mimic the effects of stimulation of the sympathetic nervous system by acting directly on adrenergic receptors or indirectly by inducing the release of transmitters at adrenergic nerve endings. The principal adrenergics are *norepinephrine* and *epinephrine.* They are used in the treatment of shock because they stimulate the heart and constrict arterioles, bringing about elevation of blood pressure. Epinephrine is used in the treatment of asthma because it dilates the bronchioles. It is also used with local anesthetics to prolong their effects. Other adrenergics are *dopamine,* the precursor of norepinephrine, and *ephedrine,* a plant alkaloid.

SYMPATHOLYTIC DRUGS (ANTIANDRENERGICS). These are adrenergic blocking agents that antagonize the effects of sympathetic stimulation, especially acting as vasodilators by relaxing smooth muscle in the arteriolar wall. Examples are *tolazoline* and *phenoxybenzamine.*

Biofeedback. *Biofeedback* is the term now applied to the voluntary control of various internal physiologic activities such as the heartbeat, blood pressure, the secretion of various glands, brain waves, and involuntary muscle contraction. Control depends on information "fed back" to the person through the use of specialized electronic instruments that are able to detect and monitor by sight or sound various physiologic activities. Through practice and training, an individual can learn to control the action of the monitor by using mental and other bodily activities. By doing so, the subject learns to control voluntarily the physiologic function of the action monitored. After a period of practicing with the monitoring equipment, the subject is usually able to bring about the desired action voluntarily, without using the device.

How and why this control is accomplished is not completely understood. Subjects learn to control various "involuntary" activities by exercise of the will, but they are not able to explain how. Biofeedback demonstrates that the power of the mind over bodily activities is much greater than has been generally recognized.

The implications of the theory of biofeedback control of bodily activities and its practical applications are almost limitless. It is used to induce relaxation for the relief of muscle tension and anxiety, to control blood flow and blood pressure, to regulate the gastrointestinal tract and the secretion of various glands, and to alleviate various disorders, such as migraine headaches, muscle spasms, psychosomatic paralysis, hypertension, and insomnia. Biofeedback confirms the inseparability of body and mind in the functioning of the body in health and disease.

Biofeedback should not be confused with "feedback control" mechanisms involved in the regulation of many physiologic activities. These are discussed in the chapter on the endocrine system, page 221.

VISCERAL AFFERENT FIBERS AND VISCERAL SENSATIONS

Sensory receptors are present in most or all of the visceral organs. These are the endings of visceral afferent fibers whose cell bodies lie in the ganglia of cranial and spinal nerves. These fibers pass in the same trunks along with autonomic fibers, but they are not considered a part of the autonomic system because this division consists only of efferent fibers passing from the central nervous system.

In the brain, visceral impulses may or may not reach cortical or conscious levels. Most remain at subconscious levels. At the conscious level, afferent fibers in the sympathetic nerves are involved mainly in the conduction of pain impulses; those in the parasympathetic nerves are concerned with the conduction of impulses that give rise to sensations of hunger, nausea, fullness of the bladder and rectum, and the like.

Visceral sensations differ from somatic sensations in that they are generally vague and poorly localized. When the abdominal cavity is opened under local anesthesia, visceral organs can be handled, cut, or even burned without producing sensations of touch or pain. At times, however, *pain* does occur in the form of cramps, a severe ache, or colic. Stimuli giving rise to these sensations are (1) *overdistention* of a viscus, as from accumulation of gas in the intestine; (2) *obstruction* of a tube, as in intestinal obstruction or the presence of a calculus in a biliary or urethral passageway; (3) *excessive contraction* or *spasm* of smooth muscle; (4) *pathologic* or *inflammatory conditions,* as in neoplasms or appendicitis; and (5) *inadequate oxygen supply,* as in ischemia.

Normally, activities of visceral organs do not produce conscious sensations; that is, one is not aware of peristaltic contractions of the intestine, the heartbeat, or changes in the diameter of blood vessels. This is probably because such activities are not of sufficient intensity to initiate impulses in sensory receptors. However, accentuated activity of an organ, such as an extremely rapid or forceful heartbeat, may register in consciousness. The occurrence of visceral pain is generally regarded as a danger signal indicative of a structural disorder or the malfunctioning of an internal organ. Visceral pain is often referred to other parts of the body. (See the section on referred pain, page 166.)

Afferent impulses arise constantly in the viscera but do not reach the conscious levels. These are of importance in the reflex control of visceral activities. The visceral afferent fibers form the first part of a reflex arc by which impulses are conducted to reflex centers in the spinal cord and brain. Among such centers are the cardiac, respiratory, vasomotor, swallowing, and vomiting centers located in the medulla, the temperature control center in the hypothalamus, and defecation, micturition, and erection centers in the spinal cord.

DISORDERS OF THE NERVOUS SYSTEM

Alzheimer's Disease. See *dementia.*

Cerebral Apoplexy (Stroke). Destruction of brain tissue or impairment of function resulting from cerebral vascular accident (CVA), which includes cerebral hemorrhage, thrombosis, embolism, or vascular insufficiency. There is usually a sudden loss of consciousness, often followed by paralysis.

Chorea. A nervous disorder characterized by involuntary, purposeless contractions of voluntary muscles.

Concussion. Condition resulting from a blow to the head, usually accompanied by transient unconsciousness.

Convulsion. Paroxysms of involuntary muscular contractions and relaxations. It may result from epilepsy, the effects of drugs such as strychnine, eclampsia, various toxemias, brain lesions, infections (meningitis, tetanus, rabies), or dietary deficiencies, especially in infants.

Dementia. An irreversible, deteriorative mental state characterized by loss of cognitive functions (reasoning power, understanding, memory). There are several types, which vary with etiology. *Alzheimer-type dementia,* which usually is associated with aging, is marked by atrophy of the cerebral cortex and development of plaques and neurofibrillary tangles.

Encephalitis. Inflammation of the brain.

Encephalocele. Protrusion of brain tissue through an opening in the skull; hernia of the brain.

Encephalomyelitis. Inflammation of the brain and spinal cord.

Fainting. See *syncope.*

Glioma. A neoplasm or tumor composed of neuroglia cells.

Hemiplegia. Paralysis of one-half of the body.

Herpes Zoster. An acute, infectious disease caused by a virus that attacks the ganglia of cranial and spinal nerves. It is characterized by a vesicular eruption and pain in the cutaneous areas supplied by the affected nerve. Also called *shingles.*

Hydrocephalus. The excessive accumulation of cerebrospinal fluid within the ventricles of the brain, resulting in expansion of these cavities and compression of the surrounding cerebral tissue (and, in neonates, enlargement of the head). It results from blockage of the cerebral aqueduct or the openings from the fourth ventricle to the subarachnoid space.

Locomotor Ataxia. See *tabes dorsalis.*

Meningitis. Inflammation of the meninges of the brain and spinal cord.

Multiple Sclerosis. See *sclerosis.*

Myelitis. Inflammation of the spinal cord; also, inflammation of the bone marrow.

Neuralgia. Pain along the course of a nerve.

Neuritis. Inflammation of a nerve.

Neuroma. A tumor on a nerve or at the end of a severed nerve.

Neuropathy. Any disorder or pathological condition that involves the nervous system.

Palsy. A synonym for *paralysis.* The term is used principally in connection with certain forms, such as *Bell's palsy,* facial paralysis due to a lesion of the facial nerve; *cerebral palsy,* paralysis resulting from defective development of the brain or injury at birth; *Erb's palsy,* paralysis of arm muscles, usually due to injuries of the brachial and cervical nerves at birth; and *shaking palsy,* paralysis agitans or *Parkinson's disease.*

Paralysis. Temporary or permanent loss of motor or sensory function. In *flacid paralysis,* there is loss of muscle tone and a reduction or loss of tendon reflexes. This results from a lesion of the spinal cord involving lower motor neurons. *Spastic paralysis* is characterized by increased muscle tone and exaggerated tendon reflexes. This is due to a lesion of upper motor neurons of the brain.

Paraplegia. Paralysis of both legs and the lower part of the body.

Paresis. Slight or incomplete paralysis; loss of muscular strength.

Parkinson's Disease. A chronic, progressive nervous disorder characterized by tremor, muscular weakness and rigidity, and a peculiar gait. Also called *paralysis agitans* and *shaking palsy.*

Psychoneurosis. A mental disorder resulting from unresolved, unconscious conflicts characterized by anxiety and disturbances in thoughts, feelings, attitudes, and behavior.

Psychosis. A major mental disorder characterized by personality disintegration, bizarre thinking and behavior, and sometimes complete loss of contact with reality. Hallucinations and delusions are common.

Quadriplegia. Paralysis of both arms and both legs.

Schizophrenia. A mental disorder characterized by disturbances in thinking, mood, and behavior. Withdrawal, aggression, and bizarre and antisocial behavior are common. Formerly called *dementia praecox.* Types include *simple, hebephrenic, paranoid,* and *catatonic schizophrenia.*

Sclerosis. An induration or hardening within the nervous system, resulting from degeneration of nervous elements. In *amyotrophic lateral sclerosis,* the anterior horn cells and the pyramidal tracts of the spinal cord are involved, resulting in *progressive muscular atrophy. Multiple* or *disseminated sclerosis* is a chronic, progressive disease of the central nervous system characterized by development of scattered patches (plaques) of demyelinized fibers.

Shingles. See *herpes zoster.*

Stroke. See *cerebral apoplexy.*

Syncope. Fainting; temporary loss of consciousness due to inadequate supply of blood to the brain. It may occur in emotional states such as excitement, fear, or grief, or it may result from hyperventilation or hypoglycemia. Fatigue, exhaustion, poor ventilation, and anemia, along with many other factors, are predisposing causes.

Syringomyelia. A disease characterized by the development of cavities within the spinal cord.

Tabes Dorsalis. A condition characterized by degeneration of the posterior columns of the spinal cord, which contain sensory fibers. It results in pain, muscular incoordination, loss of tendon reflexes, and sensory disturbances. It is due to infection by the causative organism of syphilis. Also called *locomotor ataxia.*

Tremor. An involuntary quivering or shaking movement resulting from alternate contraction and relaxation of opposing muscles. A *rest tremor* occurs when a part is at rest and disappears when active movements are attempted. An *intention tremor* occurs or is intensified when voluntary movements are attempted.

5: SENSATIONS, SENSE ORGANS, AND SENSORY RECEPTORS

The sense organs and sensory receptors and the sensations they mediate constitute the means by which an organism is made aware of its environment.

THE NATURE OF SENSATIONS

A *sensation* is a state of awareness of conditions that prevail within or outside the body or of changes in these conditions. There are four prerequisites for a sensation:

1. A stimulus capable of initiating activity within a receptor cell or neuron.
2. A receptor or sense organ that can react to the stimulus by initiation of an action potential (nerve impulse).
3. A pathway for conducting the impulse from the receptor or sense organ to the brain.
4. A region within the brain capable of translating the impulses into sensations.

Consciousness. In the cerebral cortex and to some extent the thalamus and hypothalamus, the impulses arising in the sense organs bring about specific "feelings" or sensations. These sensations are the *conscious results* of the processes that occur within the brain as a consequence of impulses received from receptors. When a person is awake, the cerebral cortex is constantly receiving countless numbers of afferent impulses. Some of them register in consciousness; others do not. A person can, within limits, select the stimuli he or she wishes to register in consciousness (for example, listening to certain sounds and disregarding others). This is the basis of *attention* and *concentration.* Some sensations can persist within the brain and be recalled to consciousness at a later time. This is the basis of *memory.*

Projection of Sensations. Sensations have their basis principally in cortical activity within the brain. One sees, hears, or suffers pain *in the brain.* This may seem contrary to personal experience, because when a part of the body is injured, the pain is localized in the injured part. Such a phenomenon is accounted for by the process called *projection:* The brain "projects" or refers the sensation to the point of stimulation or to the point of origin of the stimulus, which, in the case of auditory and visual sensations, may be outside the body. That the phenomenon is entirely mental is proved by the fact that when cortical activity ceases, as in sleep, hypnosis, or anesthesia, the ability to form and to project sensations is lost.

Classification of Sensations. Sensations are commonly classified as *exteroceptive* and *interoceptive.*

EXTEROCEPTIVE SENSATIONS. These sensations give information about the external environment. They arise from stimuli outside the body and include the sensations of touch, pressure, temperature, pain, hearing, and sight. The sense organs that mediate these sensations are called *exteroceptors,* which may be further subdivided into *contact receptors* (the stimulus is in direct contact with the sense organ) and *distance receptors* (the stimulus is of more or less remote origin).

INTEROCEPTIVE SENSATIONS. These sensations give information about the internal environment. They arise from stimuli within the body, the receptors being located principally in visceral organs. The sensations include pain, taste, and a number of indefinite sensations, such as fatigue, hunger, thirst, nausea, and suffocation, whose receptors have not been identified. The sense organs that mediate these sensations are called *visceroceptors. Proprioceptive sensations* give information about body position and movement. They include the kinesthetic (muscle, tendon, and joint) senses and the sense of equilibrium. The sense organs that mediate these sensations are called *proprioceptors.*

Modalities of Sensations. Sensations have distinct characteristics by means of which one sensation can be distinguished from another. The combinations of the characteristics that constitute a sensation are referred to as *sensory modalities.* Sensory modalities depend on the nature of the receptor or sense organ stimulated and the region of the brain where the afferent fibers terminate.

NATURE OF THE RECEPTOR OR SENSE ORGAN. Sense organs tend to be specific and to respond to a particular type of stimulus. For example, the eye normally responds to light rays but it may also respond to other types of stimuli, such as pressure or electric shock. Regardless of the type of stimulus, however, the modality of a sense organ is always fixed; that is, stimulation of a sense organ gives rise to sensations specific for that organ. This leads to the conclusion that all nerve impulses are alike in their fundamental nature; a nerve plays no role in determining the sensation or the response resulting from a stimulus. A *sensation* is determined primarily by the nature of the sense organ stimulated and the region of the brain that mediates the impulses received from it. If a response occurs, it is determined by the nature of the effector organ to which impulses are conducted.

NATURE OF THE REGION OF THE BRAIN WHERE AFFERENT FIBERS TERMINATE. In some as yet unknown way, the various sensory areas of the cerebral cortex have the ability to translate sensory impulses into specific sensations. For example, the visual center in the occipital lobe is able to produce a mental "picture" of the objects we see, a picture that in many cases can be recalled at will. A lesion of this portion of the brain may produce blindness even though all the other parts of the receptor and the afferent pathways are normal.

MODALITIES. The sensations of a given modality may differ in quality, intensity, duration, and adaptation.

Quality. Examples of quality are the colors of the light spectrum and the pitch of sound. Individual ability to distinguish between the qualities of a sensation varies. Some persons can recognize extremely slight variations in pitch, whereas others are unable to discriminate between even the most obvious variations.

Intensity. A light may seem bright or dim, a sound loud or faint, pressure upon the skin light or heavy, an odor or taste barely detectable or extremely pronounced. The intensity of a sensation depends on the frequency with which one impulse follows another, the number of receptors activated by the stimulus, or the functioning of the reticular activating system.

Duration. Every sensation lasts a definite length of time. Sometimes the duration is approximately the same as the duration of the stimulus; at other times the sensation may persist after the stimulus has been withdrawn (as when a sensation of light remains for a short period after a light has been turned off).

The persistence of a visual sensation is called an *afterimage.* Looking at the spokes of a slowly rotating wheel, a person may be able to distinguish the individual spokes. If the speed of rotation is increased slowly, the images of the individual spokes tend to become blurred and blend with one another. This occurs because the afterimage of each spoke has persisted until the sensation of the next spoke is perceived. But if the rotation is very rapid, the wheel appears to be a solid structure. Similarly, if a number of slightly varying pictures are seen in rapid succession, the afterimage of each carries over to the next so that the sensation is of one continuous picture rather than of a number of individual pictures. This is, of course, the principle underlying motion pictures.

Let us consider the sensation of touch. If a person *feels* the spokes of a rotating wheel, each spoke will be felt regardless of whether the wheel is rotating slowly or rapidly. The afterimage of the sensation of touch is of such short duration that it disappears before the next impulse is felt; consequently, there is no blending effect.

Adaptation. In some cases a sensation may disappear while the stimulus is still being applied. This phenomenon is called *adaptation.* If a hair is disturbed, a sensation is noted immediately, but this sensation disappears quickly. Similarly, if a ring is placed on the finger, its presence is quite noticeable at first, but gradually the sensation vanishes, although the stimulus may continue for days, weeks, or years. Obnoxious odors may pass unnoticed after a time; purveyors of deodorants are constantly reminding the buying public that we are rarely conscious of our own body odor.

The cause of adaptation is not precisely known. It is not due to nerve fatigue, since nerve fibers have an unlimited capacity to conduct impulses. Receptors apparently differ in their receptiveness to stimuli and in their ability to initiate and transmit action potentials. As a result, they differ in responsiveness to stimulation, some losing this capacity very rapidly, others slowly or not at all.

Although adaptation results from factors acting upon or within the receptor, a similar condition, called *habituation,* occurs as a result of the functioning of the reticular activating system in the brain. This phenomenon enables the brain to ignore monotonous, repetitive impulses that are constantly impinging upon it and to direct its attention to impulses of greater importance or interest. It is the process by which mental concentration is accomplished. A person who has the ability to concentrate may be completely oblivious to loud sounds or other sensory stimuli that would ordinarily divert attention.

CUTANEOUS SENSATIONS (SENSORY IMPRESSIONS FROM THE SKIN)

Stimulation of the skin plays a much more important role in the development and maintenance of the health of the organism than has been generally recognized. The "licking" domestic animals give their young is apparently necessary to stimulate the development and functioning of the gastrointestinal and genitourinary tracts of the young. Experiments have shown that when the young are prevented from being licked by their mothers (or any other animal), they frequently die of failure of function of these tracts. The normal licking is responsible for initiating sensory impulses that pass to the central nervous system and are there mediated to the autonomic division, from which motor impulses pass to the viscera.

It has long been known that when the newborn fails to breathe, giving it a few hearty slaps or massaging its skin will usually initiate breathing. This suggests that there is a connection between skin stimulation and the respiratory center. The reaction under a sudden cold shower of water similarly suggests such a connection.

It is also known that a breast-fed baby experiences fewer gastrointestinal disorders than a bottle-fed baby. The perioral stimulation associated with breast feeding seems to confer many physiologic benefits on the young. In humans, there is considerable evidence that a relation exists between skin stimulation and the development of the life-sustaining systems of the body.

The sensations mediated by the cutaneous sense organs or receptors are touch, pressure, cold, and heat. (Pain must also be included, but inasmuch as the sphere of pain sensations transcends the cutaneous field, pain is more fully described under a separate heading beginning on page 164).

The following table lists the principal cutaneous sensations and the receptors that mediate them.

Sensation	*Receptors*
Touch or light pressure	Meissner's corpuscles
	Merkel's discs
	Hair-root plexuses
	Free nerve endings
Deep pressure	Pacinian corpuscles
Cold	End bulbs of Krause
Heat	End organs of Ruffini
Pain	Free nerve endings
Tickling	Touch and pain endings
Itching	Pain endings

Cutaneous Receptors. The receptors in the table have a *punctiform* arrangement in the skin; that is, certain points when stimulated produce the sensation of pain, others of cold, others of pressure, and so forth. The fact that their distribution over the body surface is not uniform accounts for the considerable variation in the sensitivity of different parts of the body. This has been shown in the "two-point special discrimination test," in which the subject is tested for the ability to distinguish touch sensations coming from two points of a compass. This test shows the following order of sensitivity (from greatest to least): tip of tongue, tip of finger, side of nose, back of head, and back of neck. On the tip of the tongue, points separated by distances of less than 1.4 mm were felt as a single point; on the back of the neck, the distance was 35 to 40 mm. Touch sensitivity depends on the number of touch receptors, sensitivity being greatest where receptors are most numerous.

Classification. Cutaneous receptors are of two classes: *free* or *naked nerve endings* and *encapsulated nerve endings.*

FREE OR NAKED NERVE ENDINGS. These consist of nonmyelinated sensory fibers that divide repeatedly and end in extremely fine branches distributed among the epithelial cells. At the ends of the fibers are small knobs or expansions that may lie within or between the epithelial cells. When stimulated, these endings give rise to the sensation of pain.

Free nerve endings are not only found in the epidermis, where they are extremely abundant, but are also widely distributed in other tissues. They are present in the dermis of the skin, in mucous and serous membranes, in the cornea of the eye, in visceral organs, in the meninges, and in the pulp of teeth. They are stimulated by nearly all types of stimuli, although stretching or distension is the primary stimulus in the viscera.

Merkel's (Tactile) Discs. Some nerve fibers end in a tiny expanded disc that lies in close contact with an epithelial cell. These structures, known

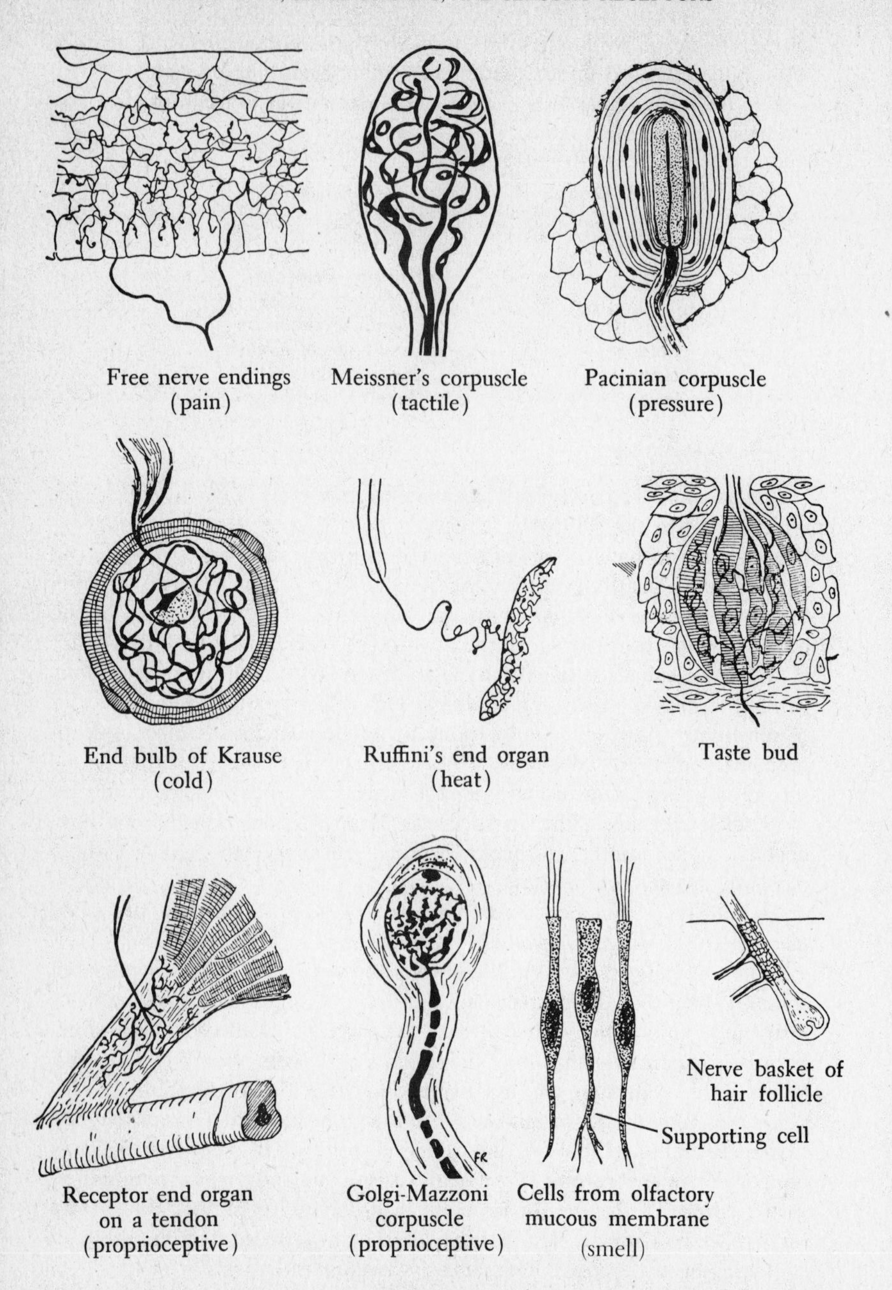

Fig. 5-1. Receptors. (Reprinted with permission of The Macmillan Company from D. C. Kimber et al., *A Textbook of Anatomy and Physiology,* 13th ed., 1955.)

as Merkel's discs or tactile discs, function as touch receptors in the epidermis. A single fiber may supply many discs.

Hair-Root Plexuses. Around the roots of hairs are found elaborate plexuses of nerve fibers ending in contact with the sheaths of the hair roots. Movement of the shaft of a hair produces a slight pressure on these endings, which gives rise to the sensation of hair movement or touch.

ENCAPSULATED NERVE ENDINGS. These structures have nerve endings surrounded by a specialized capsule of connective tissue. The capsule is usually laminated. In it are cells and fluid-filled spaces.

Among these endings are Pacinian corpuscles, end bulbs of Krause, end organs of Ruffini, and corpuscles of Meissner.

Pacinian Corpuscles. These large structures (1 to 4 mm in length) are found in the dermis and the subcutaneous layers of the skin, under mucous membranes, in mesenteries, in the pancreas, and in connective tissue in general. They mediate the sense of deep pressure. In the skin of the external genitalia and around the nipple are corpuscles of similar structure called *genital corpuscles.*

End Bulbs of Krause. These structures resemble Pacinian corpuscles except that the nerve ending is more expanded and the capsule is more diffuse. End bulbs of Krause are widely distributed in the subcutaneous connective tissue, averaging about 15 cm^2. They are the end organs for the sensation of cold.

End Organs of Ruffini. These structures are more complex than the end bulbs of Krause, although they resemble them. They are believed to mediate the sensation of heat, and possibly movement, in connective tissue. They are less numerous and more deeply situated in the tissue than are the end bulbs of Krause.

Corpuscles of Meissner. These are elongated elliptical bodies in which endings of both myelinated and unmyelinated fibers are found. The nerve fibers end in flattened processes within and between the cells of the corpuscles. They are numerous in the skin, especially on the tips of the fingers and toes and on the palms of the hands and the soles of the feet. Corpuscles of Meissner are usually found in the dermal papillae in close contact with the epidermis.

Mechanism of Cutaneous Sensations. The receptors and afferent pathways associated with the cutaneous sensations—along with certain significant phenomena—are discussed in the paragraphs that follow.

TOUCH. *Receptors.* Receptors for touch include the corpuscles of Meissner, Merkel's tactile discs, and hair-root plexuses. *Afferent Pathway.* Impulses pass through large myelinated fibers of cranial and spinal nerves. Axons of spinal sensory neurons enter the cord and pass upward and downward in the posterior funiculus. Some fibers synapse with cells in the gray matter. Axons of these secondary neurons cross to the other side of the cord and continue upward in the *ventral spinothalamic tract* to the brain. In the brain their course is not certain, but it is believed that synapses with tertiary neurons occur and that impulses reach the cerebral cortex by way of the thalamus. The touch center is in the general sensory area.

PRESSURE. *Receptors.* Receptors for pressure include Pacinian corpuscles and genital corpuscles. *Afferent Pathway.* Same as for touch.

TEMPERATURE. *Receptors.* The receptors for cold are the end bulbs of Krause; for heat, the end organs of Ruffini. *Afferent Pathway.* Impulses for the sensation of temperature pass through small, poorly myelinated fibers in cranial and spinal nerves. Impulses entering the spinal cord synapse with secondary neurons whose axons pass upward in the *lateral spinothalamic tract* along with those carrying pain impulses. On reaching the thalamus, such impulses may give rise to crude, uncritical temperature sensations. These impulses are relayed by the thalamus to the sensory or somesthetic area of the cortex.

PHENOMENA ASSOCIATED WITH TEMPERATURE SENSATIONS. Any change in temperature of the environment is a normal stimulus to the heat and cold receptors, the sensation being relative and dependent on the *degree of the change* and the *preceding condition.* To illustrate: If both hands are placed in water, one in hot and the other in cold, for a short while and then transferred to warm water, to the hand that has been in cold water, the warm water will feel hot, but to the hand that has been in hot water, the warm water will feel cold.

Temperature sensations are *poorly localized;* that is, they are felt over a considerable area around the spot that is stimulated. But *adaptation* occurs readily; for example, upon immersion of the body in water below body temperature, the water feels cold, but after remaining in it for a short time, one becomes accustomed to it. Bath water may feel almost unbearably hot at first, but again the body quickly adapts to it.

There is considerable variability in the *sensitivity of various parts of the body* to temperature changes. On exposed surfaces (face, hands, legs), the temperature sense is poorly developed. On the chest, abdomen, and anterior surface of the arm, it is highly developed. Internal organs are relatively less sensitive to extremes of temperature. Hot foods or liquids may cause pain in the mouth and even in the esophagus, but by the time they have reached the stomach, the sensation is all but lost. Hot enemas are not felt above the rectum.

SENSATIONS OF PAIN

Pain is a primitive sensation and of great importance to the state of health and even the life of an individual. As a reflection of disorder within the body, for example, it provides a warning. But whatever the stimulus for pain may be, the important fact is that the reaction of the individual, whether limited or generalized, serves either to protect the body or to withdraw it from the source of the stimulus.

Receptors. Sensations of pain are mediated by naked or free nerve endings. These endings may be stimulated by any of the usual types of stimuli (mechanical, chemical, or thermal) if the stimulus is of sufficient intensity. Pain receptors occur in most organs of the body. They are present in the skin, muscles, joints, visceral organs, cornea of the eye, and the walls of arteries. They are absent in veins, the periosteum, and cancellous bone.

Types of Pain. Two types of pain are recognized, somatic and visceral. *Somatic pain* results from stimulation of receptors located on or near the surface of the body. Pain resulting from stimulation of receptors in the skin is *superficial somatic pain;* that resulting from stimulation of receptors in underlying tissues, such as muscles, fascia, tendons, or joints, is *deep somatic pain. Visceral pain* results from stimulation of receptors in the internal or visceral organs.

Somatic pain is usually localized with a fair degree of accuracy; visceral pain is generally poorly localized and often projected or referred to a region other than the source or point of stimulation (*referred pain*).

Qualities of Pain. Terms applied to various types of pain include pricking, burning, and aching. *Pricking* is felt when the skin is cut, pinched, or penetrated by a sharp object. *Burning* is the intense, excruciating pain that results when the skin is burned or subjected to severe abrasion. *Aching* or *throbbing* is usually a persistent pain of low intensity. The quality, intensity, duration, and location of pain play an extremely important role in a physician's diagnosis of bodily dysfunction or disease.

Afferent Pathways. The path of nerve impulses for pain is through small, poorly myelinated fibers of cranial and spinal nerves. Fibers entering the spinal cord pass upward and downward only a few segments in Lissauer's tract, then pass into the posterior horn of the gray matter, where they synapse with secondary neurons whose axons pass across the cord, ascend in the *lateral spinothalamic tract,* and terminate in the thalamus. Here the axons synapse with neurons of the third order that terminate in the sensory or somesthetic areas in the cortex.

Phenomena Associated with Pain Sensations. Several phenomena associated with sensations of pain are worth noting:

1. Pain receptors may be stimulated by any type of stimulus. Overstimulation of receptors for other sensations (touch, pressure, heat, cold) may initiate pain impulses. In visceral organs, excessive distention or dilatation of an organ, excessive muscular contraction or spasm, inadequate blood supply, or the presence of chemical substances such as the products of inflammation may give rise to pain impulses.
2. There are two types of pain impulses: (a) Those that first occur following painful stimulation are strong and travel rapidly in the poorly myelinated fibers but disappear quickly after cessation of the stimulus. (b) Those that are weak travel slowly in nonmyelinated fibers and persist for

a considerable time after the stimulus is withdrawn.

3. Adaptation to pain does not readily occur.
4. In most cases, painful stimuli are harmful to the body.
5. Pain is an indicator of disease or disorder in the body. It is one of the cardinal symptoms of inflammation.

Referred Pain. Pain arising from the stimulation of pain receptors on the surface of the body can usually be fairly well localized; that is, the cerebral cortex projects the pain sensation to the point of stimulus or its vicinity. However, when pain arises from stimulation of receptors in the visceral organs, the sensation is not projected back to the point of stimulation; instead, the pain may be felt in the skin or in some surface area remote from the point of stimulation. This is known as *referred pain.* Such pain may result in a feeling of tenderness in the skin and a tensing of skeletal muscles. In general, the area to which the pain is referred is one that receives its innervation from the same segment of the cord through which the visceral organ receives its nerve supply.

An understanding of the mechanism of referred pain is a distinct advantage to the physician in the diagnosis of internal disorders. The following table shows some of the general cutaneous areas to which visceral pain is referred.

Pain in:	*May Be Referred to:*
Heart (angina pectoris)	Chest and medial surface of left arm
Liver	Region over right scapula
Stomach	Region over ensiform cartilage or upper portion of back
Kidney	Large area over lower portion of abdomen and back and over lateral and medial portions of thigh
Ovaries	Region inferior and lateral to umbilicus

Referred pain may also occur in peripheral nerves. When a nerve is stimulated somewhere along its course, the sensation is usually referred back to the region supplied by the afferent fibers customarily receiving the stimulus. In this way, pressure on the sciatic nerve in the thigh may cause a tingling feeling in the foot; similarly, irritation of the severed end of a nerve in an amputated limb may cause sensations to be referred to the part that has been cut off (known as a "phantom limb"). Pain may also be referred to a part supplied by a branch of the nerve stimulated, as in the case of an infected tooth in the lower jaw, which may cause pain to be referred to the region of the maxillary sinus, the orbital region, or the ear, all of which are supplied by branches of the trigeminal nerve.

Headache. It is curious to note that, although the pain center is in the brain, the brain tissue itself is relatively insensitive to pain either from cutting or from manipulation. Yet headaches or diffuse pain in the head region are quite common. In the majority of instances, the cause of a headache is some physiologic condition or other factor operating outside the cranial cavity. Headaches may be confined to specific lobes, or they may be quite general. They may be unilateral or bilateral. They may be continuous, intermittent, or throbbing. The range of their intensity is great: from dull pain to acute and almost unbearable pain.

Some common causes of headaches are (1) increased intracranial pressure, as from a brain abscess, brain tumor, or subdural hematoma; (2) toxic states such as those associated with nephritis or carbon monoxide poisoning; (3) inflammation of the meninges, as in meningitis, tuberculosis, or syphilis; (4) concussion or compression of the brain, as in skull fracture; (5) hypertension; (6) eye disorders such as eyestrain, iritis, or glaucoma; (7) infections of the middle ear or nasal sinuses; (8) onset of febrile diseases; and (9) emotional states such as those induced by worry, anxiety, or anger.

There are many other possible causes of headaches, most of which are associated with disturbances involving pain-sensitive structures of the head (nerves or blood vessels). Pressure on or traction applied to these structures or dilatation or constriction of blood vessels either within or outside the cranium may stimulate nerve endings, causing a headache. *Cluster* or *histamine headaches* result from dilatation of cranial blood vessels. The cause of *migraine headaches,* with their accompanying visual and digestive disturbances, is unknown, although their symptoms are related to disturbances in cranial circulation. *Tension headaches* result from sustained or excessive contraction of muscles of the head and neck with pressure on pain-sensitive blood vessels supplying the cranial structures. *Psychogenic headaches* may result from frustration, anxiety, or reactions to stressful situations. Psychotherapy in conjunction with drug therapy is often necessary in the treatment of such headaches.

Psychogenic Pain. Pain that appears to be symptomatic of a disorder in a particular organ for which no organic pathology can be found is not uncommon. If an emotional mechanism can be demonstrated, such pain is said to be *psychogenic.*

Pains of this nature are very real and not imaginary, as is commonly supposed. They frequently lead to surgery for the removal of a healthy organ through wrong diagnosis. In such cases it is not unusual that after the suspected organ has been removed, the pain manifests itself in another organ, and further unnecessary surgery may be performed.

It is now believed that psychogenic pain and the organ malfunctioning that may accompany it can be the results of the body's reaction to certain emotions, such as fear, hate, worry, or resentment, or to conflicts and frustrations related to personality. Many of these conditions have their origin in the early life of the individual, long before the onset of the organic disturbance. The treatment of such conditions usually requires the special experience and knowledge of a psychotherapist.

PROPRIOCEPTIVE SENSATIONS (KINESTHETIC SENSE)

Proprioceptive sensations provide the individual with an awareness of the activities of muscles, tendons, and joints, and, more generally, with an appreciation of position and movement (the *kinesthetic* or *muscle sense*), of importance in the maintenance of equilibrium.

Receptors. Proprioceptive receptors (*proprioceptors*) are located in muscles and at junctions of muscles with tendons or aponeuroses. They are also located in the ear and in the walls of the heart, blood vessels, and hollow organs.

NEUROMUSCULAR SPINDLES. These long, narrow complex endings consist of one or several skeletal muscle fibers enveloped in a connective-tissue capsule. The fibers within the spindle (*intrafusal fibers*) are smaller than typical muscle fibers, and each also bears a motor end plate, the end of a motor fiber. Within the capsule, the sensory nerve fibers come into close contact with the muscle fibers.

NEUROTENDINOUS SPINDLES (GOLGI TENDON ORGANS). These are similar in structure to the neuromuscular spindles, except that the sensory fibers are entwined about the collagenous fibers of the tendon instead of about muscle fibers.

Other proprioceptors are *Pacinian corpuscles,* found in tissues near joints, and the *cristae* and *maculae* of the membranous labyrinth of the inner ear (see pages 215 and 216). The cristae and maculae are concerned with maintenance of equilibrium. The receptors located in the heart, the carotid sinus, and the gastrointestinal tract are also considered to be proprioceptors. They respond to pressure changes within these organs.

Mechanism of Muscle Sensations. It is believed (1) that stretching a muscle stimulates the spiral endings about the intrafusal fibers of a neuromuscular spindle, and as a result, impulses are dispatched that result in the sensation of muscular tension; (2) that submaximal contraction is registered by the neurotendinous spindles; and (3) that maximal contraction is registered through stimulation of the branched endings of the neuromuscular spindle. Further, it is thought that the endings are stimulated by the change in shape of the muscle fibers and that the free nerve endings probably react to chemical changes taking place during muscular contractions. The hydrogen ions of lactic acid probably act as a stimulus in the *sensation of fatigue.*

Afferent Pathway. Proprioceptive impulses are conducted by fairly large, myelinated fibers in the cranial and spinal nerves. The fibers in the spinal nerves enter the cord through the posterior roots. Within the cord the axons may take one of two courses:

1. Through the *fasciculus cuneatus* and *fasciculus gracilis* of the dorsal funiculus to the corresponding nuclei of the medulla, where they synapse

with secondary neurons whose axons cross over and synapse in the thalamus with tertiary neurons. These neurons relay impulses to the somesthetic center of the cortex.

2. Through the *dorsal spinocerebellar tracts,* which are composed of axons of second-order neurons whose cell bodies lie in the gray matter of the cord. Their axons ascend in the lateral funiculus and terminate in the cerebellum; from here impulses are relayed to the red nucleus of the midbrain. Such impulses are involved in "unconscious muscle sense" and also in muscular coordination.

The Role of Proprioceptive Sensations. Through the proprioceptive or kinesthetic sense, a person is able to judge the position of various parts of the body at a given time, to realize the extent of muscular contractions, to estimate weight, and to determine the muscular effort necessary to perform work. The proper functioning of this sense is essential in the control and coordination of muscular movements and in the maintenance of muscle tone that is vital to normal posture.

Proprioceptive sensations are the key to satisfactory performance of complex or skilled activities or movements. Such activities as dressing in the dark, playing the piano, touch typing, and all athletic activities depend primarily on the ability to bring various parts of the body into definite positions at specific times. This can be accomplished only when exact information is provided by the muscle and joint receptors. The precise degree to which these sensations can be developed is illustrated by the ability of blind persons to judge distance or to recognize objects by their shape, size, or position through this proprioceptive sense. It must be admitted, however, that we are only vaguely conscious of the muscle sense. When asked to describe how we know that an arm or a leg is in a given position, we say we "feel" it, but the "feeling" is vastly different from that involving the sense of touch.

Disorders of the Proprioceptive Sense. In the condition known as *tabes dorsalis,* the fibers in the posterior funiculi of the spinal cord are destroyed, preventing proprioceptive impulses from reaching the brain. From this condition develops *locomotor ataxia,* in which there is marked disturbance in walking; patients are unable to judge the position of their legs without looking at them and are unable to stand erect with their eyes closed (*Romberg's sign*).

THIRST, HUNGER, APPETITE, AND FATIGUE

There are some sensations registered by the cerebral cortex, principally internal in origin, for which no specific end organs or receptors have been identified. Among them are thirst, hunger, appetite, and fatigue.

Thirst. Thirst is associated with reduced water content of the body. The origin of thirst may be local (inhalation of excessively dry or dusty air) or general (following hemorrhage, profuse sweating, or loss of water through diarrhea, vomiting, or excessive urination, as in diabetes). Reduced salivary secretion (as from dryness of the mouth, tongue, or pharynx) is another stimulus for the sensation of thirst.

There is evidence that cells called *osmoreceptors,* located in the supraoptic nucleus of the hypothalamus, are sensitive to changes in the osmolarity of the blood. Reduction in water content stimulates these cells, increasing secretion of the *antidiuretic hormone (ADH).* This hormone is then released from the posterior lobe of the hypophysis, and water loss through the kidney is reduced and the renin-angiotensin system activated.

Hunger. Hunger is the sensation aroused by the physical need for food. This sensation, which is projected to the region of the stomach, results from impulses initiated by the rhythmic contractions of the muscles in the wall of that organ. These contractions last about 30 sec and may occur for periods of 30 to 40 min. Thereafter there is usually a period of rest, lasting from $\frac{1}{2}$ to $2\frac{1}{2}$ hr before the hunger feelings recur.

Hunger contractions are not due to a need for food; they may occur while food is still in the intestine. Indeed, after 2 or 3 days of starvation, they disappear. Filling the stomach with food (or even indigestible bulk substances) will, however, *inhibit* the contractions. A reduction in the amount of blood sugar will *increase* hunger contractions, as seen after the injection of insulin. Vigorous exercise, smoking, or swallowing water tend to inhibit them. Emotional states (fear, anger, elation) may cause them to cease entirely.

Appetite. This sensation is similar to that of hunger but differs from it primarily in that, whereas hunger sensations are unpleasant and localized, appetite sensation is pleasant and generalized. Appetite is the desire for food and the ability to eat and enjoy it. To some degree, it is instinctive, yet both experimental animals and human beings tend to lose the desire for food and will refuse to eat it if they are fed for a protracted time on a monotonous and inadequate diet. It is curious to note that both will, if given a choice of foods, tend to select those that remedy dietary deficiencies. Humans are somewhat less efficient in this respect than are experimental animals and therefore need to plan their diet to avoid deficiencies.

Whereas hunger sensations are innate and based on a need for food, appetite is acquired and depends on previous pleasurable experiences associated with eating. Odors, the sight of food, or even the thought of food may excite this sensation. Certain specific substances in foods, called *secretogogues,* stimulate the appetite (and explain the use of so-called appetizers at the beginning of meals). Not the least of the factors involved in appetite are the appearance of foods and the environment in which they are being eaten. Metabolic conditions, too, affect the appetite. Muscular activity increases it; a good appetite is associated with general muscle tonus as well as the tonus of the stomach muscles.

Two centers in the hypothalamus control food intake: a *hunger* or *feeding center* and a *satiety center*. Starvation or lack of food causes hunger, which stimulates the feeding center, causing an individual to seek food and eat. Destruction of the center causes severe anorexia (loss of appetite). Consumption of food produces a feeling of satisfaction, which stimulates the satiety center to inhibit feeding activities. Lesions in these centers may result in abnormal eating habits. Injury to the satiety center prevents its inhibitory effect on the feeding center, resulting in overeating and obesity; injury to the feeding center may result in refusal to take in food, although the individual may be suffering from malnutrition. *Anorexia nervosa,* a self-induced aversion to food, is often the result of psychic conflicts related to body image. It may result in severe emaciation and even death by starvation.

Disorders of the appetite result from a number of factors. In diseased states appetite may be greatly reduced, as it also is when specific substances—for example, thiamine—have been omitted from the diet. Emotional disturbances have a profound effect on appetite; disgust, fear, and anger tend to reduce it, while pleasant environmental conditions and general well-being favor it. Some prolonged emotional states may lead to excessive appetite, resulting in habitual overeating. Loss of appetite is called *anorexia.*

Fatigue. The sensation of fatigue may be fairly definitely localized (as when it is experienced in a finger, arm, or leg), or it may be generalized throughout the body. Specific end organs for fatigue have not yet been identified. Fatigue may be either acute or chronic. In *acute fatigue,* the onset is rapid and the sensation is relieved by rest. Such fatigue is usually of limited duration. Overactivity and the inroads of infectious diseases are common causes. In *chronic fatigue,* there is often some disturbance of metabolism. Among the possible causes of chronic fatigue are (1) dietary deficiencies (inadequate intake of vitamins, mineral salts, proteins); (2) diseases (diabetes, anemia, tuberculosis, kidney infection, heart ailments, to name a few); (3) endocrine imbalance, such as, for example, that which accompanies hypothyroidism; and (4) abnormal physical conditions, especially those that involve the skeleton and the muscles (poor posture, "fallen arches," and others). There is also *psychogenic fatigue,* which results from personality conflicts (neuroses), frustration, and boredom.

OLFACTORY SENSE (SENSATIONS OF SMELL)

The olfactory sense is one of the basic primitive senses, involving significant individual and social consequences.

Olfactory Organ. The olfactory sense organ consists of a specialized region of nasal epithelium lying in the upper portion of each nasal fossa and covering a part of the lateral wall, the roof, and the side of the nasal septum. It covers about 500 mm^2. This epithelium is pseudostratified and contains three types of cells: *sustentacular* or *supporting cells, olfactory cells,* and *basal cells* (Fig. 5-2).

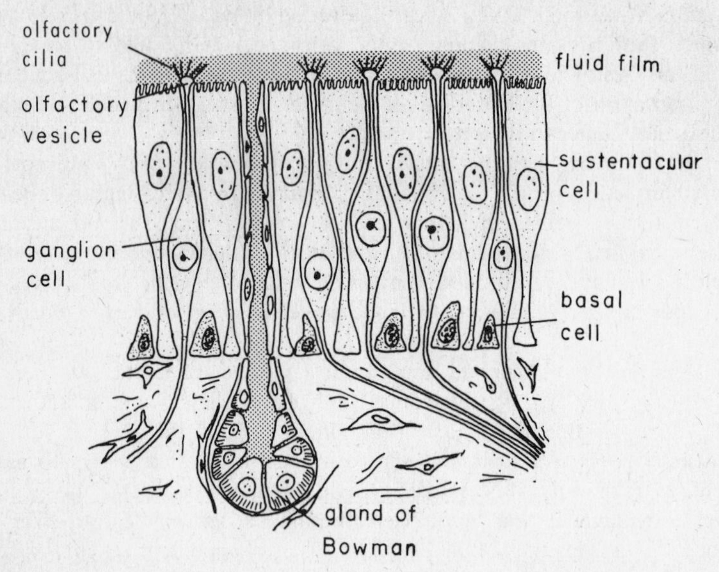

Fig. 5-2. Diagram of the olfactory epithelium. (Reprinted with permission of W. B. Saunders Co., Philadelphia, from C. R. Leeson and T. S. Leeson, *Histology,* 3d ed., 1976.)

SUSTENTACULAR OR SUPPORTING CELLS. These tall, cylindrical cells lie adjacent to the olfactory cells. Their broad free ends bear microvilli; their basal ends are narrow. They contain a yellow pigment, which gives the olfactory area a yellowish color.

OLFACTORY CELLS. These bipolar neurons lie between the sustentacular cells. They are elongated, fusiform cells with a centrally located nucleus; at the tip of its narrow, dendritic end each bears an enlargement, the *olfactory knob* or *vesicle.* This knob projects through the layer of microvilli of the sustentacular cells and bears a tuft of nonmotile *olfactory cilia* that provide the receptive surface for the stimulating molecules. The proximal end of each olfactory cell is a fine, elongated process (an unmyelinated axon) that passes through the basal portion of the epithelium, where it joins others to form a small nerve bundle. These bundles are collected into about 20 *olfactory nerves* (*fila olfactoria*), which pass through the cribriform plate to the *olfactory bulb.* The olfactory cells are the receptors for the sense of smell.

BASAL CELLS. These small pyramidal cells lie between the bases of the sustentacular cells and the olfactory cells. They are thought to be replacement cells for the supporting elements.

The olfactory epithelium is kept moist by secretions of the *olfactory glands of Bowman.* Lying in the lamina propria, these glands are branched, tubuloalveolar structures whose narrow excretory ducts open on the olfactory surface.

Mechanics of Smell. It is not known exactly how stimulation of olfactory cells is accomplished. There are two theories as to how olfactory cells respond to different types of olfactory stimulants. One, the *chemical theory,* holds that on or in the membranes of olfactory cells are receptors (chemicals) that react specifically with different types of chemical substances that act as olfactory stimulants; the other, the *physical theory,* holds that on or in the membranes of olfactory cells are physical receptors that conform in size and shape to various types of molecules, enabling these molecules to be adsorbed and exert their stimulating effect. In either case, the molecules of the stimulating substance must be water soluble in order to penetrate the mucus covering the tips of the olfactory cells and lipid or fat soluble in order to penetrate the olfactory cell membrane. Another theory holds that radiant energy given off by the stimulating molecules acts as the stimulus.

Following stimulation, impulses pass through the olfactory cells and out through axons that pass to the olfactory bulb. Within the bulb, in globular masses called *glomeruli,* these axons synapse with *mitral cells* whose axons form the olfactory tract. This tract passes into the *lateral olfactory striae* and on to the hippocampal formation, a part of the rhinencephalon or olfactory cortex. Here the impulses are interpreted as an *odor,* giving rise to the sensation of smell.

The air in the olfactory region is still or quiet air, in contrast with the air currents that in ordinary respiratory movements pass through the relatively wide nasal passageways (meatuses) lying below the inferior, middle, and superior conchae. Gaseous substances can reach the endings of olfactory cells by diffusion from the nasal cavity or the pharynx. To perceive a faint odor more distinctly, however, we open wide the external nares by a forcible, quick inspiration (sniffing), which causes air to enter the olfactory region.

Classification of Odors. The classification of odors is difficult, chiefly because of the delicate nuances associated with particular smells. Nevertheless, it is possible to recognize in a mixture of substances a number of different odors. One classification that is based principally on the mechanism by which they are perceived, lists (1) pure odors, the sensations for which arise from olfactory receptors in the nose only, (2) odors mixed with taste sensations, and (3) odors mixed with sensations arising from other receptors than those located in the nose. Pure odors have also been subdivided into nine groups: ethereal (fruits), fragrant (flowers), aromatic (benzene), ambrosial (musk), alliaceous (garlic, onions), burning (burnt wheat), goat (sweat), repulsive (garbage), and nauseating (excrement).

Special Phenomena Associated with Olfactory Sensations. When two substances are smelled simultaneously, the odor of one is recognized, then the other. Persistence, or memory of odors, is very pronounced. Once a substance has been smelled, its odor is usually recalled or recognized quite readily.

Adaptation to odors occurs rapidly. For this reason, one becomes accustomed to odors and is able to endure unpleasant ones. This also accounts for the failure of individuals to realize that gas is accumulating slowly in a room.

The sensation of smell is often confused with that of taste, since both may be aroused by the same chemical stimuli. Some food flavors depend on odors; a raw onion eaten with the nose closed can scarcely be distinguished from a piece of apple or potato.

The sensation of smell is not as highly developed in humans as in some of the lower animals, which depend on it to a greater degree in the securing of food, detection of enemies, marking of territories, and identification of young.

The nasal epithelium, in addition to receiving the endings of olfactory nerves, is innervated by sensory fibers from the trigeminal nerve, which supplies the nonolfactory nasal epithelium. These latter fibers mediate sensations of pain, cold, heat, tickling, and pressure.

The sense of smell plays an important role in *sexual selection* and *mating activities.* It is useful to a physician in the diagnosis of disease because some diseases have a characteristic odor.

The *threshold stimulus* for olfactory cells is extremely low for some substances. For example, one of the mercaptans can be detected in concentrations as low as 4.6×10^{-11} g/L of air. Other substances, such as carbon monoxide, fail to arouse any sensation.

Individuals differ widely in their ability to detect odors. In some, the sense is acute and discriminatory; in others, dull and nondiscriminatory. In women, it is more acute before and during menstruation. During pregnancy, pronounced aversions to certain odors may develop.

Inability to smell (*anosmia*) may be temporary or permanent. It may result from inflammation of the nasal mucosa or from the action of local anesthetics. *Partial anosmia* is the inability to sense certain specific odors. *Complete anosmia* may result from developmental defects or from lesions involving the olfactory receptors or pathways or the rhinencephalon. *Hyperosmia,* an abnormally acute sense of smell, may occur in certain cerebral disorders.

GUSTATORY SENSE (SENSATIONS OF TASTE)

The sense of taste is a complex sense for, although it is a chemical sense, a number of other factors play a significant role in the determination of the taste of a substance. The taste of a food is influenced by its temperature (cold, warm, or hot), its physical consistency (soft or hard, mushy or crisp), its odor (pleasant or unpleasant), and its appearance (attractive or repulsive).

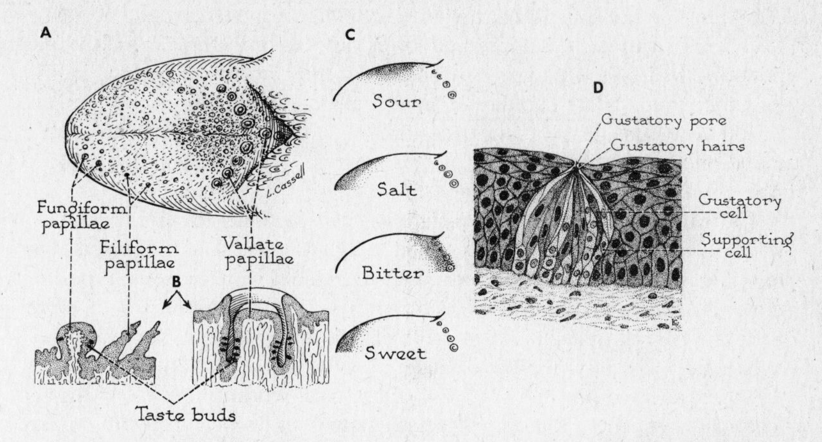

Fig. 5-3. Receptors for taste. (*A*) Dorsal surface of the tongue, showing taste papillae. (*B*) Three types of papillae in section and enlarged. (*C*) Areas of maximum sensibility to four primary taste qualities. (*D*) Vertical section of a taste bud. (Reprinted with permission of W. B. Saunders Co., Philadelphia, from B. G. King and M. J. Showers, *Human Anatomy and Physiology,* 6th ed., 1969.)

As a consequence, receptors for heat, cold, touch, smell, and sight play a role in determining whether or not a food has a pleasant or desirable taste. Taste is important in the selection of palatable and nutritious food and in the rejection of foods that might be harmful.

Receptors (Fig. 5-3). The receptors for taste are *neuroepithelial taste* or *gustatory cells* located in taste buds. A *taste bud* is an ovoid body found in the epithelium of the tongue, buccal cavity, and pharynx. Each is a barrel-shaped structure averaging 70 μm in length and 50 μm in diameter. The distal end of each bears an opening called a *taste pore.* Within each taste bud are two types of cells: supporting cells and neuroepithelial taste or gustatory cells. *Supporting cells* are spindle-shaped and enclose 4 to 20 *neuroepithelial taste* or *gustatory cells,* each of which bears at its distal end a minute *taste hair,* which projects into the taste pore. At the basal end of a taste bud is an opening through which nerve fibers enter. These terminate in close contact with taste cells. Other nerve fibers terminate between taste buds or may closely envelop them.

It is possible that there are receptor cells other than those in taste buds that can mediate taste sensations since it is known that gustatory sensations may result from stimulating areas in which there are no taste buds.

The life span of a taste cell is limited, averaging about 10 to 12 days. As older cells die, new cells migrate from the periphery of the bud inward and develop into mature sensory cells. If nerve fibers to a taste bud are destroyed, the bud degenerates.

Taste buds are, of course, most numerous on the tongue, but they are also found on the soft palate, the anterior pillar of the fauces, the posterior wall of the pharynx, the upper and lower surfaces of the epiglottis, and the posterior wall of the larynx. On the tongue they are found mostly on the fungiform and vallate papillae. A single vallate papilla may bear from 200 to 300 taste buds whose pores open into the circular furrow that surrounds the papilla. The total number of taste buds in humans is about 10,000. The number decreases in old age.

The exact mechanism by which taste cells respond to stimuli and by which action potentials are generated is not known. It is theorized that receptor sites on taste receptor cells combine with various types of molecules, thus enabling a single receptor cell to be responsive to more than one type of molecule. A nerve fiber to a taste bud may send branches to several receptor cells, or a single receptor cell may receive branches from several neurons. The intensity of the taste sensation probably results from the frequency of firing of action potentials, which depends on the concentration of the stimulating substance, but the means by which perception in qualities of taste is accomplished are unknown.

Afferent Pathway. The nerves that supply afferent fibers to taste buds are (1) the lingual branch of the chorda tympani of the facial (VII) nerve, which supplies the anterior two-thirds of the tongue; (2) the glossopharyngeal (IX) nerve, which supplies the posterior third of the tongue; and (3) fibers of the vagus (X) nerve, which supply the epiglottis and the lower pharynx. Fibers from these three nerves enter the *tractus solidarius* of the medulla oblongata and terminate in the long slender *nucleus of the tractus solidarius,* which lies in the floor of the fourth ventricle, where they synapse with second-order neurons, which pass upward to the *arcuate nucleus of the thalamus.* Here they synapse with third-order neurons whose fibers pass to the cerebral cortex, where they end in the lower portion of the postcentral gyrus of the parietal lobe.

In the nucleus of the tractus solidarius, impulses may be transmitted to the *inferior* and *superior salivary nuclei,* from which impulses are transmitted to the *submandibular* and *parotid glands,* which are reflexly stimulated to secrete saliva when food is being ingested.

Classification of Gustatory Sensations. Sensations of taste that are easily recognized fall into four primary groups: *sour, salty, bitter,* and *sweet.* There is no correlation between the taste of various substances and their chemical composition. Crystalloid substances are usually effective in arousing gustatory sensations; colloids are relatively ineffective.

Sour tastes are characteristic of acids, the sourness being attributed to the presence of hydrogen ions. *Salty tastes* are characteristic of chlorides, nitrates, sulfates, and other similar compounds. *Bitter tastes* are induced by such substances as alkaloids (quinine, morphine), some glucosides, bile salts, and many plant extractives. Most of these substances have a toxic or depressant effect on cellular activities. *Sweet tastes* are characteristic of organic substances such as sugars, alcohols, and their derivatives.

Special Phenomena Associated with Gustatory Sensations. The tongue is not equally sensitive to taste sensations at all points. Sweet tastes are registered mainly at the *tip of the tongue,* bitter at the *back* or *root,* salt at the *tip* and *sides,* and sour at the *sides.* This would indicate that there are different receptors for each type of taste, but no anatomic differences among taste buds have been discovered. Taste sensations may result from substances injected into the bloodstream, for example, nicotinic acid. They also result from electrical stimulation of the tongue.

Taste sensations of a particular substance are not always predictable. For most individuals, benzoate of soda has no taste; others may find it bitter or sweet. A condition known as *successive contrast* exists in which the taste of a sweet substance is enhanced if it is preceded by a salty or bitter substance, or vice versa. Certain tastes are *complementary* in that they tend to neutralize each other. A sweet syrup takes the sourness out of lemon juice, and vice versa.

Threshold Stimuli. The minimum concentrations of representative substances of the four main taste groups that will produce the sensation of taste are as follows: *sweet* (cane sugar)—1 part in 200; *salt* (sodium chloride)—1 part in 400; acid (hydrochloric acid)—1 part in 15,000; *bitter* (quinine)—1 part in 2 million.

Adaptation. Taste receptors adapt quickly, especially if the taste-producing substance is kept at one place in the mouth. For that reason, pleasant-tasting substances are usually moved about the mouth. Adaptation for a particular acid usually results in adaptation to other acids; however, there is no adaptation for salts and only moderate cross adaptation for bitter and sweet substances.

Variations in Taste Sensitivity. There is considerable variation among individuals to taste-producing substances. In older persons, taste sensitivity is generally reduced. It is usually altered in infectious diseases. Tastes change with age. Children often show distinct aversions to specific foods but in later life partake of these foods with relish.

Taste blindness sometimes occurs. Some individuals readily taste phenylthiocarbamide (PTC), but for others, called *nontasters,* the threshold level is markedly increased. This trait has a hereditary basis, being inherited as a Mendelian recessive. Complete loss of taste in one modality is rare. Taste sensations are sometimes accompanied by sensations of color, a condition called *pseudogeusesthesia.* A false sensation of taste in the absence of an appropriate stimulus or from no stimulus may occur in certain brain disorders.

OTHER CHEMICAL RECEPTORS

In addition to gustatory and olfactory cells that mediate the senses of taste and olfaction, the body possesses other receptors capable of responding to chemical stimuli. These include the following:

1. *Receptors in the aortic and carotid bodies.* These are stimulated by

reduced oxygen tension or increased carbon dioxide and hydrogen ion concentration in the blood.

2. *Receptors in the brain (central chemoreceptors).* These include (*a*) *receptors in the medulla oblongata* that respond to increased carbon dioxide and hydrogen ion concentration in the blood and (*b*) *receptors in the hypothalamus* that respond to changes in osmolarity and possibly to changes in the blood levels of glucose, amino acids, and fatty acids.

BARORECEPTORS

Baroreceptors, also called *pressoreceptors,* are sensory receptors in the walls of vascular structures, especially the aortic arch and carotid sinus, that are responsive to pressure. The carotid sinus is a dilated portion of the common carotid artery at its bifurcation into the internal and external carotids. Nerve endings in these structures are sensitive to changes in blood pressure and consequently play an important role in various vascular reflexes. A rise in blood pressure causes a reflex slowing in the heart rate and vasodilatation, with a resultant fall in blood pressure.

Carotid-sinus and aortic-arch baroreceptors are also involved in (1) the release of vasopressin from the posterior lobe of the hypophysis and (2) the control of the rate of respiration, especially the activity of abdominal muscles in respiratory movements.

Baroreceptors are also present in the atria of the heart and the walls of the larger arteries (common carotid, thoracic, mesenteric). They respond to changes in the volume of blood passing through them. Impulses arising from stimulation of baroreceptors do not produce conscious sensations, but they play an important role in the initiation of various respiratory and circulatory reflexes by which the body adjusts to changing internal conditions.

VISUAL SENSATIONS

Three structural units are operative in the functioning of the visual organ: the *eyeball* (fibrous tunic, vascular tunic, nervous tunic, and refractory media), wherein lie the sensory receptors of vision; the *accessory structures* (eyebrow, eyelid, conjunctiva, lacrimal apparatus, ocular muscles); and the *orbit* that encloses the eyeball.

To obtain a preliminary understanding of visual function, one has only to compare the structure of the eye with that of a camera, for each is an apparatus constructed to produce an image on a light-sensitive surface. Similarities are shown in Table 5-1.

The Eyeball (Bulb of the Eye)

The eyeball (Fig. 5-4) comprises three coats or *tunics,* which enclose the refracting media.

Coats of the Eyeball. The three tunics of the eyeball are the *fibrous*

TABLE 5-1 COMPARISON OF THE EYE WITH A CAMERA

Phenomenon	*In a Camera*	*In the Eye*
Pigment-containing wall	Box	Sclera and choroid
Light-sensitive surface	Film or plate	Retina
Light-regulating mechanism	Diaphragm	Iris
Mechanism for admitting or shutting out light	Shutter	Eyelid
Refracting media	Lenses and air	Cornea, aqueous humor, lens, vitreous body
Focusing mechanism	Adjustment of distance between lens and film	Alteration of shape of lens
Development of image	In darkroom, on film	In cortex of brain, in consciousness

tunic, composed of the *sclera* and *cornea;* the *vascular tunic* or *uvea,* composed of the *choroid, ciliary body,* and *iris;* and the *nervous tunic* or *retina.*

FIBROUS TUNIC. The fibrous tunic of the eyeball consists of the sclera and the cornea.

Sclera. The sclera is a firm, dense membrane, white in color, forming the *outermost layer* of the eye. Semirigid, it maintains the shape of the

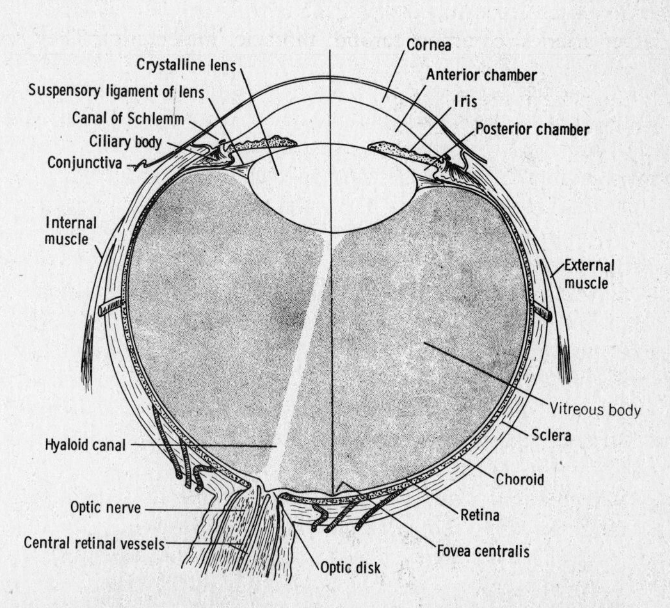

Fig. 5-4. Diagrammatic transverse section of the right eyeball. (Reprinted with permission of J. B. Lippincott Co. from E. M. Greisheimer and M. P. Wiedeman, *Physiology and Anatomy,* 9th ed., 1972.)

eyeball and protects its inner parts. To the sclera are attached the eye muscles and the surrounding fascia. Posteriorly, it is pierced by fibers of the optic nerve, which are axons of ganglion cells of the retina.

Cornea. The *most anterior* portion of the fibrous tunic, the cornea forms about one-sixth of the surface of the eyeball. It is transparent and projects anteriorly, forming a curved, convex structure. At its periphery, the outer layer of the cornea is continuous with the conjunctiva lining the space between eyelid and eyeball. The layers of the cornea are (1) the *outer epithelium,* stratified squamous in type; (2) *Bowman's membrane;* (3) the *substantia propria* (which constitutes about 90 percent of the cornea); (4) the *membrane of Descemet;* and (5) the *corneal endothelium.* Blood vessels are not found in the cornea, but it is well supplied with nerve fibers and lymph vessels. Near the corneoscleral junction, a venous sinus, the *canal of Schlemm,* passes circularly around the eye; from this arise the *anterior ciliary veins.*

VASCULAR TUNIC. The vascular tunic of the eyeball consists of the choroid, the ciliary body, and the iris.

Choroid. The choroid is the *middle layer* of the eyeball, investing about five-sixths of its posterior portion. It is a thin, dark brown, highly vascular membrane lying within the sclera, to which it is firmly attached. It contains many small arterioles and venules that terminate in a dense capillary plexus. Anteriorly, the choroid is continuous with the ciliary body and the iris.

Ciliary Body. The ciliary body extends forward from the ora serrata to the iris. It consists of three parts: the ciliary ring, the ciliary processes, and the ciliary muscle. Anterior to the *ora serrata* (the anterior border of the sensory portion of the retina) lies a circular band 3 to 4 mm wide that bears shallow, radially directed grooves; this is the *ciliary ring.* Continuing anteriorly, a circular row of medially directed ridges, about 70 in number, constitutes the *ciliary processes.* The inner surface of these ridges serves as a point of attachment for the suspensory ligament, which supports the lens. Lateral to these processes is a mass of smooth-muscle fibers, the *ciliary muscle,* which is composed principally of *meridional fibers* and *circular fibers.* The ciliary muscle is the principal agent for changing the shape of the lens in accommodation. (Accommodation is explained on pages 194–195.)

Iris. The *anteriormost* portion of the vascular coat is the *iris,* a thin, colored, circular disc suspended between the cornea and the lens; it divides the space into an *anterior chamber* and a *posterior chamber.* Its lateral margin is continuous with the ciliary body; its medial margin forms the edge of a circular, central opening, the *pupil.*

The layers of the iris are as follows: an anterior-surface *endothelium,* an anterior *stromal sheet* or *lamella* containing many *chromatophores* or pigment cells; a *vessel layer* containing many blood vessels; a *muscle layer* containing contractile elements; and a layer of *pigment epithelium* that forms the posterior surface.

The iris contains two types of contractile fibers: (1) *circular fibers* of smooth muscle that form a sphincter in the margin of the pupil and serve to *constrict* or narrow the pupil, and (2) *radially disposed fibers,* which are modified myoepithelial cells extending from the pupil to the periphery. These serve to *dilate* or enlarge the pupil. The circular fibers are supplied by parasympathetic fibers of the autonomic nervous system, the radial fibers by sympathetic fibers.

The color of the iris depends on the thickness of the stromal sheet and the pigmentation of its cells. If the layer is thick and heavily pigmented, the iris will be brown; if the layer is thin and lightly pigmented, the iris will be blue. With increasing amounts of pigment, the color becomes gray, green, or varying shades of brown. Eye color is genetically determined.

The *function of the iris* is to regulate the amount of light that can enter the eye. This is accomplished reflexly through contraction of the sphincter and dilator muscles. When a *bright* light enters the eye or the eye accommodates for near vision, the circular fibers (sphincter muscle) contract, and the pupil is reduced in size. When the light is *dim* or the eye accommodates for far vision, the radial fibers (*dilators* of the pupil) contract, and the size of the pupil increases, permitting more light to enter.

NERVOUS TUNIC OR RETINA. The retina is a delicate membrane forming the third and *innermost layer* of the eyeball. It consists of three portions: (1) the *pars optica retinae,* the light-sensitive portion, which extends forward from the optic nerve to a wavy line, the *ora serrata,* that lies a short distance behind the ciliary body; (2) the *pars ciliaries retinae,* which continues forward from the ora serrata and forms the internal covering of the ciliary processes; and (3) the *pars iridica retinae,* which continues to the margin of the pupil and forms the posterior pigmented layer of the iris. The last two portions of the retina are not sensitive to light rays.

Layers of the Retina (Fig. 5-5). The retina consists of an *outer pigmented layer* and an *inner nervous layer* made up of three sets of neurons whose cell bodies and processes are arranged to form ten layers, two of which are supporting tissues (the external and internal limiting membranes). From the surface in contact with the choroid inward, the layers are as follows:

1. *Pigment epithelium,* consisting of a single layer of cells containing melanin granules of a pigment, *fuscin.*
2. *Rods and cones,* the photoreceptors.
3. *External* or *outer limiting membrane,* formed by the inner ends of Muller's cells.
4. *Outer nuclear layer,* which contains the nuclei of the rods and cones.
5. *Outer plexiform layer,* a dense network of axons of rods and cones synapsing with dendrites of bipolar cells.
6. *Inner nuclear layer,* containing the nuclei of bipolar cells, Muller's cells, horizontal cells, and amacrine cells. *Bipolar cells* relay impulses from rods and cones to ganglion cells; *Muller's cells* are involved in supportive

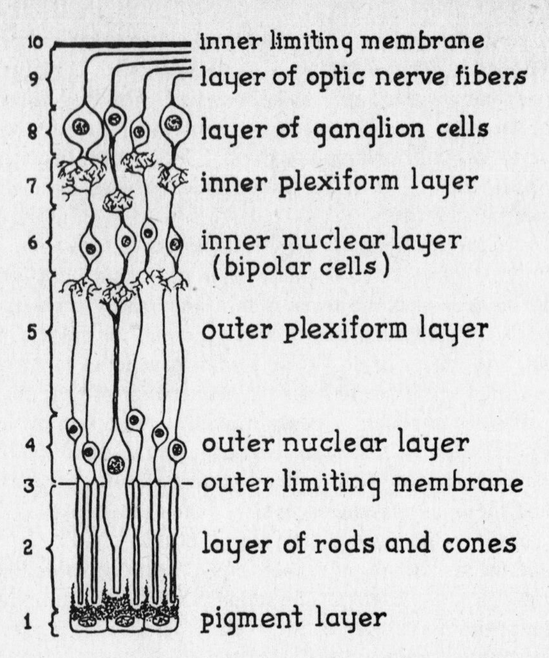

Fig. 5-5. Diagram showing the layers of the retina. (Reprinted with permission of J. B. Lippincott Co. from A. W. Ham and D. H. Cormack, *Histology,* 8th ed., 1979.)

and metabolic functions; *horizontal cells* transmit impulses to receptor cells, other horizontal cells, or bipolar cells; *amacrine cells* are true unipolar cells of unknown function.

7. *Inner plexiform layer,* where axons of bipolar cells synapse with dendrites of ganglion cells.

8. *Ganglion cell layer,* consisting principally of the cell bodies of large ganglion cells.

9. *Nerve fiber layer,* consisting principally of axons of ganglion cells that turn at right angles and follow a parallel course to their point of exit at the optic nerve.

10. *Internal* or *inner limiting membrane,* consisting of terminations of the processes of Muller's cells and forming the innermost surface of the wall of the eyeball.

Rods and Cones. The rods and cones of the retina are modified dendrites of specialized neurons that function as end organs of sight or light receptors. Light rays must pass through most of the layers of the retina to reach the rods and cones. There are an estimated 120 million rods and 5 to 7 million cones in the retina.

Rods. Rods are most numerous in the peripheral portions of the retina but are sparse in the central portion and absent entirely from the fovea. Structurally, the distal portion of each rod is divided into an inner segment and an outer segment, the two being connected by a constriction that contains a few modified cilia. The *outer segment* contains numerous, transverse membranous *discs;* the *inner segment,* the usual cell organelles. The proximal portion of the cell contains the nucleus, and at its end is an expanded *synaptic body.* The discs have a limited life span, about 10 days. As new discs are formed, the older discs migrate distally, and on reaching the tip of the cell, they are phagocytosized.

Rods contain *rhodopsin* or *visual purple,* a conjugated protein composed of a protein, *scotopsin,* and *retinene,* an aldehyde of vitamin A. When rhodopsin is bleached in bright light, the retina becomes less sensitive to light. In the presence of vitamin A, rhodopsin is quickly regenerated, and the sensitivity of the retina is restored, a process called *dark adaptation.* The rods respond to illumination of low intensity; hence, they function under conditions of poor illumination or night vision.

Cones. Cones are less numerous than rods and are concentrated in the posterior part of the retina, especially in the fovea. In basic structure they resemble rods but differ in that the outer segment is usually shorter and broader, literally cone-shaped, except for cones in the fovea, which have a long, narrow outer segment resembling that of a rod. The discs in cones are different in structure. In rods, the discs are discrete structures and expendable; in cones, the discs are infoldings of the plasma membrane and are permanent structures. In fully developed cones, no new discs are formed; however, new protein synthesized in the inner segment is continuously being incorporated into the discs, and this has a rejuvenating effect.

Cones contain one of three photopigments, each of which has the same basic composition as rhodopsin in the rods except that the protein portion, an *opsin,* differs from the scotopsin present in rhodopsin. As a result, some cones are sensitive to red light, others to blue and to green light. Various colors result from the relative numbers of the three types of cones stimulated. Cones are involved in fine discrimination and color vision.

Macula Lutea. In the center of the posterior portion of the retina is an area known as the *macula lutea,* or "yellow spot," in the center of which is a tiny depressed area, the *fovea centralis.* In this area, all layers of the retina are absent except the sensitive layer, in which *only cones,* an estimated 30,000, are present. The fovea is the area of keenest vision, being the point on which light rays are focused when the eye is directed toward an object. The foveal cones are long and closely set, and each connects with a single bipolar cell, which relays impulses by way of a ganglion cell directly to the brain.

Optic Disc. On the nasal side of the fovea is an area in which about 1 million unmyelinated axons of ganglion cells of the retina converge and make their exit through the fenestrated *lamina cribosa* of the sclera to form the *optic nerve.* This area, about 1.5 mm in diameter, is called the *optic disc* or *optic papilla.* At the center of the disc is a depression that marks the point of entrance of the *central artery,* whose branches supply the retina. This artery, which is a branch of the ophthalmic artery, enters the optic nerve a short distance behind the eyeball. Veins that accompany their corresponding arteries unite to form a *central vein,* which exits at the center of the optic disc. No rods or cones are present in the optic disc, so it is insensitive to light; it is consequently called the *blind spot.*

Refracting Media. To reach the retina, light rays must pass through four structures that constitute the *refracting media* of the eye. These are, in order, the *cornea, aqueous humor, crystalline lens,* and *vitreous body.*

CORNEA. The cornea forms the most anterior portion of the fibrous tunic. Its structure is described on page 180.

AQUEOUS HUMOR. The aqueous humor is a watery fluid that fills the anterior and posterior chambers of the eye and permeates the vitreous body. Slightly alkaline, it is similar to blood plasma except that it contains practically no proteins. Sodium chloride and considerable quantities of depolymerized hyaluronic acid are usually present, along with traces of urea and glucose. It is formed principally by diffusion from capillaries of the ciliary processes, although active secretion may be involved. From the ciliary processes it passes into the posterior chamber, then through the pupil into the anterior chamber and laterally to the *spaces of Fontana,* small spaces in the scleral meshwork at the iris angle (where the iris and cornea join). Here it filters slowly into the *canal of Schlemm,* from which aqueous veins convey it to blood-containing veins where the fluid reenters the venous circulation.

Aqueous fluid is maintained at a constant tension, the normal intraocular pressure ranging from 18 to 25 mm Hg. A complete turnover of aqueous fluid occurs every hour. If there is an interference in drainage of the fluid, intraocular pressure rises, and a condition called *glaucoma* develops.

CRYSTALLINE LENS. Immediately behind the pupil and anterior to the vitreous body lies a transparent, biconvex body, the *crystalline lens.* It consists of a *lens substance* composed of lens fibers arranged in concentric lamellae, the anterior surface of which is covered by a single-layered *lens epithelium.* The entire lens is enclosed in an elastic *lens capsule.* The crystalline lens is held in position by a series of fibers extending from the equator of the lens to the ciliary body. Collectively, these fibers form the *suspensory ligament* or *ciliary zonule.*

The lens is continually altering its shape to focus near or distant objects on the retina. With age, the lens begins to harden and becomes yellowish and more flattened, and its ability to accommodate decreases, a condition called *hyperopia* or *farsightedness.*

VITREOUS BODY. This transparent, jellylike mass fills the cavity of the eye between the lens and the retina. Its anterior surface contains a concavity, the *hyaloid fossa,* in which lies the lens. The vitreous body adheres closely to the retina and is firmly attached at the ora serrata and the optic disc. Sometimes it contains a faint canal, the *hyaloid canal,* extending between the optic disc and the lens. This canal is a vestigial remnant of the hyaloid artery in the embryo.

Accessory Structures of the Eye

The accessory structures include the eyebrows, the eyelids, the conjunctiva, the lacrimal apparatus, and the ocular muscles. Their functions are quite distinct and specialized.

Eyebrow. This thickened, movable fold of skin that covers the superior border of the orbit bears many short hairs directed laterally and serves to protect the eye.

Eyelids (Palpebrae). Two folds of the integument lie in front of the orbit of each eye. These, the eyelids, are movable structures that protect the eye from objects or from excessively bright light. They also serve to shut out light rays during sleep and to spread the lacrimal secretions over the anterior surface of the eyeball. The upper lid is the larger and more movable of the two.

The space between the free margins of the two lids is the *palpebral fissure.* When the lids are closed, it is a narrow slit; when open, it has an ovoid shape. At the two ends of the fissure are the *medial* and *lateral angles* formed by the union of the two lids, which comprise the *medial* and *lateral palpebral commissures* or *canthi.* At the medial canthus is a triangular space, the *lacrimal lake,* occupied by a reddish elevation, the *lacrimal caruncle.* The caruncle lies between two openings, the *puncta lacrimali,* each of which lies on the tip of a small elevation, the *lacrimal papilla,* located on the medial margin of each eyelid. It is through these openings that tears enter the lacrimal canaliculi or ducts.

Each eyelid (Fig. 5-6) is covered on its outer surface with *skin* that is continuous at the margin of the lid with the *palpebral conjunctiva* that lines the inner surface. Beneath the skin is a layer of loose connective tissue. In the middle of the lid is a layer of skeletal muscle, the *orbicularis oculi,* whose fibers form a sphincter that, upon contraction, closes the eyelids. Attached to the upper portion of the upper lid is the *levator palpebrae superioris,* a muscle that raises this lid, thus opening the eye. In each lid posterior to the muscle fibers is a stiff plate of dense, fibrous connective tissue, the *tarsus,* that gives form to the lid. Within its substance are the *tarsal (Meibomian) glands,* long, slender glands that open on the margins of the lids. Their secretion is sebaceous in nature and serves to lubricate the margins of the lids and to prevent the overflow of tears. At the margin of each lid are two or three rows of stiff, curved hairs called *cilia* or *eyelashes.* Just posterior to these are the openings of enlarged and modified sweat glands called *ciliary glands.*

Conjunctiva. This is a mucous membrane lining the conjunctival sac, the space between the eyelids and the eyeball. The *palpebral portion* of the

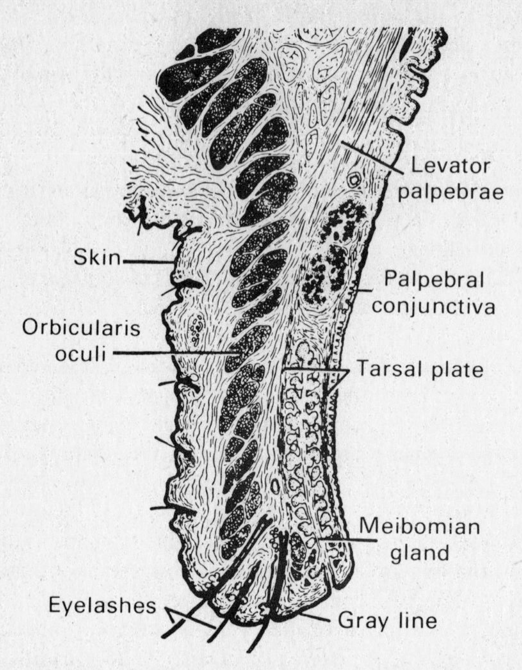

Fig. 5-6. Drawing of a perpendicular section through the upper eyelid. (After A. Duane, *Fuch's Textbook of Ophthalmology,* rev. 8th ed., Lippincott, 1924.) (Reprinted with permission of J. B. Lippincott Co. from A. W. Ham and D. H. Cormack, *Histology,* 8th ed., 1979.)

conjunctiva lines the lids, and at folds called the *superior* and *inferior fornices;* it is reflected back and is continuous with the *bulbar portion,* which covers a part of the sclera and the entire cornea. At the medial canthus, it forms a semilunar fold, the *plica semilunaris.* At the margins of the lids, it is continuous with the skin.

The conjunctiva that lines the lids is highly vascular. At the proximal region of the tarsal plate, it may possess irregular invaginations, some of which are lined with mucous cells and are regarded as glands. Over the sclera, the conjunctiva is thin, with few blood vessels; over the cornea, it is reduced to epithelium only. The epithelium over the cornea and at the margins of the lids is stratified squamous epithelium; elsewhere it is stratified columnar epithelium.

Lacrimal Apparatus (Fig. 5-7). The structures concerned with the secretion and release of tears constitute the lacrimal apparatus.

The *lacrimal glands* are two compound tubuloalveolar glands located at the upper lateral portions of the orbits. Each averages about 20 mm by 12 mm in size and consists of two parts: a *superior lacrimal gland* and an

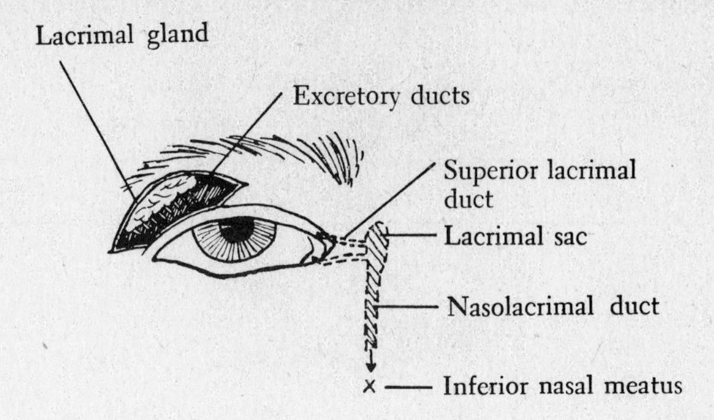

Fig. 5-7. Lacrimal apparatus. (Reprinted with permission of C. V. Mosby Company and the author from C. C. Francis, *Introduction to Human Anatomy,* 2d ed., 1954.)

inferior lacrimal gland. From the inferior portion, *excretory ducts* empty into the superior fornix of the conjunctival sac.

The *lacrimal secretion* (*tears*) passes medially and accumulates in the *lacrimal lake,* where it enters two small openings, the *puncta lacrimalia,* located on the *lacrimal papillae.* Through these it passes into two ducts, the *superior* and *inferior lacrimal ducts,* which converge medially and enter the *lacrimal sac* lying at the inner angle of the eye. From the lacrimal sac, with which it is continuous, a *nasolacrimal duct* conducts the secretion to the nasal cavity, where it opens into the inferior meatus. The secretion consists of a watery fluid containing salts and some mucin. It is sterile and slightly antiseptic. On entering the conjunctival sac, tears are spread over the surface of the eyeball by the lids. They are normally carried away through the lacrimal ducts as rapidly as they are secreted. In the event that the conjunctiva is stimulated by irritating substances (dust, bacteria, or injurious gases), or in certain mental states such as those resulting from pain, grief, or laughter, the lacrimal glands may be activated and tears secreted more rapidly than they can be carried away by the ducts. Under such conditions, the fluid accumulates and overflows the lids, this being the emotional reaction of crying or weeping. Stoppage of the flow through the nasolacrimal duct, as occurs in inflammation of the nasal cavity, may result in accumulation of tears or "watering" of the eyes. The production of tears is called *lacrimation.*

Ocular Muscles. The ocular muscles move the eyeballs and the eyelids. They include the *extrinsic muscles of the eyeball* (four rectus and two oblique) and the *levator palpebrae superioris,* which raises the upper eyelid. Any specific movement of the eyeball usually necessitates action by two or more of the extrinsic muscles. The ocular muscles are skeletal and under voluntary control. All have their origin in the vicinity of the optic foramen.

EXTRINSIC MUSCLES (Fig. 5-8). The extrinsic muscles controlling the movements of the eyeball include four rectus and two oblique muscles.

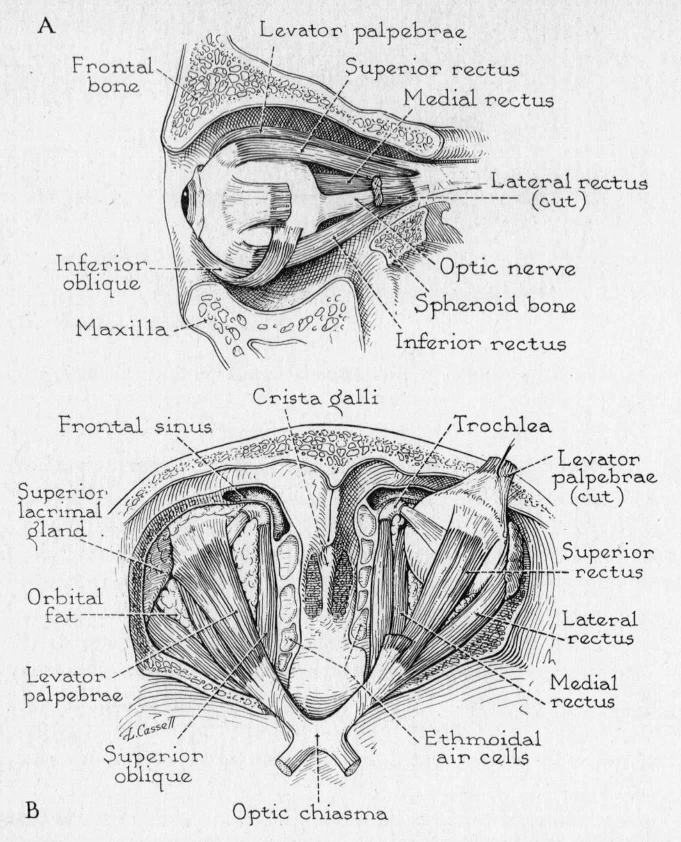

Fig. 5-8. Muscles of the orbit. (*A*) Lateral view. (*B*) Superior view. (Modified after Sobotta and Uhlenhuth; and Grant, *An Atlas of Anatomy,* Williams & Wilkins.) (Reprinted with permission of W. B. Saunders Co., Philadelphia, from B. G. King and M. J. Showers, *Human Anatomy and Physiology,* 6th ed., 1969.)

Rectus Muscles. The rectus muscles (superior, inferior, lateral, and medial) are inserted on corresponding surfaces of the eyeball. They originate at the apex of the orbit near the optic foramen and act to turn the eyeball upward, downward, laterally, and medially, as their names indicate. The superior rectus is antagonistic to the inferior rectus; the lateral rectus is antagonistic to the medial rectus.

Oblique Muscles. The *superior oblique muscle* arises at the apex of the orbit and passes anteriorly to the orbit's upper inner portion, where it becomes tendinous and passes through a fibrous loop, the *trochlea* or "pulley." Here, bending at a sharp angle, it turns laterally and posteriorly and passes to the eyeball, where it is inserted on the upper surface between the superior and lateral recti. The *inferior oblique muscle* arises from the

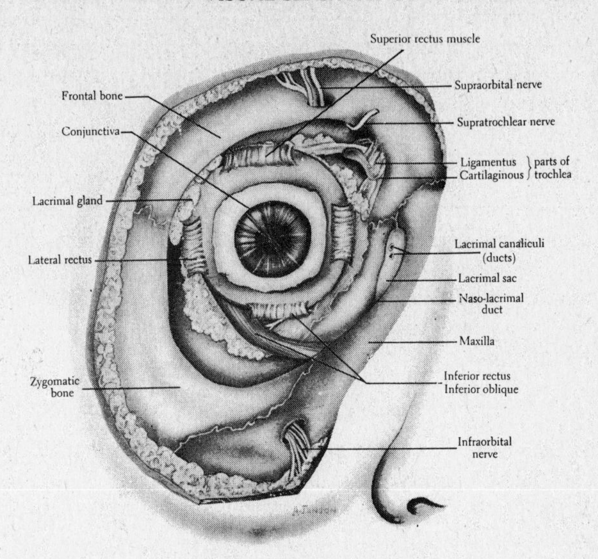

Fig. 5-9. Orbital cavity.

medial margin of the floor of the orbit. It passes laterally, upward, and backward and is inserted on the lower surface of the eyeball under the lateral rectus muscle. The oblique muscles act to rotate the eyeball on its axis.

LEVATOR PALPEBRAE SUPERIORIS. This muscle arises from the apex of the orbit and passes anteriorly to end in a wide aponeurosis that is inserted in the upper portion of the eyelid and the fornix of the conjunctiva. It elevates the upper lid, opening the eye. Paralysis of this muscle results in *ptosis* (drooping of the lid).

Movements of the Eyes. To keep the visual image focused on the fovea of the retina, four types of eye movement are utilized. These include rapid eye, smooth pursuit, divergence and convergence, and compensatory movements.

1. *Rapid eye movements* (*REM*) or *saccades* are extremely rapid, jerky movements that move the image back and forth over the receptors, thus preventing adaptation. They occur when the gaze is fixed steadily on an object and also when the eyes are closed during certain periods of sleep.
2. *Smooth pursuit movements* occur when the eyes are following a moving object.
3. *Divergence and convergence movements* occur as the eyes adjust to the distance of an object. For a near object, convergence or turning inward of the eyes is necessary; for a distant object, divergence or turning outward of the eyes occurs.

4. *Compensatory movements* for head movement are necessary to keep the image focused on the fovea when the head is moved. If the head turns to the right, the eyes must turn to the left; if the head turns to the left, the eyes must turn to the right. Up and down movements of the head similarly require reciprocal movements of the eyes.

Eye movements in the first three categories listed are regulated through neural control systems that depend on information provided by stimulation of the retina. Compensatory movements that correlate eye movements with changes in position of the head are regulated by impulses arising in receptors for position and balance located in the ampullae of the semicircular ducts of the inner ear.

The Orbit of the Eye

The cavities of the skull that enclose the eyeballs and their accessory structures are called the *orbits* (Fig. 5-9).

Walls of the Orbit. The bones comprising the walls of the orbit are as follows:

Superior wall:	Frontal, sphenoid
Inferior wall:	Maxilla, zygomatic, palatine
Lateral wall:	Sphenoid, zygomatic, frontal
Medial wall:	Maxilla, lacrimal, ethmoid, sphenoid

Note: The frontal, sphenoid, and ethmoid bones are single bones involved in the structure of *both* orbits.

Structure. The orbit is, in general, funnel- or cone-shaped, with its apex directed posteriorly, its broad end anteriorly. The rim of the orbit is formed by bony ridges of the frontal, zygomatic, and maxillary bones, which serve as protective structures. In the posterior wall, the sphenoid bone bears two openings: (1) the *optic foramen,* a round opening that is located medially and transmits the optic nerve and ophthalmic artery, and (2) the *superior orbital fissure,* an irregular slit lateral to the optic foramen that transmits the oculomotor (III), trochlear (IV), and abducent (VI) nerves that innervate the ocular muscles. It also transmits the ophthalmic branch of the facial (VII) nerve.

Fascia. This includes the structures lying within the orbit that bind together and support the eyeball and its related structures. The fascia includes:

1. The *periorbital* or *orbital periosteum* which closely invests the bone forming the walls of the orbit.
2. The *orbital septum,* a fibrous sheet that extends partially across the anterior opening of the orbit, being continuous at its margin with the periosteum.
3. The *bulbar fascia* (*capsule of Tenon*) that forms a thin, fibrous sheet enclosing all of the eyeball except the corneal portion. Its inner surface is

smooth and forms a socket, which permits limited movement of the eyeball. A continuation of the fascia encloses the optic nerve. The bulbar fascia separates the eyeball and the optic nerve from the orbital fat.

4. The *muscular fascia,* composed of thin fibrous sheets that enclose the ocular muscles; these are extensions of the sheaths of the medial and lateral rectus muscles attached to the walls of the orbits forming the *medial* and *lateral check ligaments* that *restrain* eye movements.

Orbital Fat. In addition to the eyeball and its accessory structures, the orbit contains a considerable quantity of orbital fat. This fills the spaces posterior to the eyeball and serves as a protective cushion. In starvation and in certain forms of illness, this fat is absorbed, causing a shrunken appearance of the eyes (*endophthalmos*). On the other hand, tumors, abscesses, excess development of fat, or excessive activity of the thyroid gland may cause the eyeball to protrude (*exophthalmos*). Exophthalmos may also result from the contraction of certain smooth muscles lying outside the eyeball (Müller's muscle in the floor of the orbit, a muscle in the bulbar fascia, and the superior tarsal muscle of the upper eyelid).

Orbital Circulation. The orbit of the eye contains important arteries and veins.

ARTERIES. The principal artery supplying the orbit is the *ophthalmic artery.* It is a branch of the internal carotid and enters the orbit through the optic foramen. The ophthalmic artery supplies the ocular muscles, lacrimal gland, and eyeball and sends branches to the ethmoidal cells and nasal mucous membranes. A branch of this artery, the *central retinal artery,* passes through the optic nerve to the retina.

VEINS. The *superior* and *inferior ophthalmic veins* drain the orbits. They pass through the superior orbital fissure, sometimes as a single trunk, and empty into the cavernous sinuses.

LYMPHATIC VESSELS. Lymphatic vessels and lymph nodes are lacking in the orbital structures.

Orbital Nerves. The nerves innervating the orbital structures are as follows:

1. Mixed nerves, which carry *motor* impulses to muscles and *sensory,* principally proprioceptive, impulses to the brain. These include (*a*) the *oculomotor* (III) *nerve,* which supplies the medial, superior, and inferior rectus muscles, the inferior oblique muscle, the ciliary muscle, the sphincter muscle of the iris, and the levator palpebrae superioris; (*b*) the *trochlear* (IV) *nerve,* which supplies the superior oblique muscle; and (*c*) the *abducent* (VI) *nerve,* which supplies the lateral rectus muscle.

2. The *ophthalmic branch of the trigeminal* (V) *nerve.* This supplies the cornea, ciliary body, and iris; the conjunctiva and lacrimal gland; the mucous membrane of the nasal cavity and some of the sinuses; and the skin of the eyebrow, eyelids, nose, and forehead. It is strictly a *sensory* nerve.

3. *Autonomic nerves.* The ciliary muscle and contractile fibers of the iris are innervated by fibers of the parasympathetic and sympathetic divisions of the autonomic nervous system. Postganglionic fibers of the *parasympathetic* division arise in the *ciliary ganglion,* which lies at the back part of the orbit; those of the *sympathetic* division arise from the *superior cervical ganglion.* The *lacrimal glands* are innervated by postganglionic fibers of the parasympathetic division from the sphenopalatine ganglion and by sympathetic fibers from the superior cervical ganglion.

The Physiology of Vision

The Mechanics of Vision. In the eye, receptors (rods and cones) are stimulated by light energy. This initiates nerve impulses that pass to the visual area of the cerebral cortex, where they give rise to the sensation of sight. Visual sensations are of three types: *light, color,* and *form.* Although light is the principal stimulus for light sensations, an electric shock, a blow, or even gentle pressure on the eyeball can cause a visual sensation. Such a sensation takes the form of a circle or flash of light ("seeing stars"), which may occur even in the dark or with the eyes closed. These sensations are called *phosphenes.*

LIGHT VISION. The word *light* has two meanings. Subjectively, light refers to the sensation resulting from stimulation of the rods and cones of the eye. Objectively, light or light energy consists of electromagnetic radiations of varying wavelengths, only a small range of which is capable of stimulating photoreceptors. These radiations or light rays consist of minute, discrete particles called *photons,* which possess energy and momentum and are organized into discrete packets. They move in a wavelike fashion and hence possess a wavelength and a frequency. *Wavelength* is the distance between successive wave crests; *frequency* is the number of wave peaks or cycles that pass a certain point each second. Light rays travel in a vacuum at approximately 186,000 mi/sec (just under 300,000 km/sec).

Light rays that stimulate visual receptors and give rise to visual sensations range in wavelength from 370 nm to 740 nm (nm = *nanometer,* one-billionth of a meter). These constitute the *visible spectrum.* Light rays of greater length (*infrared rays*) or lesser length (*ultraviolet rays*), which constitute "black light," do not produce visual sensations because the eye lacks receptors capable of being stimulated by them.

When light rays of the proper wavelength enter the eye, the lens focuses them on the rods and cones in the retina, where chemical changes initiate the development of action potentials. These are relayed as nerve impulses through nerve cells and fibers to the brain, where, in the cerebral cortex, they give rise to sensations of sight.

LIGHT AND DARK ADAPTATION. When a person passes from a brightly lighted room to a dark or dim room, it is impossible to distinguish objects for a time. Slowly the contents of the room begin to come into view and objects begin to take form. This process by which the eyes become adapted to vision in dimmer light is called *dark adaptation.* The change is due to (1) regeneration of visual purple in the rods, which increases their sensitivity to light rays of lower intensity, and (2) dilatation of the pupil, which permits more light to enter the eye and strike the peripheral regions of the retina where the rods predominate.

After the eyes have become dark-adapted, if the same person passes to a brightly lighted area, the light has a dazzling effect, and for a short period, vision is poor. After a few seconds, however, the eyes become adapted. This process is called *light adaptation.* The change is the result of (1) bleaching of visual purple, reducing the sensitivity of the rods, and (2) contraction of the pupil and partial closure of the eyelids, thereby reducing the amount of light entering the eye.

These two adaptive processes relate to the specialized functions of the rods and the cones. Dark adaptation makes *scotopia* or *twilight vision* possible. Light adaptation makes *photopia* or *daylight vision* possible.

Scotopia. Vision in dim light is the function of the *rods,* which have a *low* threshold of excitation; that is, they respond to light of low intensity. The rods are *not* concerned with *color* vision, their visual impulses being recorded in black or white or gray. The rods function in dim light and darkness, when color and detail are not discernible. In the rods is a reddish pigment, *rhodopsin* or *visual purple,* which is bleached upon exposure to light. Bleaching is essential for stimulation of the rods. In the dark, the pigment is resynthesized. Rhodopsin is a protein linked to a pigment of the carotene group (vitamin A). In the absence of vitamin A, rhodopsin cannot be formed, and *night blindness* or *nyctalopia* (inability to see in a dim light) results. Night blindness may also occur as a consequence of congenital lack of visual purple.

Photopia. Vision in bright light is the function of the *cones,* which have a *high* threshold of excitation; that is, light of high intensity is needed to elicit a response from them. They are the receptors in color vision and register the fine details of objects. When the eyes are directed toward an object, the retinal image falls on the *fovea centralis,* a small depressed area in the center of the macula lutea. At this point the retina is very thin and devoid of rods. This is the area of most acute vision. Surrounding the rod-free fovea, the retina contains both rods and cones, but the cones decrease progressively in number toward the periphery, where rods alone are present.

REFRACTION OF LIGHT. When light travels through a uniform medium, all the waves move at the same rate and in a straight line. However, should light rays pass from one medium to another (as from air to water), their velocity is altered, and the rays are bent or *refracted.* When the light rays pass to a medium of *greater* density, they are bent *toward* a plane perpendicular to the surface of the two media. When they pass to one of *lesser* density, they are bent *away from* this perpendicular plane. The measurement of the ability of a substance to bend a ray of light is referred to as its *refractive index.* This index is a relative measurement; air is the basis for it, the refractive index of air being assumed to be 1.0. That of water is 1.33. Light rays, on entering the eye, must pass through the cornea, aqueous humor, lens, and vitreous body to strike the retina. Accordingly, these are *refractive media.* All of them, with the exception of the lens, have approximately the same refractive index as water; that of the lens is 1.42. The formation of a sharp image on the retina depends on the proper functioning of these refractive media.

INVERSION OF THE IMAGE. Owing to the refractive power of the lens, light rays passing through it cross each other and as a consequence appear on the retina in a reversed position. The result is an *inverted image.*

Projection. When an image is formed on the retina, impulses are initiated that pass to the visual centers of the brain, where they are registered in consciousness as sight. Visual sensations thus occur within the brain. But the brain immediately projects these sensations out of the body to the sighted objects, which reflect the light rays that have stimulated the retina. In the process of projection (a mental process), an object appears as it actually is, right side up.

ACCOMMODATION. To be seen clearly, an object must form a *sharply focused* image on the retina of each eye simultaneously. The process that enables us to see objects clearly at different distances is *accommodation.* It depends primarily on the elastic properties of the crystalline lens. When the eye is focused on a distant object, the surfaces of the lens are *least convex,* and their refractive power is at a minimum; this is *far vision.* When the eyes view an object nearer than about 6 m, the lens surfaces become *more convex,* and the refractive power of the lens is at a maximum; this is *near vision.*

Mechanism of Accommodation. The crystalline lens is held in a state of tension by the pull of the suspensory ligament. Fibers of this ligament extend radially from the margin of the lens to the ciliary processes, to which they are attached. The ciliary processes are projections of the ciliary body, a part of the choroid coat of the eyeball. Intraocular pressure causes a pull to be exerted by the suspensory ligament, which reduces the curvature of the lens. This is the state of the lens when the eyes are at rest (that is, closed or focused on objects 6 m or more away). In *far vision,* the light rays that enter the pupil of the eye are nearly parallel and are brought to a focus on the retina. In *near vision,* accommodation is necessary to bring about clear vision, which is accomplished by contraction of the *ciliary muscle.* The fibers of this muscle are so arranged that when they contract, the ciliary body is pulled slightly forward. This forward movement lessens the tension on the suspensory ligament, whereupon the lens, by virtue of its elasticity, becomes more spherical. This increases the convexity of the central portion of the anterior surface of the lens, increasing its refractive power and enabling light rays from objects to be brought to a focus on the retina.

Limits of Accommodation. There are limits within which the eye can accommodate. If an object is brought toward the eyes, at a certain point the image will begin to be indistinct or blurred. The shortest distance from the eyes at which an object can be seen clearly is called the *near point of vision.* For the average young adult, this averages about 25 cm; in infants it is much less (5 to 10 cm); in older people it is much greater, because the lens gradually loses its elasticity (and consequently its power to accommodate) with advancing age. After the age of 50 years, most individuals are unable to accommodate for near vision. This condition, called *presbyopia,* can be corrected by wearing glasses with a convex curvature for close work. The farthest distance from the eyes at which an object can be seen clearly is called the *far point of vision.* This point is infinity, that is, any distance beyond 6 m.

Near vision requires not only a change in the shape of the lens but also two other correlated adjustments. These are (1) *convergence of the eyeballs,* so that the retinal images fall on identical or corresponding points to bring about single vision, and (2) *constriction of the pupil,* so that light rays pass through only the central portion of the crystalline lens, wherein lies the greatest refractive power.

Control of Accommodation. The ciliary muscle is composed of smooth muscle fibers innervated by fibers of the autonomic nervous system. Accordingly, accommodation is an involuntary reflex activity. The stimulus that initiates the reflex seems to be the contraction of the internal rectus muscles, which brings about the convergence of near vision. Voluntary control of the extrinsic muscles thereby automatically brings about coordinated activity of the ciliary muscles. The internal rectus muscles are innervated by the oculomotor (III) nerve. Postganglionic fibers supplying the ciliary muscles arise in the ciliary ganglion, which lies just behind the eye.

DEFECTS AND DISORDERS OF ACCOMMODATION

Disorders of sight involving the mechanism of accommodation are myopia, hyperopia, and astigmatism.

Myopia or *nearsightedness* arises when the anteroposterior diameter of the eyeball is longer than usual, a condition in which light rays come to a focus slightly *in front of* the retina. Myopia can be corrected by the use of concave lenses, which cause the light rays to diverge before they strike the crystalline lens of the eye. Myopic subjects who have not had this condition corrected hold objects they wish to see clearly close to their eyes; this indicates that their near point of vision is less than normal.

Hyperopia or *farsightedness* arises when the vertical diameter of the eyeball is greater than usual, a condition in which light rays come to a focus slightly *behind* the retina. Hyperopia can be corrected with convex lenses, which cause the rays to converge before they strike the crystalline lens. In hyperopic subjects, the near point of vision is more than 15 cm from the eyes. Consequently, to see any object, near or far, accommodation is necessary.

Astigmatism is due to irregularities in the curvature of the cornea or, less frequently, of the crystalline lens. This very common defect is probably present to some degree in the eyes of all persons. The curved surfaces of the cornea and the lens normally represent segments of a sphere that causes light rays to converge to a point of focus on the retina. However, if either surface is curved more vertically than horizontally, or vice versa, some parts of an object seen will be in focus and other parts will be out of focus. The result will be indistinct vision. Astigmatism is corrected with cylindrical lenses that compensate for the irregular curvature. The commonest form of astigmatism is that in which the vertical curvature is greater than the horizontal.

COLOR VISION. Colors have no objective existence; they exist only in consciousness as subjective phenomena. What corresponds to "color sensations" in our environment are electromagnetic waves. When light waves of various wavelengths strike the retina, they give rise to nervous impulses that, upon reaching the brain, are interpreted as colors.

Color vision is accomplished through the action of *cones,* of which there are three types, sensitive, respectively, to wavelengths of the three primary colors, red, green, and blue. Some cones possess a pigment that is stimulated maximally by red light, others by green light, and others by blue light. These pigments resemble the rhodopsin present in rods in that each consists of a protein, an *opsin,* and a light-sensitive *chromatophore* (retinene) that is an aldehyde of vitamin A. Light energy breaks down the pigment into its chromatophore and its opsin. This initiates an action potential, which is propagated as a nerve impulse. Following the breakdown, the pigment is reconstituted. Cones are selectively sensitive to one of the three primary colors depending on the pigment they contain. An object that appears red is red because all of the light rays striking it are absorbed except those at the wavelength of 565 nm. These red rays are reflected, and on entering the eye they strike the retina and stimulate the cone cells that contain the red-receptive pigment. Similarly, wavelengths of 540 nm and 435 nm stimulate cones containing green and blue pigments, respectively. Although each type of cone responds maximally to light rays of a particular wavelength, it can respond to other wavelengths, but not as effectively. As a consequence, color depends on the ratio of the number of cones stimulated maximally, moderately, minimally, or not at all. The result is that subjectively we can recognize every shade of color between the ends of the visible spectrum.

There are approximately 150 identifiable colors or hues. Two colors that, when combined, cancel each other and produce the sensation of white are called *complementary colors*—for example, red and greenish blue. *Primary colors* are those that in various combinations can produce all other colors. *White* is the color resulting from reflection of all rays of the spectrum; *black* is the total absence of light.

The foregoing theory of color vision, in which perception of color is based on the presence of three types of color receptors sensitive to rays of all wavelengths, but preferentially sensitive to one region of the spectrum, is known as the *trichromatic theory.* Certain phenomena of color vision, however, cannot be readily explained by this theory. Light rays of certain wavelengths, when mixed together, tend to neutralize each other. To explain this, an *opponent-process theory* has been proposed, in which it is assumed that an inhibitory mechanism, either in the bipolar-ganglion cell pathway or in the thalamus, can turn off certain stimuli. Color sensations are considered to depend on competitive interactions within two pairs of color receptors, one sensitive to red-green stimuli, the other to blue-yellow stimuli.

Cones and Light. When the image of a well-lighted object falls on the fovea or the part of the retina immediately surrounding it, all the spectral colors of the light emitted by the object can be distinguished. The cones in this area are especially adapted for responding to light of all wavelengths. In areas a short distance from the fovea, yellows and blues can be distinguished, but not reds and greens. In the extreme peripheral regions of the retina, color sensations are not elicited, only light and shades. Color discrimination is correlated to the distribution of the cones in the retina, the cones being most numerous in the fovea, reduced in number or absent in the peripheral regions.

Color Blindness. Color blindness is the inability to distinguish between colors. The condition may exist in a mild degree (*color weakness*), or it may be complete (*achromatopia*). The latter state, in which an individual is totally insensitive to color, is rare and is believed to be due to an absence of cones or their inability to function. In *trichomats,* all three cone systems are present and functioning, but one is weak. In *dichromats,* only two cone systems are functioning, so individuals tend to confuse reds and greens or yellows and blues. In *monochromats,* only one cone system is functioning, and the individual is completely color-blind. A color-blind individual is often unaware of the visual defect until tested and informed of the result.

About 8 percent of all males, but less than 1 percent of all females, are color-blind to some extent. This is because the condition is inherited and the gene is sex-linked, its transmission being similar to that in hemophilia.

AFTERIMAGES. A visual sensation may persist for a short time beyond the period during which the stimulus is applied. If, for example, a person looks at a bright light for a moment and then closes the eyes or turns them toward a dark surface, the sensation of light will continue for a noticeable period, then gradually fade away. By the same token, if a person looks at a bright-colored object and then looks at a dark surface, an image of the object in the first color will persist. The persistence of a visual sensation after the stimulus has ceased is called an *afterimage.* If the bright parts of the image remain bright, it is called a *positive afterimage.* The occurrence of afterimages enables visual sensations to appear to be continuous even though the stimuli may be intermittent. Motion pictures take advantage of this phenomenon. Still pictures are projected with sufficient rapidity (16 per second) for the afterimage of each picture to persist until the next picture is seen, with the resulting illusion of a continuous picture.

The length of time required for a light stimulus to evoke a sensation is extremely short. A light flash of adequate intensity can be distinguished, though it may persist for only 1/8,000,000 of a second. Positive afterimages last only a fraction of a second.

If a person looks at a *colored* object for a few seconds and then directs the eyes to a sheet of white paper, the image of the object will appear on the paper, but in the color that is complementary to the first; that is to say, if the object is yellow, the afterimage will be seen in blue. This is called a *negative afterimage.* Should the object be black, the negative afterimage will be white (and vice versa).

When a person looks at a green object for a protracted time and then at a red object, the color of the latter will be intensified; this phenomenon is called *successive contrast.*

Visual Acuity. The degree of sharpness or distinctness of vision is called *visual acuity.* It is measured either by (1) the smallness of an object that can be seen clearly at a standard distance or (2) the greatest distance at which a standard-sized object can be seen clearly. A common type of test for visual acuity is that based on a prepared chart of test letters (*Snellen test*). The degree of acuity is indicated by fractions such as 20/20, 20/10, 20/40, which express the subject's visual acuity as compared with that of a normal individual. A *normal eye* is one by which block letters of an established size can be distinguished clearly at a distance of 20 ft (6 m).

Visual Acuity (Army Standard)

20/40	One-half normal. Subject must stand 10 ft from chart to read letters normally read at 20 ft.
20/30	One-third normal. Subject must stand one-third nearer to read letters.
20/20	Normal vision.
20/15	Better than normal. Subject can read letters at one-third greater distance than normal.
20/10	Twice normal. Subject can read letters at twice the normal distance.

Visual acuity is also measured by a grating test pattern that employs lines oriented in various directions or by use of a Landolt ring. In the latter test, rings of various sizes with a gap are presented to the subject; the ring with the smallest size for which the gap can be specifically located determines the visual acuity.

Visual acuity is greatest in the *fovea,* where only cones are present and where they are most concentrated. It decreases gradually toward the periphery of the retina. For an object to be seen in three dimensions, three separate receptors must be stimulated; a line may be seen when only two receptors are stimulated. The smallest discernible image on the fovea has a diameter of 0.004 mm (4 μm). Images on the peripheral portions of the retina must be larger than this to be discerned.

Visual acuity varies greatly among individuals due to differences in structure of the retina, sensitivity of the rods and cones, and other factors, such as brightness and illumination of the stimulus, where the image falls on the retina, whether the eye is stationary or moving, the condition of the refracting media, and the length of time the subject is exposed to the stimulus.

VISUAL FIELDS. The part of the outside world seen by one eye constitutes the *visual field* of that eye. In humans, the visual fields of the two eyes overlap, with the result that most objects within the field of vision are seen by both eyes. This is termed *binocular vision.* The movements of the eyes are controlled reflexly in such a manner that both eyes are directed toward the same object; consequently, an image of the object is formed in each eye. The places where these two images of a given object lie are called the *corresponding points* of the two retinas.

The term *visual field* is also used to describe the surface of the retina that is stimulated by light rays coming from the external visual field. The diagnosis of certain diseases of the retina and the optic pathways can be made by an analysis of changes that have occurred in the visual fields. These can be mapped with the aid of an instrument called a *perimeter.* The area of the visual field of each eye is reduced somewhat by the eyebrow, nose, and cheek, which prevent some light rays from entering the eye.

BINOCULAR OR STEREOSCOPIC VISION. The ability to judge distance or to perceive depth is due largely to binocular vision because the two retinal images are slightly dissimilar, owing to the fact that the two eyes view the object from different angles. The two images give rise to impulses that are fused in the brain into a single composite image. Distance is judged by the degree of convergence of the eyeballs; the nearer the object, the greater the degree of convergence.

Several other factors are also of importance in the estimation of depth or distance:

1. *Size of the image on the visual field.* The nearer the object, the larger the image formed on the retina.
2. *Relative distinctness of detail.* In near objects, color, form, and fine details are distinct; in distant objects, they are less so.
3. *Perspective.* Objects that are aligned (such as railroad tracks) appear to come together in the distance. This effect is employed in subjective judgment of distance. In a drawing, the illusion of distance is created by introducing lines that converge toward a point in the background.
4. *Parallax.* This is the apparent relative movement of adjacent objects owing to changes in the observer's position. As one moves forward, near objects appear to move in the opposite direction and distant objects in the same direction. In finer perceptions (as in the reading of a thermometer and many other scales), the same effect is produced by mere movement of

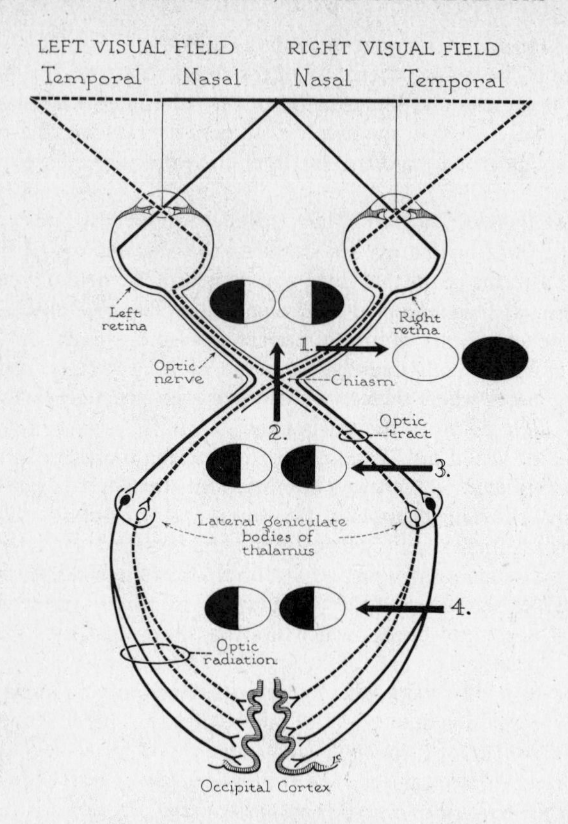

Fig. 5-10. The visual pathway. Arrows indicate possible sites of injury of the pathway. Black portion of the oval suggests the visual-field loss resulting from the injury. (Reprinted with permission of W. B. Saunders Co., Philadelphia, from B. G. King and M. J. Showers, *Human Anatomy and Physiology,* 6th ed., 1969.)

the head or eyes. Because the eyes are more or less constantly in motion, the resulting parallax is highly useful in depth perception.

5. *Blocking out of distant objects by near objects.* Regardless of their relative sizes, near objects can block out distant objects. A coin held close to the eye can block out a huge building several hundred meters away. Judgment in such cases depends on prior knowledge of the comparative true sizes of the objects.

VISUAL PATHWAY (Fig. 5-10). The visual pathway comprises the structures through which visual impulses pass in their course from the sensory receptors to the sensory areas of the brain. From the rods and cones, impulses are transmitted through bipolar neurons to ganglion cells whose cell bodies lie in the retina and whose axons leave the eye through the optic nerve. As the axons pass through the *optic chiasma,* a crossing of the optic nerves, some fibers cross to the opposite side (*decussation*), while others remain uncrossed (*semidecussation*). Passing through the optic chiasma, the fibers, now constituting the *optic tract,* enter the brain and end in the *lateral geniculate body* of the thalamus. Here they synapse with neurons whose axons pass through the *internal capsule* to the *visual centers* located in the cortex of the occipital lobes of the cerebrum. The nerve fibers that pass from the lateral geniculate body to the occipital cortex are called the *optic radiation.*

In traveling from the rods and cones to the occipital cortex, visual impulses must pass through at least four neurons: (1) a rod or cone cell, (2) a bipolar cell, (3) a ganglion cell, and (4) a neuron with its cell body in the thalamus. The cell bodies of the first three lie in the retina.

Cones in the fovea have a direct route to the cortex; that is, each bipolar cell receives impulses from one or a few cones, and the ganglion cell receives impulses from only one or a few bipolar cells. As a result, precise and detailed information is relayed from cones in the fovea and registered in the brain, thus providing acute vision. In other areas of the retina, many rod and cone cells may synapse with a single bipolar cell and several bipolar cells with a single ganglion cell. As a result, stimulation of peripheral areas of the retina provides less detailed information about color and form but more information about light intensity. There are approximately 1 million fibers in the optic nerve and an estimated 125 million rods and cones in the retina. It is obvious that for the major portion of the retina, there must be a considerable degree of convergence of impulses from rods and cones through bipolar cells on to ganglion cells whose fibers comprise the optic nerve.

The visual field (Fig. 5-10) of each eye is divided into two regions: the *medial* or *nasal* half and the *lateral* or *temporal* half. For each eye, light rays from an object in the nasal half of the visual field fall on the temporal half of the retina; those from objects in the temporal half of the visual field fall on the nasal half of the retina. It should be noted that in the optic chiasma, the nerve fibers from the nasal halves of the retinas *cross* and continue on to the thalamus; the nerve fibers from the temporal halves of the retinas *do not cross* but continue directly to the thalamus. In this way, the visual center in the occipital cortex on each side receives impulses from the nasal half of the retina of one eye and the temporal half of the retina of the other eye. Accordingly, a lesion or injury to the *retina* or the *optic nerve* of one eye results in interference with (or loss of) vision in *one eye only,* but owing to the peculiar course of the nerve fibers in the optic pathway, a lesion or injury to the nerve fibers in the *optic chiasma, optic tract, optic radiation,* or *visual center of the cortex* causes blindness in half of *each retina,* a condition known as *hemianopsia* or half-blindness.

DISORDERS OF VISION

Blepharitis. Inflammation of the eyelids, especially their margins, involving the hair follicles and sebaceous glands.

Blindness. Blindness or loss of sight sensation may be partial or complete. It may involve one eye, both eyes, or parts of both eyes. Blindness may result from (1) loss of transparency of any of the refractory media; (2) disease, injury, or abnormalities of the retina; (3) disease, injury, or abnormalities of the optic pathway; and (4) lesions or malfunction of the visual center of the brain.

LOSS OF TRANSPARENCY OF REFRACTING MEDIA. The parts most frequently involved in this form of blindness are the cornea and the crystalline lens.

A diseased or injured *cornea* forms new vascular tissue known as a *pannus.* Such a cornea is usually cloudy and has an uneven surface covered with a thin film of blood vessels; the film may cover a portion of the cornea or all of it. This condition is seen in trachoma, eczema, and granular conjunctivitis. Another form is gonorrheal conjunctivitis in newborn infants (hence the required use of silver nitrate or an antibiotic as a prophylactic measure).

Loss of transparency of the crystalline lens is known as *cataract.* It is due to degenerative changes in the lens proteins that lead to opacity. A cataract usually begins in the center of the lens and proceeds peripherally to involve the capsule. It is most common in older persons (senile cataract). The condition can be corrected by surgically removing the opaque lens and substituting for it a plastic lens or corrective eyeglasses to compensate for the loss of refractive power. Cataract may also be congenital, infantile, traumatic, or occupational. It is frequently seen in association with diabetes. Its incidence is high in tropical areas, where it is thought to be due to excessive exposure to ultraviolet rays.

Sometimes minute semiopaque bodies are present in the vitreous body. Usually they are of no significance, but they can create a disturbance of vision manifested as the appearance of flying specks (*muscae volitantes*) when the eyes are closed or when the eyes are directed toward a clear space. These are caused by the presence in the vitreous body of epithelial cells or remnants of embryonic structures that cast shadows on the retina.

DISEASES, INJURIES, OR ABNORMALITIES OF THE RETINA. Injury to the optic nerve or the optic tract results in degeneration of the nerve cells of the retina and consequent blindness. Injury to the eyeball may cause the retina to become detached from the choroid coat. Pathological changes in the retina commonly occur in diabetes mellitus. *Retinitis pigmentosa* is an inherited disease that leads to blindness; *retrolental fibroplasia* is an oxygen-induced degeneration of the retina that occurs in premature infants.

Glaucoma is a common cause of blindness. In this condition, intraocular pressure (normally 25 mm Hg) increases markedly due to either (1) an increase in production of intraocular fluid or (2) blockage of the draining canal (*of Schlemm*) located at the angle of the anterior chamber. This leads to pathological changes in the retina and the optic nerve. There are two types: *primary glaucoma,* of unknown cause, which is not preceded by other ocular disease, and *secondary glaucoma,* which follows some other eye disease.

DISEASE, INJURY, OR ABNORMALITIES OF THE OPTIC PATHWAY. If a lesion involves the optic nerve, blindness of the corresponding eye results. If a lesion involves the nerve fibers between the optic chiasma and the occipital cortex, partial blindness of both eyes results; this is due to the decussation of some of the fibers in the optic chiasma. Brain tumors that involve the optic pathway may cause blindness.

LESIONS OR MALFUNCTION OF THE VISUAL CENTER OF THE BRAIN. Lesions from blows, fractures, brain tumors, and other causes may involve the visual center in the occipital lobe. If the lesion is unilateral, partial blindness results, involving the halves of the eyes corresponding to the side of the injury; if bilateral, total blindness results.

Other types of blindness that sometimes occur are *color blindness* (inability to distinguish one or more colors), *night blindness* (imperfect vision at night or in dim light), *psychic blindness* (failure of a person to recognize what is seen, such as may occur in migraine headaches, hysteria, or certain toxic states), and *sun blindness* (retinal injury caused by gazing at the sun).

Chalazion. A cyst or small, hard tumor that develops on the eyelid. It is formed by a distention of the tarsal or Meibomian glands.

Conjunctivitis. Inflammation of the conjunctiva, the membrane that lines the eyelids and is reflected over the anterior surface of the eyeball.

Diplopia (Double Vision). In this condition, the optical axes of the two eyes cannot be directed simultaneously at the same object. In diplopia, the light rays from the object do not fall on corresponding points of the two retinas, giving rise to double images. (See also *strabismus* or *squint.*)

Endophthalmos. Recession of the eyeball into the orbit due to absorption of orbital fat. This condition may occur in emaciation from malnutrition, disease, or old age.

Exophthalmos. Abnormal protrusion of the eyeballs, which may be caused by abnormal deposition of fat behind the eyeball or a widening of the palpebral fissure. The condition is a common symptom of hyperthyroidism, which results in edema and lymphoid infiltration of the orbit. Other causes include inflammation of the orbit and enlargement of the eyeball.

Nystagmus. Involuntary to-and-fro movements (quickly forward, slowly backward) of the eyes following rotation of the head.

Scotoma. An islandlike blind spot in the visual field; it may be temporary or permanent. It may follow excessive use of alcohol or tobacco or overexposure to light. It often accompanies migraine. Scotomas may arise from an injured or diseased retina or from conditions involving the optic tract or the brain.

Strabismus or squint. A condition in which the visual axes of the two eyes cannot be directed toward the visual object due to inequality of the opposing muscles. If the eyes converge, it is known as *esotropia* or *cross-eye;* if the eyes diverge, it is known as *exotropia* or *walleye.*

Stye (Hordeolum). An inflammation of a sebaceous gland in the edge of the eyelid.

Trachoma. Chronic contagious granular conjunctivitis, caused by a strain of *Chlamydia.* It may lead to deformities of the eyelids with resultant visual disability or even blindness.

Tunnel Vision. A concentric narrowing of the field of vision that makes the external world appear as seen through a tunnel. Vision is so restricted as to require gross movements of the head and the eyes. Its cause is unknown. This condition is a common symptom of hysteria.

Ulcers. Ulcers may occur on the cornea and are commonly the result of injury to or section of the trigeminal nerve. Because this nerve contains sensory fibers that are involved in eye reflexes, the presence of foreign bodies may not be detected or protective mechanisms brought into play to prevent injury to the cornea.

PRACTICAL CONSIDERATIONS

Eye Examination. Examination of the eyes is of importance not only for the detection of disorders involving the eyes themselves but because pathological conditions not associated with the eyes or visual sense can often be diagnosed through observation of structural or circulatory changes within the eyes. Examination of the eyes may be performed by an ophthalmologist or optometrist. An *ophthalmologist,* also called an *oculist,* is a physician who specializes in diseases and disorders of the eyes, and is trained in the diagnosis and treatment of various pathological conditions and in the correction of disorders of vision. An *optometrist* is a nonmedical person trained in optometry and licensed to examine and test eyes. An optometrist's work is principally measuring the powers of vision and prescribing appropriate lenses or exercises for their correction. Optometrists are not allowed to prescribe drugs or to perform surgery. An *optician* is a person skilled in the making of optical instruments or lenses or one who dispenses them. An optician grinds lenses prescribed by an ophthalmologist or optometrist, sets them into frames, and fits glasses.

Instruments used in the examination of the eyes include the ophthalmoscope and the retinoscope. An *ophthalmoscope* enables the examiner to look into the interior of the eye and distinguish details of the retina. The instrument consists of a small mirror with a central opening through which the examiner views the subject's eye, which is illuminated by light rays from the mirror. A *retinoscope* is an instrument used to determine the refractive power of the eye, especially in subjects who are unable to converse and thus unable to be tested by the examiner, such as a small child or a severely retarded individual.

Details of the structure of the retina can be observed by use of an *ophthalmoscope.* Blood vessels can be clearly seen and changes symptomatic of certain diseases noted. The condition of the optic disc (the point where the optic nerve enters the retina) is of importance. Excessive internal pressure, as in glaucoma, causes this disc to be pushed backward and to have a "cupped" appearance; excessive external pressure, as from a brain tumor, causes it to be pushed forward, creating the condition known as *choked disc.*

Drugs Affecting the Pupil. A *mydriatic,* such as atropine, epinephrine, or cocaine, *dilates* the pupil. A *miotic,* such as pilocarpine, physostigmine, or morphine, *constricts* the pupil. The pinpoint pupil characteristic of morphine addiction is well known.

THE SENSATIONS OF HEARING AND EQUILIBRIUM

The sensation of hearing (the *auditory sense*) and that of equilibrium (the *labyrinthine sense*) are presented together because of their close *anatomical* association. Although the external and middle ears are concerned exclusively with hearing, parts of the internal ear (especially the membranous labyrinth, hence *labyrinthine* sense) play a role in the sense of balance.

Structure of the Ear

The ear, sense organ of hearing and of equilibrium, consists of three divisions: the *external ear,* the *middle ear,* and the *internal ear* (Fig. 5-11).

External Ear. The external ear consists of the *pinna* and the *external acoustic (auditory) meatus.*

The *pinna* or *auricle* is the projecting portion located on the side of the head. It consists of supporting elastic *cartilage* and six poorly developed intrinsic muscles covered by skin. The prominent in-curved rim is called the *helix;* a second curved ridge, the *antihelix,* lies within and parallels the helix. Enclosed by the antihelix is a deep cavity, the *concha,* which leads to the external acoustic meatus. Projecting over the anterior part of the meatus is a small, pointed prominence, the *tragus.* The lowermost portion of the pinna is the *lobule,* which may be free or attached. The lobule, which lacks cartilage, contains fibrous and adipose tissue. Three small ligaments and three small extrinsic muscles attach the ear to the head.

The *external acoustic meatus,* a canal about 2.5 cm in length, leads from the pinna to the middle ear. It has an outer *cartilaginous portion* and an inner *osseous portion.* The skin of the cartilaginous portion is thick and contains many fine hairs and sebaceous glands; the latter are lacking in the osseous portion. Also present are tubular *ceruminous glands* (actually modified sweat glands) whose secretion, together with that of the sebaceous glands, forms *cerumen* or *earwax,* which lubricates the ear canal.

Middle Ear. The middle ear includes the *tympanic membrane,* the *tympanic cavity* and the *auditory ossicles* enclosed by it, the *tympanic antrum,* and the *auditory tube.*

TYMPANIC MEMBRANE. This membrane, also called the *eardrum* or *drum membrane,* is a semitransparent structure that forms a partition between the external auditory meatus and the tympanic cavity. It is somewhat elliptical in shape, measuring 9 to 10 mm in length at its greatest diameter. Its external surface is slightly concave owing to the pull exerted on it by the malleus (one of the ear bones), which is attached to its inner surface. Histologically it consists of two layers of collagen fibers covered externally by stratified epithelium and internally by a mucous membrane consisting of a single layer of low cuboidal cells.

Technically, the "eardrum" or "drum" is the tympanum or tympanic

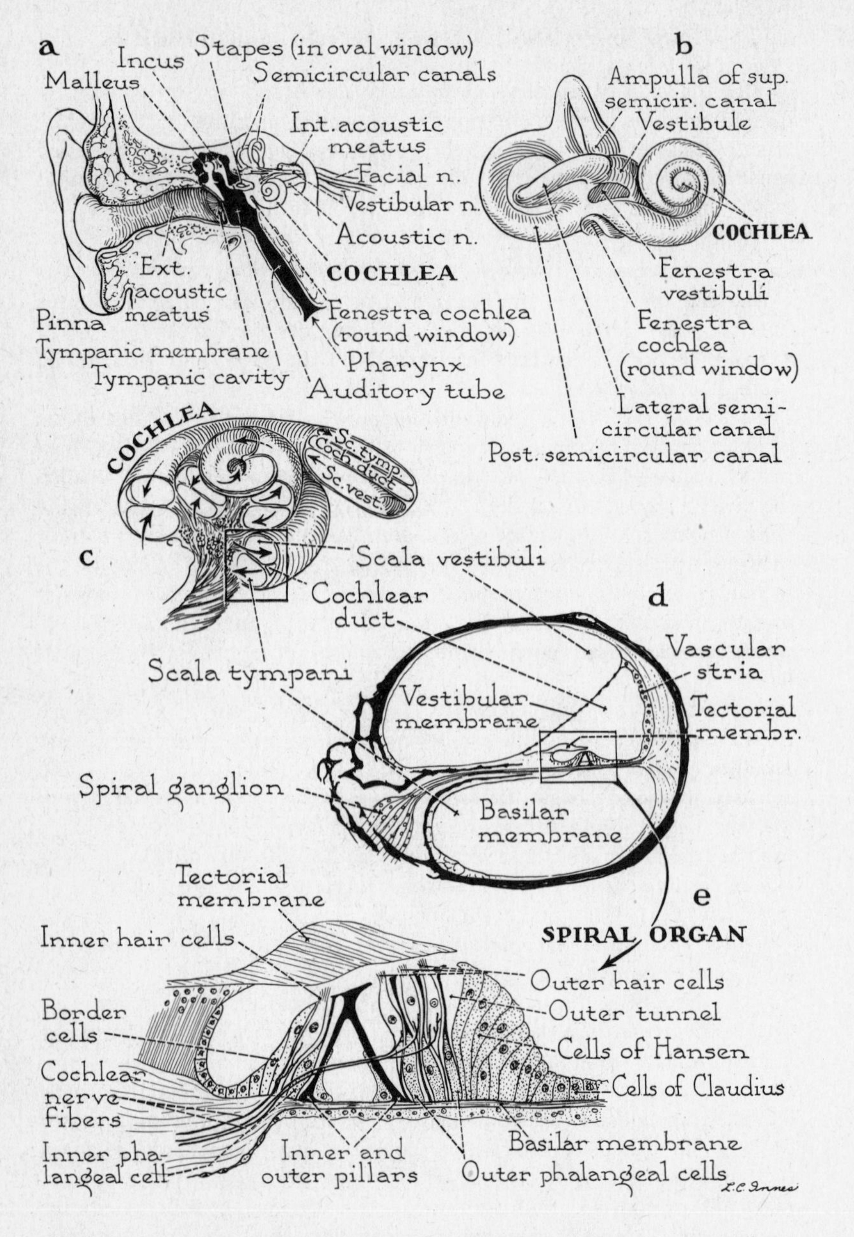

Fig. 5-11. External, middle, and internal ear. (Reprinted with permission of W. B. Saunders Co., Philadelphia, from B. G. King and M. J. Showers, *Human Anatomy and Physiology,* 6th ed., 1969).

cavity. However, in most medical texts, the term *eardrum* is used as a synonym for tympanic membrane or drum membrane. In this text it will be so used.

TYMPANIC CAVITY. The *tympanic cavity* or *tympanum,* is a small air-filled cavity within the petrous portion of the temporal bone. It occupies the space between the external ear and the internal ear and communicates with the pharynx by means of the auditory tube. It is lined with mucous membrane, which also covers the surfaces of the ossicles and the two tympanic membranes. The epithelium of the middle ear is continuous with the epithelium of the pharynx by way of the auditory tube and with that of the mastoid cells by way of the tympanic antrum. This explains why respiratory infections frequently involve the middle ear and may spread to the mastoid cells, causing mastoiditis.

AUDITORY OSSICLES. Extending across the tympanic cavity is a series of three small bones, the *auditory ossicles.* Named for their shape, they are the *malleus* (hammer), the *incus* (anvil), and the *stapes* (stirrup). The malleus is attached to the tympanic membrane, and its head articulates with the base of the anvil. A process of the anvil articulates with the stirrup, the footplate of which lies in a small oval opening, the *oval window* (*fenestra vestibuli*), located on the medial wall of the tympanic cavity. The joints between these ossicles are diarthroidal joints with synovial cavities. Two small muscles act on the ossicles. The *tensor tympani* (length, 25 mm) is attached to the malleus and serves to tighten the tympanic membrane; the *stapedius* (length, 6.3 mm) is the smallest skeletal muscle of the body and acts on the stapes. These muscles protect the inner ear from excessively loud sounds.

Directly below the fenestra vestibuli is a smaller opening, the *round window* (*fenestra cochleae*), which is closed by a thin membrane. The oval and round windows connect the middle ear with the vestibule of the internal ear.

TYMPANIC ANTRUM. An irregular chamber about the size of a small bean, the tympanic antrum is connected with the upper part of the *epitympanic recess* or *attic.* Into it open numerous *mastoid cells,* small irregular spaces that honeycomb the mastoid portion of the temporal bone. They vary greatly in number in individuals.

AUDITORY TUBE. Better known as the *eustachian tube,* the *pharyngotympanic* or *auditory tube* consists of an *osseous portion* contained within the temporal bone and a *cartilaginous portion* opening into the nasopharynx. This tube is lined with mucous membrane, which is continuous with that of the pharynx and the middle ear. The auditory tube opens and closes with swallowing and yawning movements, which action equalizes the air pressure on the two sides of the tympanic membrane.

Internal Ear. The internal ear consists of a complicated series of canals located within the petrous portion of the temporal bone and containing the *sensory receptors for the senses of hearing and equilibrium.* Because the inner ear consists of complex communicating passages, the term *labyrinth* is applied to it. It is divided into two parts: the *osseous* or *bony labyrinth* and the *membranous labyrinth.*

OSSEOUS LABYRINTH. The osseous labyrinth consists of three regions: the *vestibule,* the *semicircular canals,* and the *cochlea.* The osseous labyrinth is lined with a thin layer of periosteum and contains a fluid, the *perilymph,* that surrounds the membranous labyrinth. The membranous labyrinth is enclosed within the osseous labyrinth.

The *vestibule* is the ovoid central portion. Extending upward and posteriorly from it are the three *semicircular canals,* each of which forms a curved arc lying in a plane approximately at right angles with the other two. On the basis of their position, they are designated *superior, posterior,* and *lateral canals.* They open into the vestibule by five openings. One end of each canal is enlarged to form an *ampulla.*

Opening from the anterior surface and lying in front of the vestibule is a spiral structure, the *cochlea,* named for its resemblance to a snail's shell. It consists of a spiral canal about 30 mm in length, which turns about $2\frac{1}{4}$ times around a short, conical bony structure, the *modiolus.* Projecting upward from the modiolus, a thin bony plate, the *spiral lamina,* extends about halfway into the cochlea, dividing the cochlear canal into two passageways, an upper *scala vestibuli* and a lower *scala tympani.* These two scalae join at the *apex* or *cupula* of the cochlea. The *base* of the cochlea adjoins the medial wall of the vestibule, into which the scala vestibuli opens. The scala tympani ends at the fenestra cochleae. The perilymph of the vestibule is continuous with that of the scala vestibuli.

MEMBRANOUS LABYRINTH (Fig. 5-12). The membranous labyrinth is a series of interconnected sacs and tubes lying within the osseous labyrinth. Although it is smaller than the osseous labyrinth, it has the same general form. The space between these labyrinths is filled with *perilymph,* except where fibrous bands connect them. The membranous labyrinth is lined with epithelium and contains a fluid, the *endolymph.*

Within the bony cochlea and the semicircular canals of the osseous labyrinth lie the *cochlear duct* (*scala media*) and the *semicircular ducts,* which have almost identical shapes. Within the vestibule, the membranous portion is divided into two sacs (the *utricle* and the *saccule*), which are connected with each other by a small duct. From the posterior wall of the utricle extend the semicircular ducts, each of which has an *ampulla.* The saccule is slightly smaller than the utricle, to which it is attached. From its anterior surface extends the cochlear duct. Also connected with the saccule is a small, short *endolymphatic duct* that ends blindly in the *endolymphatic sac.*

An axial section of the cochlea (Fig. 5-11) reveals its detailed structure. Five sections are seen, each composed of three canals, the scala vestibuli, the scala tympani, and, between them, the cochlear duct. An *osseous spiral lamina* and *basilar membrane* separate the scala tympani from the cochlear

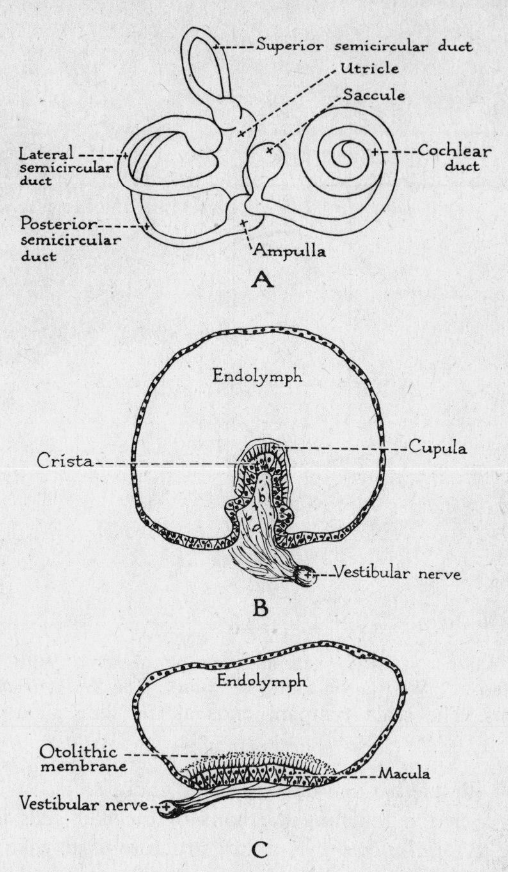

Fig. 5-12. (*A*) Membranous labyrinth. (*B*) Crista. (*C*) Macula. (Reprinted with permission of W. B. Saunders Co., Philadelphia, from B. G. King and M. J. Showers, *Human Anatomy and Physiology,* 6th ed., 1969.)

duct and scala vestibuli; the *vestibular membrane* separates the cochlear duct from the scala vestibuli. In the floor of the cochlear duct is the spiral *organ of Corti.*

The spiral organ of Corti (Fig. 5-11), which lies on the basilar membrane, extends from the round window to the tip of the cochlea. It consists of *hair cells* and *supporting cells* arranged in a complicated manner. The hair cells comprise a single layer of *inner hair cells* and three or four rows of *outer hair cells.* The inner hair cells are ovoid and completely enclosed within their supporting phalangeal cells; the outer hair cells are cylindrical, each with a rounded basal end that rests in the concave end of an outer phalangeal cell. Each hair cell bears on its apical end a number of minute, nonmotile *hairs* or *stereocilia.* The basal ends of the hair cells are in

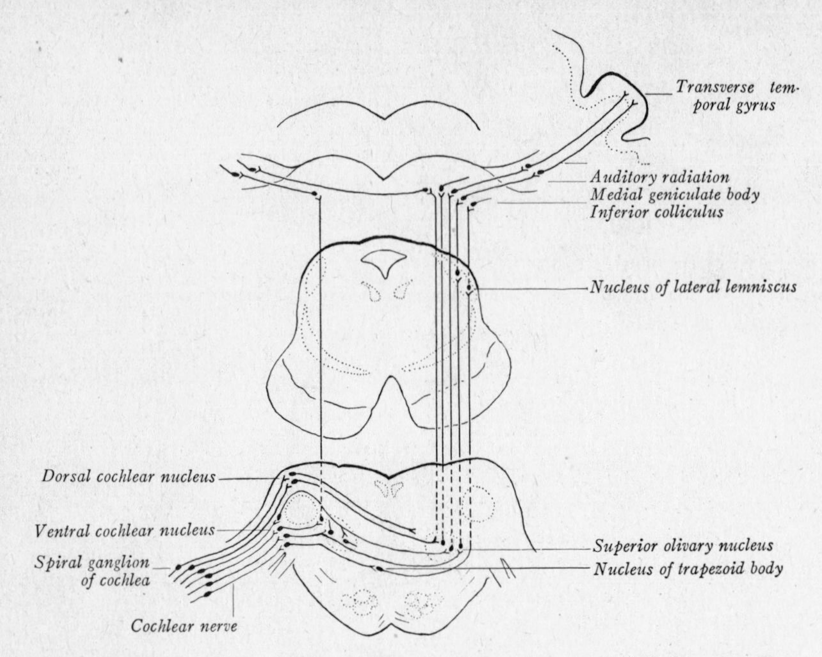

Fig. 5-13. The auditory pathway. (Reprinted with permission of W. B. Saunders Co., Philadelphia, from S. W. Ranson and S. L. Clark, *The Anatomy of the Nervous System,* 1959.)

contact with both afferent and efferent nerve endings.

Projecting over and touching the hairs of the hair cells is a shelflike structure, the *tectorial membrane,* a thin structure of jellylike consistency. Beneath the spiral organ is the *basilar membrane,* composed principally of bundles of minute collagenous fibers; the latter sometimes are called *auditory strings.* The basilar membrane, about 30 mm long, extends from the round window to the apex of the cochlea. It increases gradually in width from base to apex, even though the cochlea as a whole becomes narrower. Its width at the round window is 0.16 mm; at the tip of the cochlea, 0.52 mm.

Auditory Pathways (Fig. 5-13). Auditory impulses from the hair cells of the cochlea to the cerebral cortex pass through three or four neurons. Fibers in contact with the hair cells are the dendrites of *bipolar cells* whose cell bodies lie in the *spiral ganglion.* Axons of these cells form the cochlear branch of the vestibulocochlear (VIII) nerve, which enters the cranial cavity through the internal acoustic meatus. On entering the medulla oblongata, these fibers terminate in the *dorsal* or *ventral cochlear nuclei,* where they synapse with second-order neurons, some of which pass to the *superior olivary nuclei* of the medulla; others pass upward and terminate in the *nuclei* of the *lateral lemnisci* or in the *inferior colliculus* of the *midbrain.* At each of these levels, fibers may decussate to the opposite side. From the inferior colliculus of each side, tertiary fibers ascend to the *medial geniculate nucleus* of the thalamus. From the thalamus, fibers pass by way of the internal capsule to the *auditory region* or *hearing center* of the cerebral cortex, located in the *temporal lobe.* Fibers carrying impulses from sounds of high frequencies end deep within the cortex; those carrying impulses of low frequencies end near the surface. There is evidence that impulses of loud sounds may reach the cortex by way of the reticular activating system. In addition to a hearing center, there is a *sound memory center,* which stores sounds from birth. Destruction of this center, also located in the temporal lobe, results in *acoustic agnosia,* the inability to recognize the significance of sounds.

In addition to afferent pathways, there are *efferent pathways* through which impulses pass from the cortex to the hair cells of the cochlea and muscles of the middle ear. Cortical fibers terminating in the olivary and cochlear nuclei of the medulla and fibers terminating in the cochlea and middle ear have been identified. It is postulated that through these fibers, the brain can partially control or eliminate sounds that have no immediate significance but can respond to sounds of critical importance; for example, a mother can be oblivious to sounds of children playing but will react immediately to the cry of her baby.

The Physiology of Hearing

Sound. *Objectively,* sound is a form of energy consisting of "waves" of molecules caused by a vibrating body and transmitted through a medium such as air, water, or a solid material. In physics, the study of sound is called *acoustics. Subjectively,* sound is the sensation occurring within the mind of the listener that results from the stimulation of auditory receptors. *Hearing* is the process by which the energy of sound waves transmitted to the auditory receptors in the ear results in the production of the sensation of sound and ultimately in the identification of particular sounds.

Sound waves travel at approximately the following rates: through air at 20°C, 331.5 m (1125 ft)/sec; through water, 1500 m (4856 ft)/sec; through iron, 3240 m (10,627 ft)/sec. Sound waves result from alternate regions of compression and rarefaction in the molecules of the transmitting medium initiated by the vibrating object. Sound waves may be refracted, diffracted, or absorbed by objects in their path.

Process of Hearing. The events that occur in the process of hearing sounds transmitted through the air are as follows:

1. Sound waves are directed by the pinna into the external acoustic meatus.
2. The tympanic membrane or eardrum is set in motion by the vibrations of the sound waves.

3. The auditory ossicles or ear bones (malleus, incus, stapes) act as a series of levers to transmit the vibrations across the middle ear to the oval window. Through increased leverage and the larger size of the tympanic membrane as compared to the oval window (about 20 times larger), the force of the vibrations is amplified about 90 times.

4. Movements of the footplate of the stapes at the oval window set up vibrations within the perilymph of the scala vestibuli of the cochlea. These vibrations may (*a*) pass through the vestibular membrane to the endolymph of the cochlea and then through the basilar membrane to the perilymph of the scala tympani, or (*b*) pass up the scala vestibuli to the apex of the cochlea and then down the scala tympani. In either case the vibrations are dissipated at the round window.

5. High tones cause the basilar membrane, which is narrow and stiff at its end near the stirrup, to vibrate; low tones cause the end of the membrane to vibrate near the apex of the cochlea, where it is wider and more elastic. The hair cells of the organ of Corti which rest on the basilar membrane, are moved by the vibrations. As a result, the hairs of the hair cells, which are in contact with the tectorial membrane, are deflected by a shearing action. This deflection in some unknown manner initiates action potentials in the hair cells, which are transmitted through neurons of the auditory pathway to the cerebral cortex, where they are interpreted as sounds.

Bone Conduction. Sound may also result from vibrations reaching the receptor cells of the organ of Corti through bone. The vibrations of a tuning fork placed against the skull are perceived much more readily than when the fork is held away from the ear. Much of the sound of a person's own voice and such sounds as the clicking of the teeth are the result of vibrations transmitted through the skull bones. Air and bone stimulation result in the same types of movement of the basilar membrane and involve the same auditory pathways.

Theories of Hearing. Sound waves have their origin in a vibrating body, such as, for example, a musical instrument, human vocal cords, metal striking against metal, and the sound-producing parts of insects. These vibrations vary in *amplitude* and *frequency*. Sound has three qualities that can be perceived by the human ear: pitch, loudness, and quality. *Pitch* depends on the frequency of vibrations. The human ear is capable of hearing sounds ranging in frequency from 16 or 20 Hz to 20,000 Hz (Hz = hertz, cycles per second). Sounds of greater or fewer vibrations (higher or lower frequency) are inaudible. The *loudness* or *intensity* of a sound depends on the amplitude of the vibrations. It is measured in *decibels* (db). The human voice at a conversational level at a distance of 60 cm has an intensity of 50 db; a pneumatic drill, 120 db. The *quality* or *timbre* of a sound depends on the *overtones* produced along with the fundamental tones. *Noise* is sound produced by fundamental pitch complicated by irregularities in the sound waves.

The mechanism by which variations in frequency of sound waves striking the ear give rise to nervous impulses that the brain interprets as differences in pitch is not known with certainty. Two theories have been proposed to account for this; they are the *resonance theory* and the *telephone theory.*

RESONANCE THEORY. The *resonance theory* (*Helmholtz's theory*) postulates that the cochlea serves as an analyzer of sound. The basilar membrane, with its 24,000 fibers, is assumed to be a resonating structure. In much the same way that sound produced by a horn or the human voice near a piano will cause a string of the piano tuned to the same number of vibrations to vibrate ("sympathetic resonance"), so the fibers of the basilar membrane are thought to react to the vibrations of sound waves. Vibrations of different frequencies cause different regions of the basilar membrane to vibrate; the short fibers at the base respond to high frequencies and the long ones at the apex to low frequencies. The hair cells in the organ of Corti overlying the fibers that are set into sympathetic vibration are stimulated, and impulses are carried to the brain, where they are interpreted.

TELEPHONE THEORY. The *telephone theory* postulates that the ear functions somewhat like a telephone system. The transmitter converts sound waves into electrical impulses of the same frequency. These impulses, when transmitted over a wire, set up vibrations in the receiver that reproduce the original sound. In a similar way, it is believed that the basilar membrane vibrates as a whole and that the frequency of its vibrations stimulates the hair cells, initiating impulses that on reaching the auditory center are interpreted and serve as the basis of pitch discrimination.

There is considerable evidence to support both theories, but there are many facts that both leave unexplained. In spite of intensive research on hearing, the mechanism by which the hair cells in the organ of Corti respond to the vibrations of the basilar membrane and initiate action potentials is still unknown.

Practical Considerations of Hearing

Audiometry. Relative hearing acuity can be tested by determining the distance at which a sound (such as the tick of a watch or the vibration of a tuning fork) can be heard from the subject's ear in comparison with a standard set for normal individuals. A more precise means is through the use of the *audiometer,* an instrument that produces electrical tones that can be altered in pitch and intensity (amplitude). Results are plotted on a chart called an *audiogram.* Other methods used to test acuteness of hearing include bone conduction tests and the testing of acoustic reflexes. Acoustic reflexes include the *cochleopalpebral reflex* and *middle-ear muscle reflexes.* The cochleopalpebral reflex is the sudden contraction of the orbicularis oculi muscles upon hearing a sudden, unexpected sound. Reflex contractions of the tensor tympani and stapedius muscles can be recorded by direct observation through a perforated tympanic membrane or indirectly by electromyography.

Hearing Aids. For persons with conduction deafness, a hearing aid may markedly improve hearing. A *hearing aid* is essentially a miniature public-address system in which sound waves are converted into electrical energy, which is amplified and then converted back to sound. The resulting increase in volume may be effective, but sound distortion may occur. If structures of the inner ear are defective or if deafness is of central origin, however, a hearing aid is of little or no value.

Localization of Sound. Like the sensation of sight, the sensation of sound is projected externally to the source of the stimulus. The ability to determine the exact source of a sound is, however, rather limited. When the sound waves strike both ears simultaneously, it is difficult to locate the source, but if the vibrations strike one ear before the other, the direction can be determined with fair accuracy. It is for this reason that a person turns the head to hear more clearly.

DISORDERS OF HEARING

Deafness. Deafness is the loss or impairment of the ability to hear sounds ordinarily heard by the average individual. It may be partial or complete, temporary or permanent, unilateral or bilateral. Three types of deafness are generally recognized: *transmission deafness, nerve deafness,* and *central deafness.*

TRANSMISSION OR CONDUCTION DEAFNESS. Any condition that interferes with or prevents the passage of sound waves through the external or middle ear brings about transmission deafness. Some common causes are (1) presence of an obstruction in the external auditory meatus, (2) severe rupture or abnormal thickening of the eardrum, (3) inflammation of the middle ear (otitis media) or destruction of its structures, (4) ankylosis (stiffness) of the auditory ossicles, (5) fixation of the footplate of the stapes, and (6) blockage of the auditory tube, which may be temporary (from respiratory infections) or permanent (from adhesions or presence of adenoid tissue over the nasopharyngeal opening).

NERVE DEAFNESS. This deafness, also called *sensory* or *perceptive deafness,* results from conditions involving the cochlear structures or the auditory nerve. In this type of deafness, a person may be deaf only to tones of high frequencies or of low frequencies. It is sometimes seen as occupational deafness, occurring in persons working where there are very loud noises. It involves lesions of the basilar membrane. Degeneration of cochlear and nerve structures, producing nerve deafness, is common in advanced age. Some drugs, for example, streptomycin, may cause degeneration of the hair cells. Cochlear dysfunction may also result from such diseases as measles, mumps, meningitis, and diabetes.

CENTRAL DEAFNESS. Pathological states of the auditory pathway or the auditory area of the cerebral cortex may give rise to central deafness. The causes include tumors, abscesses, injuries from blows or from skull fractures, cerebral vascular accidents, and the effects of certain drugs or poisons.

OTHER DISORDERS. The following are common disorders of hearing:

Otitis. Inflammation of the ear is designated in accordance with the part of the ear in which it is located: otitis externa, otitis media, otitis interna. Inflammation of the middle ear commonly results from infectious microorganisms that enter by way of the auditory tube. It is also a common complication of measles, scarlet fever, pneumonia, and influenza.

Otosclerosis. This is a chronic, progressive condition in which the bone in the region of the oval window becomes spongy and then hardens, resulting in the stapes's becoming fixed or immovable. It is a common cause of deafness in young adults. By an operation called *stapes mobilization,* the stapes can often be freed and hearing restored. Another procedure, called *fenestration,* the making of a new opening into the vestibule, is sometimes employed when the stapes cannot be freed. Sometimes the stapes is removed and replaced by a tissue graft.

Tinnitus. This is the perception of sound in the absence of an acoustic stimulus, usually manifested by a ringing or buzzing sound in the ears. It may be due to impacted wax, otitis media, otosclerosis, interruption in normal blood supply to the cochlea, the effects of certain drugs (for example, salicylates and quinine), or an increase in endolymphatic pressure. The last is a symptom of *Ménière's disease,* which also generally involves the nonauditory labyrinth and leads to vertigo accompanied by nausea.

Cauliflower Ear. An enlarged and grossly disfigured ear, commonly seen in boxers and wrestlers, develops from calcified hematoma between the perichondrium and the cartilage of the pinna.

Structure and Functions of the Apparatus of Equilibrium

The *semicircular ducts, utricle,* and *saccule* are the sense organs of equilibrium and position. The utricle, into which the semicircular ducts open, and the saccule lie within the vestibule. These sense organs function in the maintenance of equilibrium and the coordination of body movements. They also provide awareness of the position of the body in space. It must be recognized, however, that other senses (sight, touch, muscle sense) assist in these functions.

Functions of the Semicircular Ducts. The three membranous semicircular ducts lie within corresponding bony semicircular canals. The ducts are filled with endolymph, and each lies in a plane approximately perpendicular to the other canals. In the ampulla, a dilated portion of each duct, there is a small elevation, the *crista.* Each crista consists of a group of hair cells covered by a mass of gelatinous material called a *cupula.* The hair cells are the sensory receptors. Two types exist, one flask-shaped, the other cylindrical. The latter receives both afferent and efferent nerve fibers. Movement of the head brings about a movement of the endolymph, which stimulates the hair cells. When the head is turned, the entire vestibular apparatus including the cristae turns with it. The fluid endolymph, which fills the semicircular ducts, however, because of inertia, retains its initial position. The movement of the cupula against the relatively stationary endolymph causes the hair cells embedded in the gelatinous cupula to bend. This bending or distortion of the hair cells gives rise to action potentials, which are transmitted by nerve fibers in the vestibular branch of the vestibulocochlear (VIII) nerve.

The semicircular ducts are the sense organs of *dynamic equilibrium* and are primarily involved in righting reflexes. They are also involved in reflex control of eye movement so that regardless of the position of the head, the eyes retain a relatively fixed position with reference to the object viewed. As the position of the head is changed, afferent impulses from the cristae are relayed through the vestibular nuclei to the ocular muscles; contraction occurs to maintain the eyes directed on the object. If the head is turned to the right, the eyes are rotated to the left, and vice versa; if the head is turned upward, the eyes are rotated downward, and vice versa. If a person with head erect is rotated rapidly and the rotation is abruptly stopped, involuntary to-and-fro movements of the eyes occur for a time afterward. This movement of the eyes is called *nystagmus.*

Functions of the Utricle and the Saccule. The utricle and the saccule each contain within their walls areas of sensory neuroepithelium called the *macula utriculi* and *macula sacculi.* These maculae are oval structures measuring about 2 by 3 mm. Their epithelium contains two types of cells: *sustentacular* or *supporting cells* and *hair cells.* Over the surface of the maculae is a platelike, gelatinous layer, the *otolithic membrane,* into which the hairs of the hair cells penetrate. In the upper portion of the "jelly" are found many minute crystalline bodies called *otoliths* or *otoconia,* which are made up of calcium carbonate and a protein.

Changes in the position of the head or the body shift the position of the otoliths, which stimulate the hair cells of the maculae. This provides information concerning the position of the head with reference to the force of gravity. Impulses from the maculae are transmitted to the brain through the vestibular branch of the vestibulocochlear (VIII) nerve. The utricle is the primary sense organ for *static equilibrium.* It is essential for the maintenance of posture and for certain kinds of equilibrium reactions, namely, righting and attitudinal reflexes. Although the maculae in the saccule are similar in structure to those of the utricle, the exact role of the saccule in equilibrium reactions is not clearly understood.

Labyrinthine Pathway. The cristae of the ampullae of the semicircular ducts and the maculae of the utricle and the saccule are innervated by the superior and inferior divisions of the *vestibular nerve,* a branch of the vestibulocochlear (VIII) nerve. Fibers of these nerves are dendrites or peripheral processes of neurons whose cell bodies lie in the vestibular ganglion that is located in the internal auditory meatus of the temporal bone. Axons or central processes of these neurons pass through the vestibulocochlear nerve to the vestibular nuclei of the medulla, where synapses are made with (1) neurons whose fibers pass upward to the motor nuclei of the cranial nerves that innervate eye muscles or (2) neurons whose fibers pass through descending tracts of the spinal cord. These fibers synapse with motor neurons innervating muscles of the neck, trunk, and limbs. Through these two pathways, ocular and postural reflexes are mediated. The cortical connections of the vestibular nerve fibers have not been definitely determined.

DISORDERS OF THE LABYRINTHINE SENSE

Disorders of the sense of equilibrium are vertigo, dizziness, and motion sickness.

Vertigo. The sensation that the outside world is revolving about the subject is called *objective vertigo;* the associated sensation that one is revolving in space is called *subjective vertigo.* Use of this term as a synonym for *dizziness* is erroneous. Vertigo is a result of a disorder of the labyrinthine apparatus (semicircular ducts, utricle, saccule) or associated structures (vestibulocochlear nerve, vestibular nuclei, and reflex centers within the central nervous system). These disorders may result from or be affected by many various conditions of the body, including (1) pathogenic conditions of the ear, including Ménière's disease; (2) the toxic effects of various drugs (alcohol, salicylates, streptomycin, and others); (3) various cardiovascular disturbances within the ear (arteriosclerosis, ischemia); (4) ocular disturbances; (5) infectious diseases (influenza, encephalitis, measles, mumps); (6) neoplasms or lesions of the brain or ear; and (7) miscellaneous causes, such as psychogenic disorders, epilepsy, diabetes, and multiple sclerosis.

Dizziness. This is the sensation of unsteadiness with a feeling of movement (whirling) within the head. Also called *giddiness* or *lightheadedness.*

Motion Sickness. A feeling of malaise, the principal feature of which is nausea, arises from the effects of motion created by various means of transportation. The susceptibility of individuals varies greatly, but the cause is the same: stimulation of the vertical canals. The condition is variously referred to as *seasickness, airsickness, carsickness, trainsickness,* and so on.

6: THE ENDOCRINE SYSTEM

The regulation, control, and integration of bodily activities are accomplished in two ways: (1) through nervous impulses conducted by the nervous system and (2) through chemical substances or *hormones* carried by the blood and lymph. The organs that secrete hormones are called *endocrine glands* or glands of internal secretion. Collectively, these glands comprise the *endocrine system* (Fig. 6-1). Included in this system are the hypophysis (pituitary gland), thyroid, parathyroid, adrenal glands, pancreatic islets of Langerhans, gastrointestinal mucosa, testes, ovaries, and placenta. To these organs can be added the thymus gland, pineal body, skin, and kidneys, for these organs have been shown to produce substances that can be classified as hormones.

Endocrine glands are *ductless,* their secretions being discharged into the blood or lymph, by which they are transported to all parts of the body. In this respect they are differentiated from *exocrine glands,* such as salivary or sweat glands, whose products are discharged through ducts that open onto a surface.

HORMONES

A *hormone* is a chemical substance produced by an organ or a tissue that has a specific effect on tissues that are more or less remote from its place of origin. Sometimes this effect is general, affecting the body as a whole; in other cases the effects are limited, their action being restricted to a specific structure designated a *target organ* or *tissue.*

Internal secretions or *endocrines* may have either an excitatory or an inhibitory effect. Originally, the term *hormone* (literally, "exciting") was applied only to those having excitatory effects, while those having inhibitory effects were called *chalones.* The latter term, however, was not generally adopted, and *hormone* has come to be applied to any internal secretion regardless of its effects.

Neurohormones and Neurohumors. Certain nerve cells have the ability to function both as nerve cells transmitting impulses and as gland cells secreting chemical messengers. These cells, called *neurosecretory cells,* include the cells in the hypothalamus that produce oxytocin and vasopressin. Because these hormones are produced by nerve cells, they are called *neurohormones.* Nerve cells also produce substances at the tips of axons that function as *transmitter agents* at synapses or at effector endings. These agents, called *neurohumors,* are produced in minute quantities and act for a limited period of time. The principal neurohumors are *acetylcholine* and

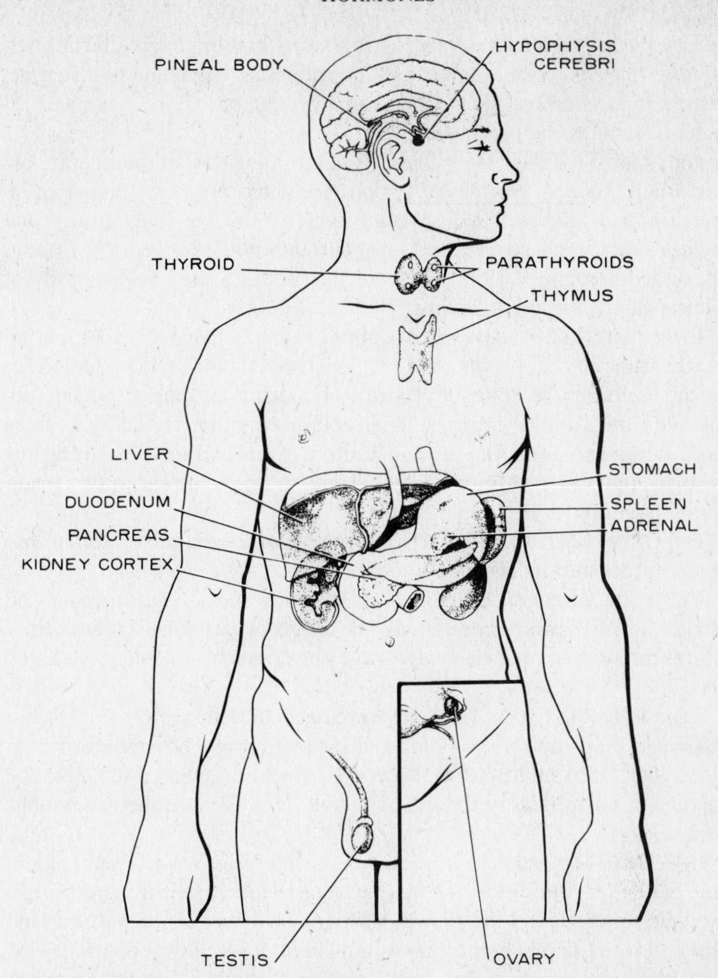

Fig. 6-1. Diagram showing the location of the endocrine glands. (From *General Endocrinology,* 6th ed., by C. Donnell Turner, Ph.D., and Joseph T. Bagnara, Ph.D. Copyright © 1976 by W. B. Saunders Co. Reprinted by permission of Holt, Rinehart and Winston, CBS College Publishing.)

norepinephrine. Norepinephrine, incidentally, is also a true hormone secreted by the adrenal medulla. Other neurohumors include *serotonin* and *dopamine.*

Chemical Nature of Hormones. Hormones vary greatly in their chemical composition. They may be (1) *modified amino acids,* such as norepinephrine and thyroxine; (2) *small peptides,* such as vasopressin and oxytocin; (3) *larger peptides,* such as gastrin and glucagon; (4) *proteins,* such as insulin, the parathyroid hormone, and various growth hormones;

(5) *glycoproteins,* such as the follicle-stimulating and luteinizing hormones; (6) *steroids,* such as the lipid-soluble sex hormones, estrogen and testosterone, (7) *mucopolysaccharides,* such as various hypothalamic hormones; and (8) *fatty acids,* such as the prostaglandins.

Methods of Studying Hormones. Endocrinology is of comparatively recent origin. Indeed, only in 1905 was the term *hormone* first used in connection with *secretin,* a substance secreted by the duodenum that stimulates pancreatic secretion, and most current knowledge about hormones has been acquired since 1920. Some of the methods used in the study of endocrine glands are the following:

1. EXPERIMENTAL REMOVAL. An organ can be identified as having an endocrine function if (*a*) its removal or inactivation by irradiation or chemical inhibitors in either immature or mature animals produces observable or measurable changes in structure or function and (*b*) these changes disappear upon the administration of an extract containing the active principle obtained from normal glands or upon successful transplantation of glandular tissue.

2. INJECTION METHOD. The active principle of the gland is injected into normal animals and its effects observed.

3. CLINICAL METHOD. By clinical observation, bodily dysfunction and glandular disorders can be correlated. The effects of glandular abnormalities resulting from disease, accident, defective development, or heredity can be noted.

4. ANALYTIC METHOD. Tests are employed to determine the presence or absence of hormones in the blood, urine, saliva, or body tissues.

5. ISOTOPIC TRACER METHOD. Radioactive isotopes are used to determine the uptake of hormones by the various tissues and to trace their metabolic pathways.

6. IN VITRO METHODS. Through (*a*) the perfusion of excised organs, (*b*) the study of thin slices of living organs immersed in appropriate physiologic solutions, (*c*) the growth of living cells in a tissue culture, and (*d*) the study of tissue homogenates prepared from hormone-producing tissues, much information has been obtained about chemical reactions involved in the secretion of hormones and their degradation.

Factors Controlling the Secretion of Hormones. Two principal classes of factors that regulate hormone secretion by endocrine glands are nervous factors and chemical factors.

NERVOUS FACTORS. Some endocrine tissues are innervated by the autonomic nervous system, and their activity is controlled by nerve impulses that may either stimulate or inhibit glandular secretion. The adrenal medulla receives fibers from the sympathetic division of this system; the secretion of epinephrine increases or decreases with the system's activity. Nerve impulses to the hypothalamus influence its activity.

CHEMICAL FACTORS. Chemical substances carried by the blood are of primary importance in the regulation of the secretory activity of endocrine glands. These substances may be hormones produced by other glands or nonhormonal substances. The secretory activity of the thyroid gland, the adrenal cortex, and the gonads depends on certain hormones secreted by the hypophysis. In turn, the secretion of these hormones by the hypophysis is influenced by other hormones produced by the hypothalamus. The production of insulin by the islets of Langerhans in the pancreas depends primarily on the blood sugar level. An increase in blood sugar increases secretory activity; a decrease diminishes it. The presence of fatty or acid substances in the duodenum induces the secretion of *secretin* by the duodenal mucosa.

NEUROHUMORAL CONTROL. Some endocrine secretion involves both neural and endocrine control. For example, the "letdown" of milk in a lactating female mammal results from a *neuroendocrine reflex* initiated by the sucking of the young. Impulses pass from the nipple through nerves to the hypothalamus. These stimulate secretory cells (neurons) to secrete a hormone, *oxytocin.* This hormone passes to the posterior lobe of the pituitary, which releases it into the blood. On reaching the mammary gland, oxytocin causes the myoepithelial cells surrounding the secretory alveoli to contract, resulting in the discharge of milk.

FEEDBACK CONTROL. Since the function of most endocrine secretion is to maintain a state of constancy (homeostasis) in bodily activities, various mechanisms exist to hold the amount of hormone secreted within physiologic limits. This is accomplished through the operation of feedback systems. In a *negative feedback system,* for example, when the blood calcium level decreases, the secretion of parathyroid hormone is stimulated, and this restores the blood calcium through the resorption of bone, decreased excretion of calcium by the kidney, and increased absorption of calcium by the intestine. When blood calcium level returns to normal, parathyroid secretion ceases. A *positive feedback system* is illustrated by the secretion of insulin to regulate the blood sugar level. When there is an increase in blood sugar following a meal, the islet cells, which possess some sort of sensing mechanism, respond and increase their insulin output. Insulin acts on liver cells to store more sugar and on other target cells to increase their utilization of sugar. As a consequence, blood sugar level is reduced, and the secretion of insulin ceases. Should the blood sugar level fall below normal, other islet cells are stimulated to secrete glucagon, which acts on target organs to raise the blood sugar level.

Mode of Action. A hormone may exert its effect within cells by (1) acting as an enzyme, (2) inducing the biosynthesis of enzymes, (3) altering the activity of a critical enzyme in a metabolic pathway, or (4) altering the membrane transport of a substance. Hormones do not initiate new activities within cells but act to regulate the rate at which cellular processes occur.

Three theories have been proposed to account for the specificity of action of hormones on target tissues or organs: the *receptor hypothesis,* the *second-messenger theory,* and the *hormone-gene theory.*

RECEPTOR HYPOTHESIS. Here it is assumed that cells of target tissues possess *receptors* or *receptor sites* whose molecular structure is such that they accept or bind with specific hormones. These sites may be on the plasma membrane or within the cytoplasm of the target cell.

SECOND-MESSENGER THEORY. In this theory it is assumed that the hormone, called the *first messenger,* reacts with the receptor on the surface of a target cell. A *second messenger,* thought to be cyclic adenosine monophosphate (cAMP), is then elaborated and mediates the appropriate response within the target cell. It is postulated that prostaglandins may have a modulating effect on the hormonal control of cellular metabolism.

HORMONE-GENE THEORY. This theory assumes that after entering the target cell, the hormone travels to and enters the nucleus, where it acts on specific genes, either activating or suppressing them. This can alter the rate of protein synthesis or the activation of cellular enzymes.

Hormone Transport. Hormone molecules may exist free in the blood, or they may be bound to a plasma protein. The free form is readily available for utilization by target cells, whereas the bound form serves as a reserve source and protects the organism by limiting the amount available to target cells. Thyroxin combines in the blood with a *thyroxin-binding globulin* (TBG). Sex hormones, corticosteroids, and the human growth hormone also combine with various plasma globulins. The combination of hormones with carrier proteins forms a convenient transport mechanism and also serves to prevent small, molecular-sized proteins from passing through the blood-brain and placental barriers (with possible detrimental effects to the brain or developing fetus, respectively) or the glomerular membranes of the kidney (with loss due to excretion).

Fate of Hormones. Hormones are secreted at varying rates, depending on the age and sex of the individual and environmental factors, such as temperature, pH of reacting tissues, electrolyte balance, and presence or absence of other hormones. With the exception of thyroxin and ovarian estrogens, hormones are not stored, except temporarily when bound to a transporting agent. Their life is short and their turnover rapid.

Most hormones disappear from the circulation a short time after secretion, being inactivated and metabolized in target organs, the liver, or the kidney. Transformation processes in the liver include oxidation, reduction, deamination, methylation, and conjugation. Degradation products are passed through the bile into the intestine, where they are excreted with the feces or reabsorbed. In the kidney, complete hormones, such as human chorionic gonadotrophin and insulin, may be excreted. In addition, degradation products of various hormones, especially steroids, formed in the kidney, liver, and various other organs are excreted in the urine.

Assay of Hormones. Because they exist in such low concentrations, determination of the presence of a hormone in a fluid, its potency, length of life, and other characteristics is difficult. If the molecular structure of a hormone is known, highly sensitive physiochemical methods can be employed for its detection. Other methods employed include bioassay and radioimmunoassay techniques. In *bioassay,* the effects of a hormone on an organism or a living tissue are noted. The effects of insulin on glucose uptake in rat muscle, androgens in promoting growth of a capon's comb, and estrogens in inducing the cornification of the vaginal epithelium in rodents are examples. *Radioimmunoassay techniques* employ various radiochemical procedures to detect certain immunologic reactions. These techniques are extremely sensitive and have a high degree of accuracy. They are used to measure the blood insulin concentration and to detect the presence of human chorionic gonadotrophin in pregnancy urine.

Some General Characteristics of Hormones. The following are some characteristics that apply to hormones in general.

1. Hormones are potent substances. Very minute amounts injected into the bloodstream of an animal may produce marked effects. For example, the injection of 0.001 mg of epinephrine (adrenaline) into a cat produces noticeable effects on the heart and the blood vessels.
2. Most hormones are secreted in small amounts spontaneously; however, the rate of secretion can be altered (increased or decreased) through the action of nervous or endocrine controlling mechanisms.
3. The production and release of hormones by the secreting cells is an active energy-consuming process requiring ATP.
4. The secretion and release of many hormones depend on the potassium and calcium content of the extracellular fluid.
5. Some hormones (insulin, parathyroid hormone, ACTH, and other peptides) are first synthesized in the form of more complex precursors or *prohormones.* A prohormone is then converted into the active hormone by the enzymatic removal of certain peptides prior to its release from the secreting cell.
6. Complex hormone interrelationships may exist in which the effect of a hormone on a target organ depends on the previous action of one or more hormones. Several hormones are essential for the functioning of the mammary glands. The action of a hormone may be increased (*potentiated*) or decreased (*limited*) by another hormone. When a hormone action requires the previous action of another hormone, a condition of *permissiveness* exists.
7. In general, the basic fundamental processes of the body are under hormonal control; growth, development, maturation, and reproduction are among them. Furthermore, the rates of various physiologic processes, their rhythmic variations, the rate of energy expenditure—in short, the basic life processes—are all regulated by hormones. These substances also have a profound effect on the functioning of the nervous system. Much of a person's behavior and most of the traits that collectively constitute personality depend on the normal functioning of the endocrine glands.

PITUITARY GLAND (HYPOPHYSIS CEREBRI)

The *pituitary gland* or *hypophysis cerebri* is a rounded body attached to the base of the brain by a thin *infundibular stalk*—a downward extension of the floor of the third ventricle. It lies in the sella turcica of the sphenoid bone. The hypophysis averages 1.3 by 1.0 by 0.5 cm and weighs about 0.5 g.

Gross Structure of the Pituitary. The pituitary gland consists of two primary parts, which have different embryonic origins (Fig. 6-2). The *adenohypophysis* or glandular portion, which comprises the anterior and intermediate lobes, develops from an evagination (Rathke's pouch) of the ectoderm of the embryonic mouth cavity or stomodeum; the *neurohypophysis* or posterior lobe arises from neural ectoderm which develops as a downgrowth of the diencephalon.

Blood Supply of the Pituitary. The pituitary gland receives blood through the *superior* and *inferior hypophyseal* arteries, which arise from the internal carotid arteries and the circulus arteriosus. Branches of the superior arteries terminate in the anterior lobe and in a capillary plexus in the median eminence located at the upper end of the infundibular stalk. Venous capillaries from the median eminence converge to form veins of the *hypothalamohypophyseal portal system,* which conveys blood to the anterior lobe.

The posterior lobe receives blood from the inferior hypophyseal artery, which also supplies the anterior lobe. Veins from both lobes empty into venous sinuses, which empty into the internal jugular vein.

Nerve Supply of the Pituitary. Very few nerve fibers end in the anterior lobe, and those present are thought to be vasomotor and to play no role in the secretion of hormones. A large number of nerve fibers pass through the infundibular stalk and terminate in the posterior lobe. These fibers are grouped into two tracts, the *supraopticohypophyseal* and *tuberohypophyseal tracts.* They are the axons of neurons whose cell bodies lie in nuclei in the hypothalamus. They convey secretory products (neurohormones) to the posterior lobe for release into the bloodstream.

Microscopic Structure of the Pituitary. The anterior lobe contains two types of cells, *chromophils* (cells whose cytoplasm takes a stain or color) and *chromophobes* (cells whose cytoplasm does not stain). Chromophils include two types, *basophils* (cells whose granules are stained with basic stains) and *acidophils* (cells whose granules are stained with acid stains). Efforts have been made to identify the cells responsible for the secretion of specific hormones by the use of various staining techniques, electron microscopy, and immunofluorescence techniques. It is now thought that with one or two exceptions, each hormone is secreted by a specific type of cell.

The *posterior* or *neural lobe* consists of branching cells called *pituicytes,*

The Hypophysis

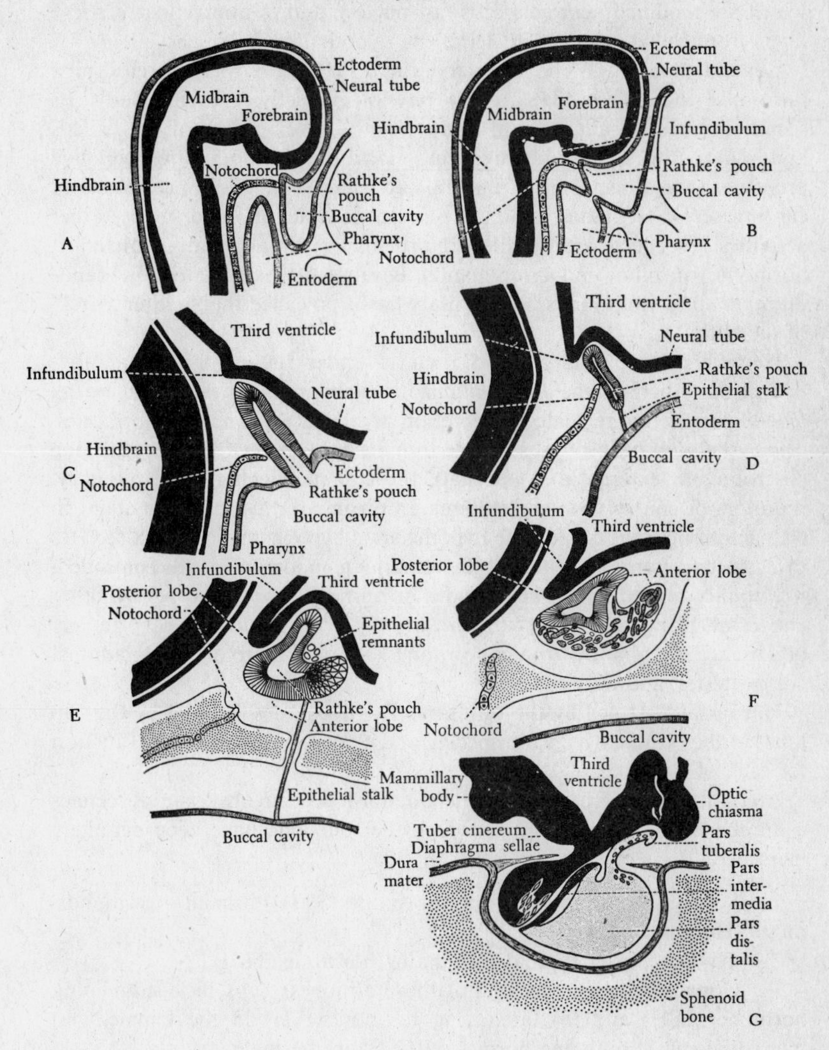

Fig. 6-2. Development of the pituitary gland. (*A*) Rathke's pouch, originating from the roof of the visceral cavity. (*B*) Infundibulum, originating in the forebrain from the floor of the third ventricle. (*C*) Rathke's pouch, fully developed, opening as a craniobuccal duct into the buccal cavity. (*D*) Rathke's pouch, now a closed epithelial sac. Epithelial stalk still connects it with the buccal cavity. (*E*) Residual epithelial stalk traversing the sphenoid bone is still present. (*F*) Epithelial stalk has disappeared. Anterior lobe undergoes differentiation. (*G*) Relationship of adult pituitary to adjacent structures. (After J. H. Globus.) (Reprinted with permission of Blakiston Division, McGraw-Hill Book Company, from *Morris' Human Anatomy,* 11th ed., ed. by J. P. Schaeffer, 1953.)

which are modified neuroglia cells (astrocytes), and of unmyelinated nerve fibers from cells located in nuclei of the hypothalamus.

General Remarks. The pituitary gland or hypophysis secretes nine hormones that affect nearly every physiologic activity of the body. Its secretions regulate and control body growth, development of the gonads and maturation of reproductive cells, lactation and other reproductive processes, general metabolism, mineral and water metabolism, pigmentation, cardiovascular activities, smooth muscle contraction, and many other activities. Its effects are manifested principally through the action of its hormones on other endocrine glands. Because of its regulatory influence on other endocrine glands, the pituitary has been called the "master gland" of the body.

However, the pituitary gland itself is under the control of another endocrine structure, the *hypothalamus,* which produces *hormones* or *releasing factors* that regulate the secretion and release of pituitary hormones. The hypothalamus, being susceptible to influences arising from the changing environment (external and internal), through its effects on the pituitary, monitors or controls the adjustments an organism must make to adapt to changing conditions of life. The hypothalamic-hypophyseal complex secretes more specific hormones than all the remaining endocrine glands combined.

Pituitary hormones are all peptides or proteins. Those that act on other endocrine glands are called *trophic* or *tropic* hormones.* The principal organs acted upon by tropic hormones are the thyroid gland, adrenal cortex, testes, and ovaries.

Hormones Secreted by the Pituitary. The hormones secreted by the two lobes of the pituitary body are specific and clearly differentiated in function for each of the lobes.

ANTERIOR-LOBE HORMONES. The anterior lobe of the hypophysis secretes a number of hormones, among which the following have been definitely identified:

1. Somatotropin, somatotropic hormone (STH), human growth hormone (HGH).
2. Thyrotropin or thyroid-stimulating hormone (TSH).
3. Gonadotropins (gonad-stimulating hormones): (*a*) follicle-stimulating hormone (FSH) and (*b*) luteinizing hormone (LH) in the female, and interstitial cell–stimulating hormone (ICSH) in the male.
4. Prolactin (lactogenic hormone, mammotropin).

* There is a lack of uniformity in the spelling of the suffix used in designating stimulating hormones. In a few research articles and textbooks, the spelling is *-trophic;* in the remainder, *-tropic.* Etymologically, *-trophy* or *-trophic* is a combining form derived from the Greek *trophein,* meaning "to feed, to nourish." It is used in words such as *atrophy* (lack of growth) or *hypertrophy* (excessive growth). The ending *-tropic* is a combining form derived from the Greek word *tropos,* meaning "a turn or turning." Because the suffix *tropic* has come to be generally accepted by endocrinologists and is now widely used, it will be employed in this book.

5. Adrenocorticotropin (adrenal cortex–stimulating hormone, ACTH).

Somatotropin (STH) or Growth Hormone. This hormone, a protein, regulates the growth of the skeleton and body growth in general. In addition to promoting the growth of cartilage and bone and the development of associated tissues, it plays an important role in the metabolism of proteins, carbohydrates, and fats. Somatotropin promotes the uptake of amino acids and the synthesis of proteins as well as the utilization of glucose by adipose tissue and muscle cells. It induces a positive nitrogen balance and facilitates the action of pancreatic, adrenocortical, and thyroidal hormones on carbohydrate metabolism. Somatotropin stimulates the production of lymphocytes by the thymus and other lymphocyte-producing organs.

Somatotropin and prolactin (lactogenic hormone) are so similar in their chemical structure and biological properties that it was not certain until recently that they are separate hormones. Somatotropin produces its effects on tissues, especially cartilage growth, by stimulating the liver and other tissues to produce a peptide, *somatomedin,* that stimulates DNA and RNA synthesis and collagen formation.

Thyrotropin or Thyroid-Stimulating Hormone (TSH). This hormone, a glycoprotein, regulates the development and functioning of the thyroid gland. In its absence, the thyroid atrophies. When TSH is present in excess, the thyroid undergoes hypertrophy and hyperplasia. Thyrotropin is the primary factor in the control of iodine uptake and its utilization by the thyroid.

Gonadotropic Hormones. These hormones that stimulate the gonads include the follicle-stimulating hormone, luteinizing hormone, and interstitial cell–stimulating hormone.

Follicle-Stimulating Hormone (FSH). This hormone, a glycoprotein, is essential for the development of ovarian follicles, the maturation and (with LH) the discharge of the ovum, and the secretion of estrogens by the follicle. It is produced during the first half of the menstrual cycle (Fig. 7-7). In the male, FSH stimulates the production and maturation of spermatozoa within the germinal epithelium lining the seminiferous tubules.

Luteinizing Hormone (LH). This hormone, a glycoprotein, acts synergistically with FSH in the female to stimulate the secretion of estrogen by developing follicles and to induce ovulation. It is essential for the formation of the corpus luteum in the ruptured follicle and, with prolactin, stimulates the secretion of progesterone and estrogens by the corpora lutea. In the male, LH acts on the interstitial cells (of Leydig) of the testes, which is why it is called the *interstitial cell–stimulating hormone* (*ICSH*). It is essential for the development of the interstitial cells and the secretion of androgens by these cells.

Prolactin (PRL). This hormone, also called *lactogenic hormone,* is a protein similar in structure and chemical properties to the human growth hormone (HGH) or somatotropin (STH). In many of its metabolic actions, it mimics the activity of HGH, including stimulation of growth of the skin and its derivatives, especially the glands and hair. One of its primary actions is stimulation of the development of the mammary glands in the later stages of pregnancy and the initiation and maintenance of milk secretion following birth of the young. In some animals, prolactin is essential for the maintenance of the corpus luteum, a luteotrophic action, but this has not been demonstrated in humans. The action of prolactin in the male is uncertain, but there is evidence that it works synergistically with testosterone to promote protein synthesis.

Adrenocorticotropin (ACTH). This hormone, a peptide, is essential for the growth and development of the adrenal cortex and for the secretion of cortical hormones, especially glucocorticoids. Its action on the secreting cells of the adrenal gland is primarily on the plasma membrane. Extra-adrenal actions include promotion of lipolysis in adipose tissue, increased glucose and amino acid uptake by muscle cells, and increased secretion of insulin by pancreatic islet cells and of somatotrophin by the pituitary.

INTERMEDIATE-LOBE HORMONE. Also developing from the adenohypophysis is the *intermediate lobe,* which secretes a single hormone, the *melanocyte-stimulating hormone* (MSH). This hormone, a peptide, in lower animals promotes the synthesis of pigments and their dispersal in chromatophores. In rats, it has a favorable effect on learning and memory. In humans, it promotes the synthesis of melanin in the melanocytes of the epidermis, and there is evidence that it improves visual retention and reduces anxiety. It also enhances attention.

Control of Secretion of Hormones of the Adenohypophysis. The control of the secretion of hormones liberated from the anterior and intermediate lobes of the hypophysis resides in the *hypothalamus,* which lies immediately above the pituitary gland. The hypothalamus, a neuroendocrine structure, produces a number of hormones that regulate the functioning of the anterior and intermediate lobes. (See "Hormones of the Hypothalamus," page 230.) These neurohormones, called *releasing hormones* or *releasing factors,* are discharged into capillaries in the region of the median eminence (Fig. 6-3). From here they are transported a very short distance by hypothalamohypophyseal vessels (*hypophyseal portal veins*) to the secreting cells of the anterior and intermediate lobes. Of the factors released, there is a stimulating hormone for each of the hormones produced in these lobes. For three of the hormones (GH, PRL, and MSH), a factor that inhibits secretion is also produced.

POSTERIOR-LOBE HORMONES. The posterior lobe or neurohypophysis liberates two hormones into the bloodstream but does not secrete them. These hormones, called *neurohormones,* are called vasopressin and oxytocin.

Vasopressin. Vasopressin, also called the *antidiuretic hormone* (*ADH*), is a polypeptide secreted by the neuroendocrine cells of the hypothalamus, especially those located in the supraoptic nuclei (see Fig. 6-3). The secretory

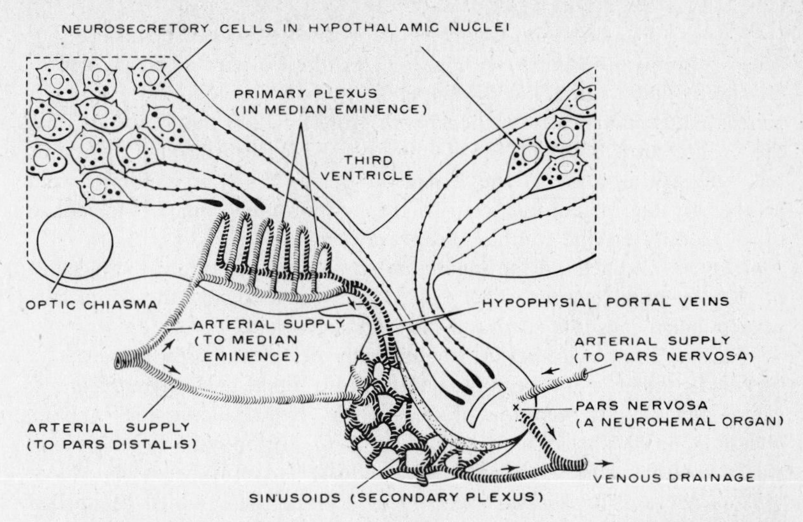

Fig. 6-3. Diagram showing the anatomic relationship between the hypothalamus and the pituitary gland. (From *General Endocrinology,* 6th ed., by C. Donnell Turner, Ph.D., and Joseph T. Bagnara, Ph.D. Copyright © 1976 by W. B. Saunders Co. Reprinted by permission of Holt, Rinehart and Winston, CBS College Publishing.)

granules, combined with *neurophysin,* a carrier protein, travel down the axons of the hypothalamohypophyseal tract in the infundibular stalk to the posterior lobe of the pituitary, where they are stored and later released into the bloodstream.

Vasopressin has two physiologic actions in the body:

1. It acts on the smooth muscle cells in the walls of blood vessels. Its *vasopressor effect* elevates blood pressure.
2. It acts on the kidneys to prevent excessive excretion of water, an *antidiuretic action.* This effect is brought about by its action on the distal convoluted tubules, increasing their permeability to water and thus promoting its reabsorption. In the absence of vasopressin, large quantities of a very dilute urine are excreted, as in *diabetes insipidus.*

The secretion of vasopressin depends on the volume of extracellular fluid and osmotic pressure of the blood plasma. *Osmoreceptors* in the hypothalamus respond to a decrease in water content and an increase in sodium-ion concentration in and osmotic pressure of the blood plasma. These act on the secretory neurons in the hypothalamus, resulting in the secretion and release of vasopressin as required. Baroreceptors in the carotid sinus are also involved in the production and release of vasopressin. When these receptors are stimulated by stretching or by an increase in blood pressure, vasopressin release is inhibited.

Oxytocin. This hormone, a polypeptide, is also secreted by neurosecretory cells of the hypothalamus (see Fig. 6-3). Combined with a carrier, neurophysin, oxytocin passes through axons to the posterior lobe, from which it is released. Oxytocin acts on the smooth muscle cells of the uterus, causing their contraction, which aids in (1) the passage of sperm through the uterus following coitus and (2) the expulsion of the fetus at childbirth and postpartum uterine contractions. It is sometimes used clinically to induce or to intensify uterine contractions at childbirth.

Oxytocin also acts on the mammary gland, bringing about the ejection of milk or milk letdown. This results from its action on the contractile myoepithelial cells that surround the secretory alveoli.

The secretion of oxytocin results from nerve impulses initiated by sucking by the newborn. Impulses pass from the nipple to the hypothalamus, where they initiate secretion of oxytocin by hypothalamic cells and its release from the posterior or neural lobe of the hypophysis. The hormone passes by way of the circulatory system to the mammary glands, where milk is released. This *neuroendocrine reflex* may be initiated by such a conditioned stimulus as the cry of an infant.

HYPOTHALAMUS

The hypothalamus is a bilaterally symmetrical structure which forms the walls and floor of the third ventricle of the brain. Its structure is described on pages 105–106. Since it produces a number of releasing factors and hormones, it may properly be considered an endocrine gland.

Hormones of the Hypothalamus. The hormones or factors released by the hypothalamus (Fig. 6-4) that affect the anterior and intermediate lobes of the pituitary are as follows.

Stimulating Factors or Hormones:

1. Somatotropin releasing factor or hormone (SRF or SRH)* and growth hormone releasing factor or hormone (GRF or GRH).
2. Corticotropin releasing factor or hormone (CRF or CRH).
3. Follicle-stimulating hormone releasing factor (FSH-RF).
4. Luteinizing hormone releasing factor (LH-RF or LRF). In the male, this stimulates the secretion of the interstitial cell–stimulating hormone (ICSH).
5. Prolactin releasing factor (PRF; vasoactive intestinal peptide, VIP).
6. Melanocyte-stimulating hormone release factor (MRF).

Inhibiting Factors:

1. Growth hormone release-inhibiting factor (GIF) and somatostatin.

* Sometimes the word *hormone* is substituted for *factor,* in which case the letter *H* is substituted for *F* in the abbreviation.

2. Prolactin release-inhibiting factor (PIF).
3. Melanocyte-stimulating factor inhibiting factor (MIF).

In addition to the hormones listed, the hypothalamus produces two hormones, *vasopressin* and *oxytocin*, which are released from the posterior lobe of the pituitary. These hormones were described earlier.

THYROID GLAND

The thyroid gland (Fig. 6-5) is a bilobed structure lying in the midportion of the neck, just anterior to the upper portion of the trachea and slightly below the larynx. Each of its two lobes, connected across the midline by a narrow isthmus, is about 5 by 3 by 2 cm. The entire gland weighs about 30 g. The size and structure of the thyroid gland are, however, extremely variable, depending on a large number of influences, among them age, sex, body temperature, and diet. The iodine content of the diet plays a dominant role. The gland tends to be slightly larger in females; during pregnancy its size increases.

Microscopic Structure of the Thyroid (Fig. 6-6). The thyroid gland consists of a large number of closed, spherical or ovoid vesicles called *follicles*, surrounded by a fibrous capsule that is continuous with the cervical fascia. Each follicle is lined with a single layer of cuboidal or low columnar epithelial cells and encloses a homogeneous, gelatinous substance called *colloid*. This substance has a high iodine content and contains the active principle secreted by the gland.

The cells of the follicles are of two types, follicular and parafollicular. (1) *Follicular cells* make up the main portion of the wall of a follicle and are concerned with the production and resorption of the colloid and its contained hormones. (2) *Parafollicular cells*, also called *C cells*, lie at the bases of follicular cells and are not in contact with the colloid. They secrete calcitonin.

Blood and Lymph Supply of the Thyroid. The thyroid gland, a highly vascular organ, receives its blood supply from two *superior thyroid arteries* (branches of the external carotids) and two *inferior thyroid arteries* (branches of the subclavian arteries). An inconstant artery, the *thyroidea ima*, from the brachiocephalic artery, when present appears to compensate for the absence of or deficiency in one or the other thyroid arteries. In proportion to its size, and with the possible exception of the adrenals, more blood flows through this gland than through any other organ of the body.

From the thyroid gland, the blood drains into the *superior, middle*, and *inferior thyroid veins*, which empty into the *internal jugular* and *brachiocephalic veins*. The gland is well supplied with *lymphatic vessels*, which drain into the main lymph ducts.

Nerve Supply of the Thyroid. This gland receives sympathetic fibers from the superior and inferior sympathetic ganglia. Parasympathetic (vagal)

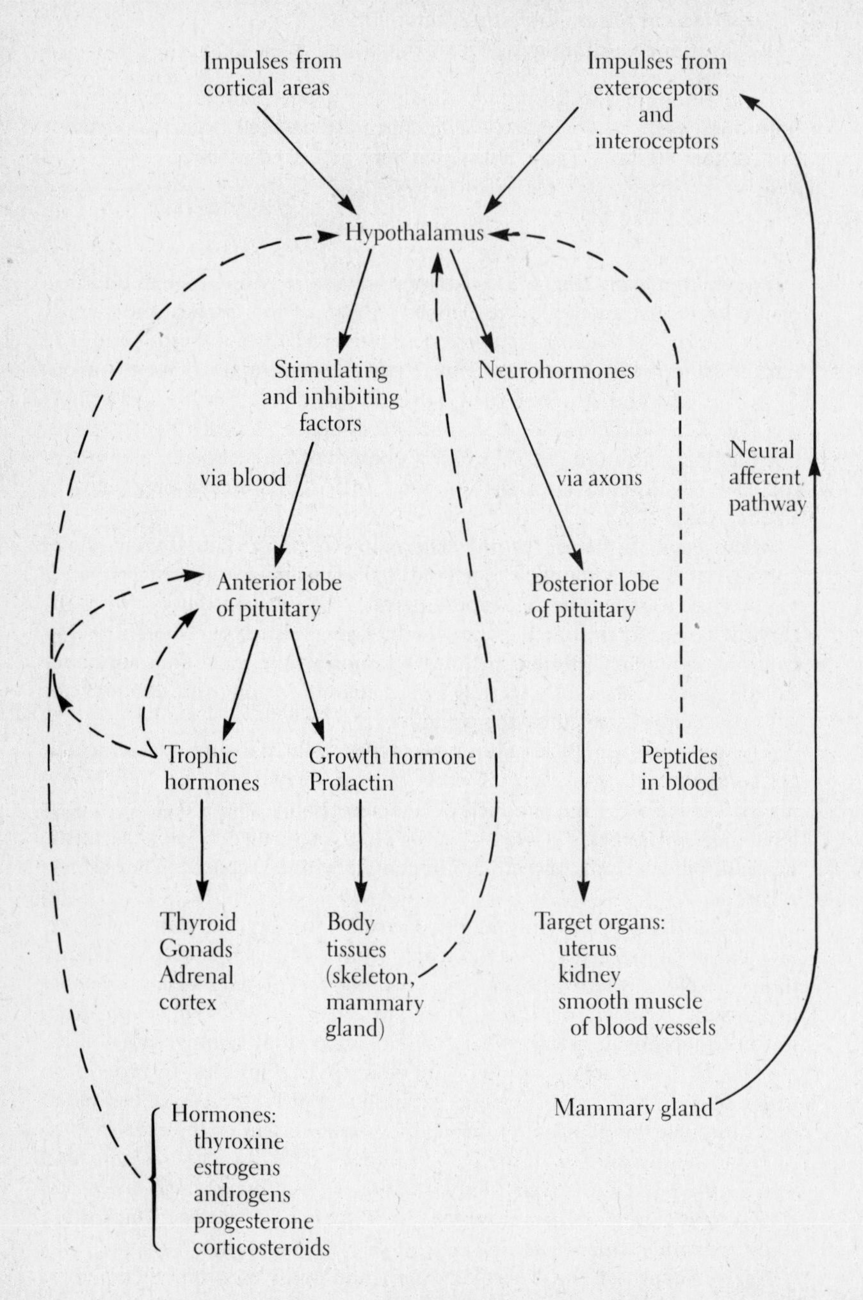

Fig. 6-4. Hypothalamic-hypophyseal relationships.

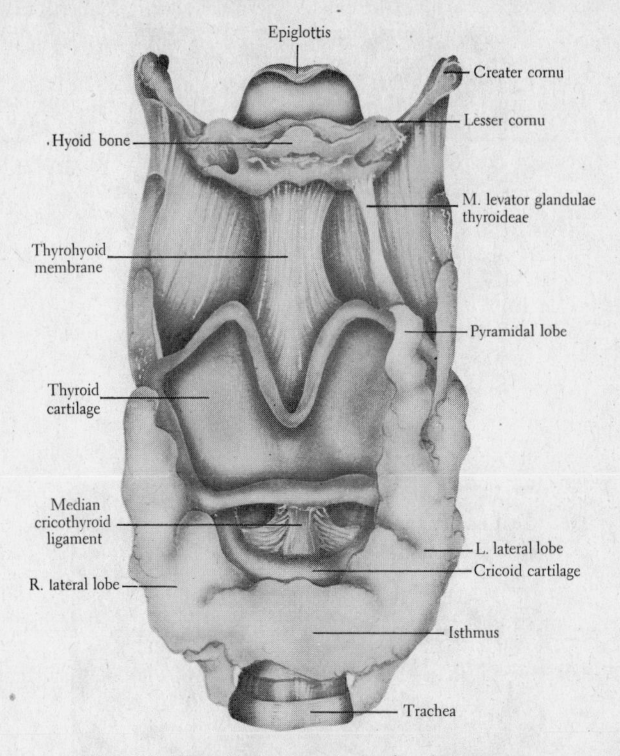

Fig. 6-5. Thyroid gland and associated structures.

fibers also innervate the gland. The fibers end principally in the walls of the blood vessels and are vasomotor in their effect. Whether secretory fibers exist is not certain. The thyroid gland has the ability to secrete hormones even if its normal nerve supply is interrupted or if it is transplanted to another site.

Embryonic Origin of the Thyroid. The thyroid gland is endodermal in origin, arising as a median ventral outgrowth of the floor of the pharynx. It grows caudally as a hollow tube, its distal end thickening and forming a solid mass of epithelium, which differentiates into cords; these cords eventually give rise to the follicles. The connection with the pharynx disappears, though it sometimes persists as the *thyroglossal duct.* The foramen cecum at the base of the tongue indicates the point of embryonic origin.

Hormones of the Thyroid. Two primary hormones are produced by the thyroid gland: (1) *thyroxine* and related iodine-containing compounds and (2) *calcitonin.* Thyroxine is concerned with the regulation of bodily metabolism, growth, and development, calcitonin with the lowering of blood calcium.

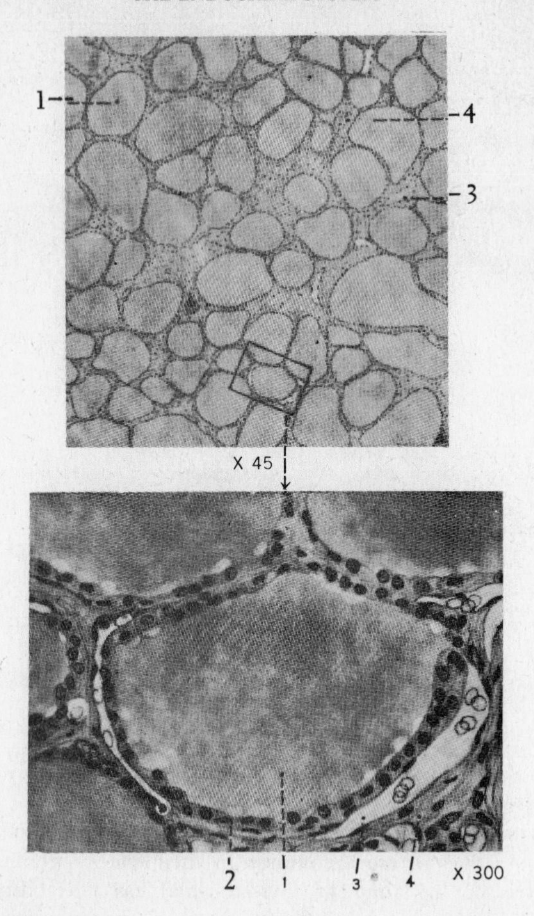

Fig. 6-6. Section through the thyroid. (*1*) Follicle containing colloid. (*2*) Cuboidal follicular epithelium. (*3*) Blood vessel. (*4*) Intercellular connective tissue. (From *Atlas of Human Anatomy,* Barnes & Noble, Inc., 1961.)

THYROXINE. An iodinated glycoprotein, *thyroglobulin,* is present in the colloid stored in the thyroid follicles. From it, by hydrolysis, a number of iodinated amino acids called *thyronines* are formed. Of these, two comprise the active circulating thyroid hormones, namely *3,5,3-triiodothyronine* (T_3) and *thyroxine* (T_4). The former is the more potent of the two. These hormones are transported in the blood in combination with certain proteins. Thyroxine is bound to a globulin called *thyroxine-binding globulin* (*TBG*). It is the principal circulating hormone and considered to be the most effective.

Effects of Thyroxine. The regulatory effects of thyroid hormones fall into two categories: those involving metabolism and those involving growth and development.

Metabolic Effects. Thyroxine acts to control the basal metabolic rate (BMR). An excess of the hormone increases the respiratory rate and oxygen consumption, with increased heat production (*calorigenesis*). Carbohydrate and fat metabolism are accelerated. A decrease in thyroxine secretion produces the opposite effects. In general, thyroxine promotes catabolic reactions involved in the release of energy.

Effects on Growth and Development. In the organism as a whole, thyroxine promotes the synthesis of new proteins and is essential for normal growth and development as well as for the development of the central nervous system. A deficiency in early life results in *cretinism,* a condition in which physical and mental growth are severely retarded.

Tests for Thyroid Function. Tests to determine thyroid function include (1) determination of metabolic rate, (2) determination of the amount of protein-bound iodine (PBI) or butanol-extractable iodine (BEI), (3) measurement of iodine uptake by the thyroid, and (4) determination of plasma TSH.

Fate of Thyroid Hormones. The iodine is removed from iodine-containing hormones by the action of enzymes present in the liver. Some of the free iodine is excreted in the urine, but most is circulated and reincorporated into thyroid hormones. The remaining amino acids pass into the intestine in the bile, and from there they may be reabsorbed or lost in the feces. Some of the hormones may be degraded in the cells of the tissues within which they act.

Regulation of the Thyroid Gland. The secretion of iodinated hormones is primarily under the control of thyrotrophin (TSH) secreted by the anterior pituitary, which is in turn controlled by the thyrotrophin releasing factor from the hypothalamus. In the absence of TSH, there is reduced synthesis and storage of the thyroid hormones, reduced blood supply and oxygen consumption, decreased iodine uptake, and general atrophy of the gland.

A negative feedback mechanism exists by which an increase in thyroxine inhibits the release of hypothalamic TSH. Nervous stimuli resulting from such environmental factors as cold, emotional disturbances, and physical stress (trauma or hemorrhage) may also trigger the release of hypothalamic releasing factors.

A number of other factors may affect the rate of thyroxine secretion. Diet plays an important role. Since *iodine* is a basic constituent of the thyroid hormones, a deficiency leads to enlargement of the gland, as occurs in *goiter.* Other substances in food can interfere with the utilization of iodine by thyroid cells and produce goiter. Such substances, called *goitrogens,* include thiocyanates, nitrates, and perchlorates. Some drugs, such as salicylates, interfere with the binding of thyroid hormones to blood proteins, thus impairing their utilization.

A number of pathogenic conditions may result from the malfunctioning of the thyroid gland. These are discussed in the section on dysfunction of the endocrine glands at the end of this chapter.

CALCITONIN. Calcitonin is a polypeptide secreted by the parafollicular or C cells of the thyroid gland, principally in response to elevated serum calcium. It acts to lower blood calcium, hence its effects are antagonistic to those of the parathyroid hormone. It promotes the action of osteoblasts in bone formation and reduces the number of osteoclasts. It decreases the absorption of calcium by the gastrointestinal tract and increases the excretion of calcium and phosphate ions by the kidneys. All of these actions act to reduce the level of calcium in the blood.

PARATHYROID GLANDS

The parathyroid glands, usually four in number, lie within the capsule of the thyroid gland. They are generally ovoid in shape and are found along the posterior medial surface of each lobe of the thyroid gland. Each parathyroid averages 5 by 5 by 3 mm in size. Total weight averages about 0.5 g.

Microscopic Structure of the Parathyroids (Fig. 6-7). Each parathyroid gland consists of densely packed cells within a connective-tissue framework of reticular fibers; between the cells, capillaries and sinusoids form an anastomosing network. In some instances, the cells form compact masses; in others, they have a cordlike arrangement. Two types of cells are present: *principal* or *chief cells,* which have a clear, nongranular cytoplasm, and *oxyphil cells,* larger in size and containing granules that stain with acid stains. The principal or chief cells are regarded as the primary hormone-secreting cells.

Blood, Lymph, and Nerve Supply of the Parathyroids. The blood and nerve supply is the same as that of the thyroid gland. Lymphatic drainage is into the vessels that drain the thyroid.

Embryonic Origin of the Parathyroids. The parathyroids develop from dorsal portions of the third and fourth *pharyngeal pouches.* The fact that these pouches also give rise to the thymus gland probably accounts for the occasional presence of extra masses of parathyroid tissue in the connective tissue of the neck, near (or sometimes within) the thymus.

Parathyroid Hormone and Its Functions. The parathyroid hormone (PTH), a polypeptide, functions in the maintenance of a proper level of blood calcium essential for the maximal functioning of the tissues dependent on calcium. Calcium plays a critical role in many physiologic activities including nerve and muscle irritability, coagulation of the blood, cell permeability, formation of bone, mediation of hormone action, utilization of ATP, and acid-base balance. Blood-plasma calcium averages 9 to 11

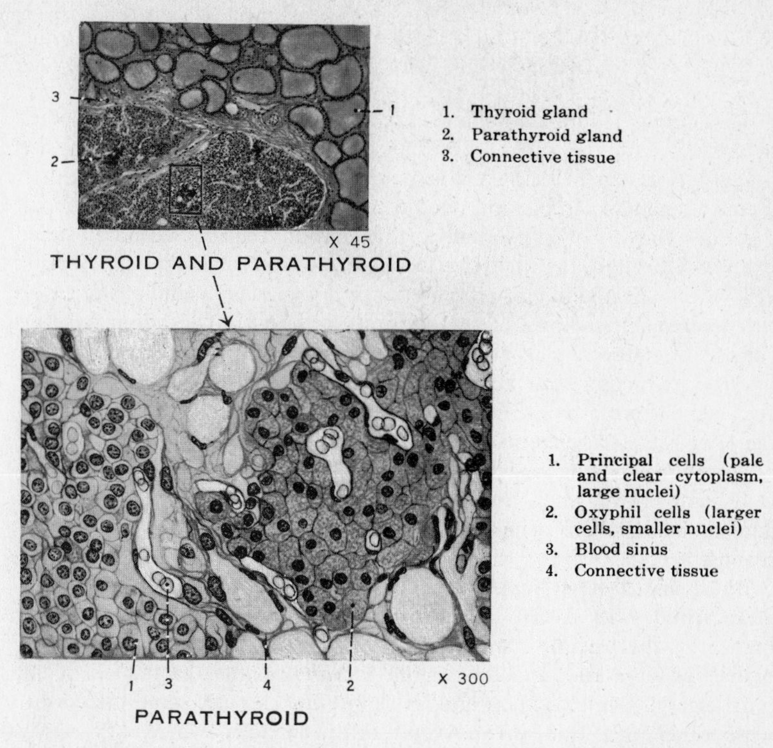

Fig. 6-7. Section through the parathyroid. (From *Atlas of Human Anatomy,* Barnes & Noble, Inc., 1961.)

mg/100 ml of blood. About half is bound to proteins and is nondiffusible; most of the remainder is present as free calcium ions.

A sudden elevation in blood calcium may lead to alterations in heart rhythm and possible cardiac arrest. Sustained elevation results in deposition of calcium in blood vessels or in tissues. A lowering of plasma calcium increases neuromuscular irritability, resulting in involuntary tremors and muscle spasms. In children, there is impaired development of bones and teeth; in adults, softening of bone.

Parathyroid hormone increases in the blood when calcium levels are low and decreases when calcium levels are high. The primary target organs for PTH are bone, the kidneys, and the intestine. The hormone increases the reabsorption of bone by increasing the number of osteoclasts involved in demineralization of bone. This releases calcium into circulation. In the kidneys, resorption of calcium is increased, while that of phosphate is decreased. In the intestine, the absorption of calcium is increased because of the synergistic action of PTH with vitamin D.

ADRENAL (SUPRARENAL) GLANDS

The adrenal (suprarenal) glands are two encapsulated, flattened bodies in contact with the superior medial surface of each kidney. They are roughly triangular in shape and somewhat variable in size, averaging 3 to 5 cm in length, 4 cm in width, and 5 cm in thickness. Each weighs about 5 g. Each consists of an external cortical portion, the *cortex,* and an inner medullary portion, the *medulla.* The outer portion of the cortex is bright yellow in color; the inner portion, reddish brown. The medulla is grayish.

Microscopic Structure of the Adrenals (Fig. 6-8). The adrenal *cortex* consists of epithelial cells arranged in three zones: the outer *zona glomerulosa,* the middle *zona fasiculata,* and the inner *zona reticularis.* The cortical cells are arranged in more or less parallel cords perpendicular to the long axis of the capsule. The *medulla* consists largely of chromaffin cells forming irregular masses separated by sinusoidal vessels. Chromaffin cells stain brown when treated with chromic acid. Also present in the adrenal medulla are sympathetic ganglion cells, occurring singly or in groups.

Blood and Lymph Supply of the Adrenals. Both adrenals are highly vascularized. They receive blood through the *superior suprarenal artery* (a branch of the inferior phrenic artery), the *middle suprarenal artery* (a branch of the aorta), and the *inferior suprarenal artery* (a branch of the renal artery). Some of these arteries supply the capsule, some the cortex, while others pass through the cortex to the medulla, where they empty into sinusoids that form an anastomosing network. Blood is drained into the *central veins* of the medulla, which lead to the *suprarenal vein.* On the right side, the suprarenal vein empties into the *inferior vena cava;* on the left side, the *left renal vein.*

Lymphatic vessels are present in the capsule and the medulla. They lead to vessels that drain into the lymph nodes along the abdominal aorta or into the mediastinal lymph nodes.

Nerve Supply of the Adrenals. The adrenal glands are innervated principally by autonomic fibers from the celiac plexus of the sympathetic division. The fibers are preganglionic and pass without interruption through the splanchnic nerves to the capsules of the adrenals, where they form extensive plexuses. Branches from these plexuses pass through the cortex to the medulla, where some fibers end in contact with chromaffin cells of the medulla, which are homologues of sympathetic postganglionic cells. The transmitter agent for these fibers is acetylcholine. The adrenal glands also receive parasympathetic fibers from the vagus nerve, but their function is unknown.

Embryonic Origin of the Adrenals. The adrenal glands have a double origin. The cortex develops from the coelomic mesoderm in the immediate vicinity of the embryonic urogenital organs. The medulla develops from

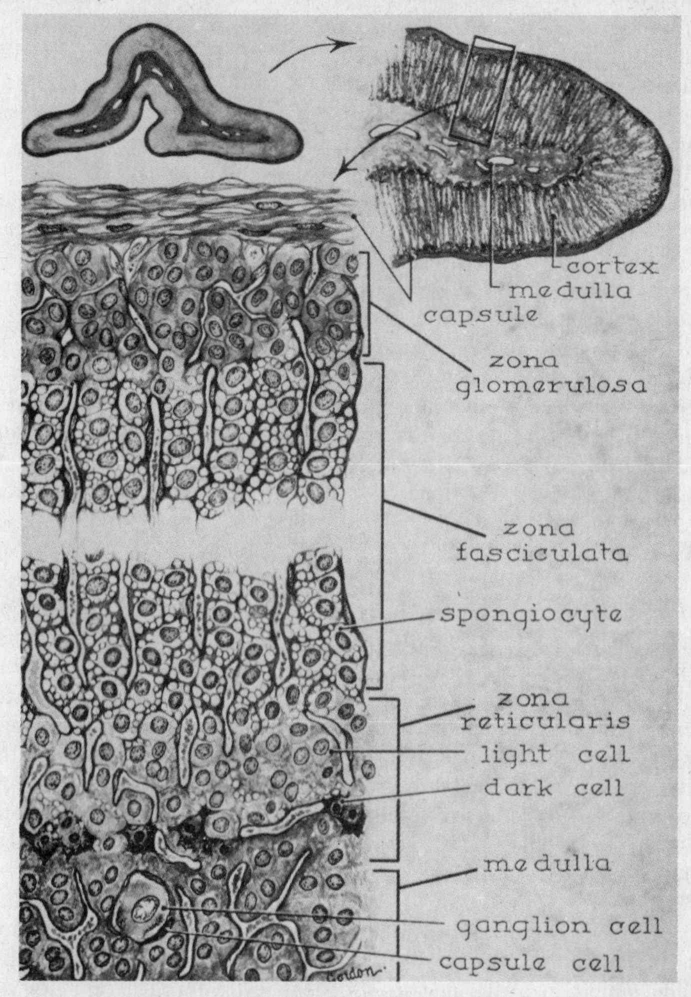

Fig. 6-8. Composite drawing showing the representative regions of an adrenal gland. (Reprinted with permission of J. B. Lippincott Co. from A. W. Ham and D. H. Cormack, *Histology,* 8th ed., 1979.)

ectoderm. The chromaffin cells of the medulla are derived from neural crest cells, which also give rise to the ganglion cells of the sympathetic nervous system. These cells migrate into the primordium of the cortex and acquire a central position.

Function of the Adrenals. The adrenal glands play an important role in the regulation of many activities of the body. Some of the known functions of the adrenals are the following:

CORTICAL FUNCTIONS. The adrenal cortex produces a number of hormones necessary for the maintenance of life. They are involved in the regulation of salt balance, water and electrolyte balance, carbohydrate and protein metabolism, and reproductive activities. They also have anti-inflammatory and immunosuppressive actions.

MEDULLARY FUNCTIONS. Through its hormones epinephrine and nor-epinephrine, the adrenal medulla duplicates or augments activities resulting from stimulation of the sympathetic nervous system, a *sympathomimetic action*. Many structures not innervated by sympathetic nerves may also be affected. Metabolic rate may be increased and carbohydrate and protein metabolism stimulated. These responses are those involved in "fight or flight" responses to emotion or stress.

Hormones of the Adrenals. Many steroids have been extracted from the adrenal cortex. The most important ones are discussed here. The adrenal medulla secretes three hormones: epinephrine, norepinephrine, and dopamine. Their actions are also discussed here.

ADRENAL CORTICAL HORMONES. All hormones secreted by the adrenal cortex are steroids synthesized from cholesterol, which the adrenal can either synthesize or take up from circulating blood. They include glucocorticoids, mineralocorticoids, and sex hormones.

Glucocorticoids. These include a number of adrenal steroids that profoundly affect carbohydrate and protein metabolism. Corticoids belonging to this group include *hydrocortisone (cortisol), cortisone, corticosterone,* and *11-dehydrocorticosterone.* Of these, hydrocortisone has the greatest potency. Glucocorticoids promote activities involved in the release of energy. These include release of proteins from tissue cells and their transport to and deamination in the liver. Glucocorticoids promote gluconeogenesis and storage of glycogen in the liver, facilitate mobilization and oxidation of lipids, and inhibit amino acid uptake and protein synthesis in the tissues.

When produced in excessive amounts or administered therapeutically, glucocorticoids exert anti-inflammatory and immunosuppressive actions. Anti-inflammatory actions, which counteract the effects of inflammation, include inhibition of the migration of leukocytes through capillary walls into the tissues, inhibition of the proliferation of fibroblasts and the formation of granulomas, and the stabilization of lysosomes that damage adjoining tissue when released from injured cells.

Immunosuppressive actions include reduction in the size of the thymus, lymph nodes, and spleen; a decrease in number of circulating lymphocytes; and decreased antibody production.

After release from the adrenal gland, most corticoids are transported in the blood in combination with a *corticosteroid-binding globulin* (CBG), also called *transcortin.* Corticoids are rapidly catabolized in the liver and excreted in the urine.

Mineralocorticoids. These include steroids that have a specific effect on the transmission of ions across cell membranes, with special reference to the retention of sodium and the excretion of potassium. The principal steroid involved is *aldosterone,* although others, such as deoxycorticosterone (DOC), may be involved to a limited extent.

Aldosterone acts on cells in the distal convoluted tubules of the kidney, promoting the reabsorption of sodium and the excretion of potassium in the urine. In the absence or reduced secretion of aldosterone, there is a loss of sodium chloride and retention of potassium. This seriously alters water and electrolyte balance in the body fluids, leading to increased excretion of water, reduced blood volume, and eventual circulatory failure.

Sex Hormones. Sex hormones secreted by the adrenal cortex include androgens, estrogens, and progesterone. All originate from *cholesterol,* which is converted into *pregnenolone* and then through various biosynthetic pathways into the final products. The principal androgen is *testosterone;* its effects, however, are weak compared with testicular testosterone. In certain pathologic conditions, excess amounts may be produced, which in a female may lead to virilization. In a pregnant woman, a female fetus may be affected by abnormal development of the urogenital system, resulting in pseudohermaphroditism. Estrogens secreted by the adrenal cortex include *estrone, estradiol,* and *progesterone.* Their effects are usually inconsequential, except in pathologic cases, where excess production may lead to feminization of males.

REGULATION OF ADRENAL CORTICAL SECRETION. The secretion of corticoids, especially glucocorticoids, is regulated principally by the *adrenocorticotropic hormone* (ACTH) secreted by the anterior lobe of the pituitary. The secretion of ACTH depends in turn on the release from the hypothalamus of a corticotropin releasing factor (CRF). Feedback mechanisms exist whereby an increase in hydrocortisone suppresses ACTH secretion; a decrease results in increased ACTH production. The production of ACTH is subject to diurnal rhythms; levels are higher in the morning than in the evening. Stress is a powerful stimulus to ACTH production. Stressful situations, as when the body is subjected to severe trauma, pyogenic infections, various inflammatory conditions, hypoglycemia, or intoxications, induce ACTH secretion and thus an increase in corticosteroid secretion.

Aldosterone secretion depends on a number of factors. Of primary importance is the renin-angiotensin system, which responds to changes in blood volume. A reduction in blood volume or a decrease in sodium concentration causes the juxtaglomerular apparatus of the kidney to produce renin. Renin acts on angiotensinogen, a blood globulin, to form angiotensin I, which is converted into angiotensin II, which stimulates the adrenal cortex to secrete aldosterone.

Certain hormones may influence the secretion of aldosterone. Among these are the adrenocorticotropic hormone (ACTH) and somatotropin, which stimulate aldosterone secretion. Aldosterone production varies with sodium ion and potassium ion concentrations in the blood plasma.

ADRENAL MEDULLARY HORMONES. The two principal hormones secreted by the medulla are epinephrine (adrenaline) and norepinephrine. A third hormone, dopamine, is also produced. All are catecholamines. *Epinephrine* is synthesized in the adrenal medulla and is also produced in small quantities in the brain. *Norepinephrine* is produced in the adrenal medulla but is also synthesized in the brain and in the terminations of all postganglionic neurons of the sympathetic nervous system, where it serves as a transmitter agent. *Dopamine* is produced in both the adrenal medulla and the brain. It is also a neurotransmitter. The three hormones are synthesized from phenylalanine and tyrosine. Dopa is formed first; then dopamine, epinephrine, and norepinephrine are synthesized, in that order.

The medullary hormones are produced by chromaffin cells. These cells are derived from neural crest cells and are consequently homologous with postganglionic sympathetic ganglion cells. Their secretion is controlled by sympathetic impulses transmitted to the adrenals through preganglionic neurons of the splanchnic nerve.

Effects of Medullary Hormones.

Epinephrine. The principal medullary hormone increases the rate and force of the heartbeat, raising systolic but not diastolic pressure. Peripheral dilatation occurs, and peripheral resistance is reduced; however, skin capillaries are constricted. Blood flow to muscles, the liver, and the brain is markedly increased; flow to the kidneys is decreased. The respiratory rate is increased, and respiratory passageways are dilated. Epinephrine promotes carbohydrate metabolism, including glycogenolysis, in the liver. Blood sugar levels are increased. Other effects include an increase in coagulability of the blood, dilatation of the pupils of the eyes, decrease in the tone of muscles of the gastrointestinal tract, erection of the hairs resulting from contraction of piloerector muscles, and activation of sweat glands.

Norepinephrine. Many of the effects of norepinephrine are the same as listed for epinephrine. Both stimulate the heart, increasing systolic pressure, thus promoting blood flow to skeletal muscles. However, norepinephrine reduces blood flow to the skin, mucous membranes, and kidneys. Both stimulate bodily metabolism and elevate body temperature. Both promote carbohydrate metabolism and lipolysis.

Dopamine. Dopamine functions primarily as a neurotransmitter. It acts in the central nervous system, especially in the regulation of hypothalamic secretion. The amounts produced by the adrenal medulla are small, and their physiologic significance is unknown.

Control of Medullary Hormone Secretion. In general, the effects of the release of medullary hormones are similar to those resulting from the stimulation of the sympathetic nervous system. This occurs in *emergency situations* that necessitate the quick mobilization and release of energy as occurs in "fight or flight" situations or in response to stress. After the

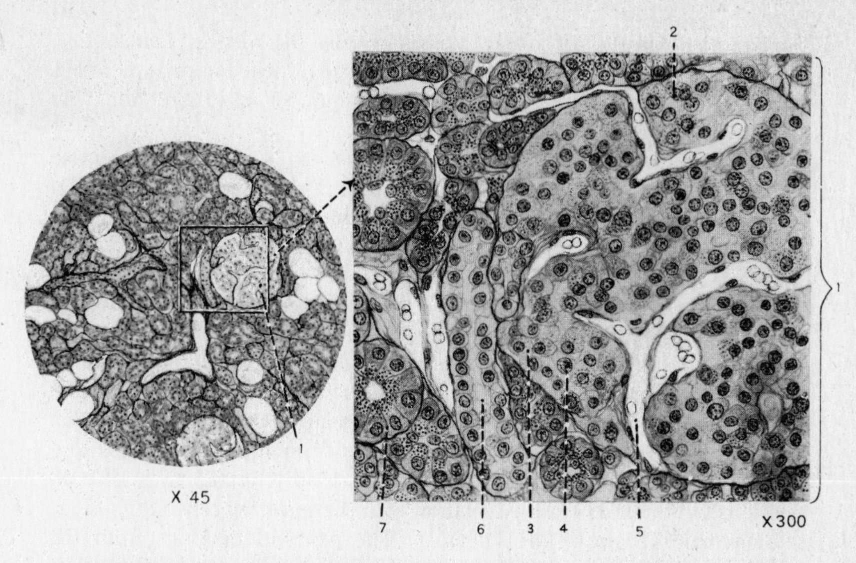

Fig. 6-9. Portion of the pancreas. (*1*) Islet of Langerhans. (*2*) A cells. (*3*) B cells. (*4*) D cells. (*5*) Blood vessel. (*6*) Pancreatic duct. (*7*) Pancreatic acinar cells. (From *Atlas of Human Anatomy,* Barnes & Noble, Inc., 1961.)

immediate response brought about by sympathetic nervous impulses, the discharge of hormones into the blood prolongs the effects. Because of the widespread effects and the close interrelationships between the adrenal medulla and the sympathetic nervous system, the two are referred to as the *sympathicoadrenal* system.

The secretion of medullary hormones is regulated by autonomic nerve impulses mediated primarily by the hypothalamus. The adrenal medulla is especially responsive to heightened emotional activity, as in anger, fear, rage, the effects of noxious stimuli, and various stressful situations such as those resulting from trauma, burns, childbirth, hypoxia, and extreme cold. All these increase the rate of catecholamine synthesis.

Medullary hormones are stored in cytoplasmic granules in cells of the adrenal medulla and sympathetic nerve endings. After release at effector sites, they are quickly inactivated by enzymes in the blood plasma, rebound to cytoplasmic granules, or excreted.

PANCREAS (ISLETS OF LANGERHANS)

The pancreas, a double-functioning gland, produces an exocrine secretion that is discharged into the duodenum and an endocrine secretion that is discharged into the bloodstream.

Microscopic Structure of the Islets (Fig. 6-9). The endocrine portion of the pancreas consists of isolated masses of cells, the *islets of Langerhans,* that lie scattered throughout the acinar tissue. They comprise about 1 percent of the mass of the pancreas and may number up to 2 million. The islets are most numerous in the tail portion.

The islets contain four types of cells, *alpha* or *A cells,* which secrete glucagon; *beta* or *B cells,* which secrete insulin; and C cells and delta (D) cells of unknown function. The B cells are the most abundant. The cytoplasm of each type of cell contains secretory granules that stain selectively.

Blood, Lymph, and Nerve Supply of the Pancreas. The pancreas receives blood through branches of the splenic, superior mesenteric, and hepatic arteries, and the blood is returned through the splenic and superior mesenteric veins to the portal vein. The islets are highly vascularized. *Lymphatic vessels* drain into the celiac nodes along the celiac artery. The pancreas is innervated by sympathetic fibers from the celiac plexus and parasympathetic fibers from the vagus nerve. Unmyelinated fibers of both divisions terminate in close association with the secretory cells of the islets.

Embryonic Origin of the Pancreas. The pancreas develops from the endoderm of the foregut in the region of the liver diverticulum. It arises as two portions: a dorsal and a ventral. The dorsal portion arises from the wall of the duodenum, the ventral from the embryonic bile duct. From the former, the tail, which contains most of the islets, develops; the latter gives rise to the head portion, and its duct becomes the functional pancreatic duct. The acini and the islets develop from anastomosing tubules that connect with the pancreatic duct, but the tubules connected with islets never become patent or functional.

Endocrine Function of the Pancreas. Through its hormones *insulin* and *glucagon,* in conjunction with the hormones of other glands (adrenal cortex, adrenal medulla, thyroid, and pituitary), the pancreas regulates the metabolism of carbohydrates. This includes the utilization of glucose by tissue cells, glycogen formation in the liver, and conversion of glycogen into glucose. Indirectly, the metabolism of fats and amino acids is influenced.

Hormones of the Pancreatic Islets. The pancreatic islets secrete two hormones, insulin and glucagon. Both are polypeptides.

INSULIN.

Action of Insulin. Insulin facilitates the utilization of glucose by increasing glucose oxidation and glycogenesis. It prevents glycogenolysis or the breakdown of glycogen stored in muscle and liver cells. By preventing an increase in blood sugar level, it acts as a *hypoglycemic agent.* Insulin promotes the uptake and utilization of glucose by adipose tissue (lipogenesis), and it is antagonistic to fat degradation by inhibiting lipase activity. Insulin promotes protein metabolism by facilitating the uptake of amino acids and the synthesis of proteins.

The exact mode of action of insulin is not known, but it is thought that it acts primarily on the cell membranes of target tissues, enhancing permeability and thus facilitating the transport of amino acids and glucose. Insulin binds tightly to specific receptors in the plasma membrane of target cells, as shown by the use of radioactive insulin. The mode of action of insulin within target cells in the promotion of metabolic reactions is unknown, but it is probably through the activation of enzyme systems.

Regulation of Insulin Secretion. Insulin is formed within secreting cells from a precursor, *proinsulin.* The primary factor stimulating insulin secretion is an increase in blood glucose level. Several hormones indirectly stimulate insulin secretion. These include somatotropin, thyroxine, glucocorticoids, and sex hormones. Certain drugs, such as tolbutamide (Orinase), stimulate insulin secretion. Secretion may also result from parasympathetic nerve stimulation. Secretion is inhibited or reduced by low blood sugar, low blood calcium, or the action of epinephrine, norepinephrine, or somatostatin. Certain drugs, such as *alloxan,* destroy insulin-producing cells. The life of insulin is short; it is degraded in all the tissues of the body, especially the liver and kidneys.

GLUCAGON. *Glucagon,* a hyperglycemic factor (HGF) secreted by the A cells of the pancreatic islets, acts antagonistically to insulin, its action being to increase blood sugar level. This is accomplished by increasing the rate of conversion of liver glycogen to glucose. Glucagon acts on enzymes that bring about the breakdown of glycogen (glycogenolysis) and at the same time inhibits the synthesis of glycogen (glycogenesis). Glucagon also promotes the synthesis of glucose from pyruvate and lactate and stimulates the formation of glucose from amino acids. Through these actions, glucagon acts to prevent abnormally low blood sugar (*hypoglycemia*), with its serious physiologic effects. Glucagon is also a lipolytic agent. The principal factor that stimulates the secretion of glucagon is low blood sugar.

GASTROINTESTINAL TRACT

The gastrointestinal tract elaborates several hormones that play an important role in the activities of the pancreas, liver, gallbladder, and stomach. Among the hormones produced are *secretin* and *cholecystokinin-pancreozymin,* secreted by the duodenal mucosa, and *gastrin,* secreted by the mucosa of the stomach. All these hormones are polypeptides.

Secretin. This hormone acts on the acinar cells of the pancreas and brings about the secretion of pancreatic juice with a high water and carbonate content. It also stimulates water and bicarbonate secretion by the liver, increasing the flow of bile. It inhibits smooth muscle contraction in the stomach and intestines. Secretin is produced by the mucosal cells of the duodenum and upper jejunum. Acid chyme from the stomach is the primary stimulus for its formation and release.

Cholecystokinin-Pancreozymin. This is a single hormone with two separate and distinct actions. It stimulates (1) contraction of the gallbladder and (2) secretion of pancreatic juice rich in enzymes. The latter action was formerly attributed to a separate hormone, *pancreozymin,* but studies of purified extracts revealed that pancreozymin and cholecystokinin are identical.

Cholecystokinin-pancreozymin is a powerful stimulant of the gallbladder, initiating contractions that bring about the discharge of bile. It also relaxes the sphincter of Oddi in the bile duct. It stimulates muscle contractions in the small and large intestine and promotes blood flow to the gastrointestinal tract. It stimulates the release of insulin and acts on the stomach, stimulating the secretion of hydrochloric acid and pepsin but inhibiting gastric motility. It has been found in brain tissue and is thought to play a role in the induction of satiation. The secretion and release of cholecystokinin-pancreozymin from the intestinal mucosa results from the presence of acid chyme in the duodenum.

Gastrin. Gastrin is a polypeptide hormone (or one of a group of hormones) secreted by mucosal cells in the pyloric antrum of the stomach in response to the presence of secretagogues in food, vagal stimulation, or mechanical distention. It enters the bloodstream and is transported to cells in the fundic region, where it stimulates the secretion of all constituents of gastric juice (hydrochloric acid, pepsin, and the intrinsic factor). It has a wide range of additional actions, including stimulation of pancreatic secretion, the release of secretin from the duodenum, the release of insulin from the pancreas, inhibition of water and electrolyte absorption in the small intestine, stimulation of smooth muscle contraction in the esophageal sphincter, stomach, and other organs, and stimulation of blood flow to the stomach. Gastrin is also secreted by the duodenal mucosa.

Other Gastrointestinal Hormones. A number of other substances considered to be hormonal in nature have been identified as having their origin in the intestinal mucosa. Among them are *gastric-inhibitory peptide* (GIP), *vasoactive intestinal peptide* (VIP), *enterogastrone, motilin, chymodenin, substance P, bombesin,* and *somatostatin.* All are peptides. The evidence pertaining to them is, however, inadequate, and experimental results have in some cases proved too conflicting to merit including them, at this time, in the category of true hormones.

PINEAL GLAND

The *pineal gland* or *body,* also called the *epiphysis,* is a cone-shaped structure that projects upward from the posterior end of the roof of the diencephalon. It averages 5 to 8 mm in length and 5 mm in width and is enclosed by a capsule formed from the pia mater. Well supplied with blood vessels, its principal nerve supply is by sympathetic fibers from the superior cervical ganglion. It develops from birth to puberty, following which gradual involution occurs, and calcified material (*brain sand*) may accumulate in it. Functional activity, however, continues. The principal cells in the pineal gland are *pinealocytes,* of neuroectodermal origin, characterized by numerous microtubules. *Glia cells,* of mesodermal origin, are also present. These cells possess bulbous processes that terminate on pinealocytes or other glia cells.

Until recently, the pineal gland was considered a vestigial organ, a rudimentary third eye. However, evidence accumulated in recent years indicates that it may be an endocrine organ of importance in the regulation of the functioning of the gonads. Two substances of possible endocrine function are produced by the pineal gland: melatonin and serotonin. *Melatonin* and the enzymes that catalyze it are abundant in the pineal glands of most vertebrates. Melatonin is also present in blood plasma and in various body tissues. *Serotonin* is synthesized not only in the pineal gland but also in various other body tissues. Serotonin can be converted into melatonin.

Melatonin (MLT) is considered the primary pineal hormone. Observations in many animals confirm that melatonin production follows a circadian rhythm, more being produced at night than during the day; thus blood plasma levels of melatonin are higher at night. During the day, nervous impulses initiated by light are transmitted indirectly to the pineal gland through sympathetic nerves. These inhibit melatonin production. Melatonin production also depends on norepinephrine production, which is inhibited by light.

The role of melatonin in human physiology is not well established, but it is thought that it acts on the gonads to delay maturation and functional activity. It also acts on the adrenal cortex and the thyroid and is thought to influence such cyclic reproductive activities as ovarian and uterine changes during the menstrual cycle. The onset of puberty may depend on melatonin, since pineal tumors that inhibit the production of melatonin result in precocious puberty, and excessive production of melatonin inhibits gonadal development and functioning.

THYMUS GLAND

The *thymus gland* is a lymphatic and endocrine organ located in the upper portion of the mediastinal cavity anterior to and superior to the heart. Its structure, its relationship to the development of lymphatic organs and lymphocytes, and its role in the functioning of the immune system of the body are covered in Chapter 10, Vol. 1, pages 344–346.

Hormones of the Thymus Gland. Recent research has revealed that the thymus gland produces a number of humoral factors that play an important role in the production of lymphocytes and their acquisition of immunologic competence. The following hormones have been identified.

HOMEOSTATIC THYMIC HORMONE. Maintains immunologic competence of bone marrow cells.

THYMIC FACTOR. Stimulates intrathymic differentiation of T cells, and later stages of T-cell maturation; also increases thymic antigen activity and reactivity of marrow cells.

THYMIC REPLACING FACTOR. Restores immune responsiveness of T cell–deprived organs.

LYMPHOCYTOSIS STIMULATING HORMONES (LSH_R AND LSH_r). Stimulate the production of lymphocytes. Both LSH_R and LSH_r enhance the formation of plaque-forming cells and antibody responses.

THYMIC HUMORAL FACTOR. A polypeptide; increases mitosis; concentrations are diminished in old age, by certain diseases, by toxicities, and by radiation.

THYMOSIN. A protein; enhances maturation of T cells. Highly specific for growth and differentiation of thymus-dependent lymphoid cells. Several of its fractions stimulate secretion of the luteinizing hormone-releasing factor (LHRF). Low levels of thymosin result in Di George syndrome, characterized by T-cell deficiency. Thymosin blood levels are high in childhood, but begin to fall in the third and fourth decades and are low in the aged. Repeated injections of thymosin in patients with T-cell deficiency have produced dramatic improvements. Injection of thymosin in children with genetic forms of immunodeficiency disorder and in patients with lymphocytopenia secondary to irradiation or drug treatment for cancer have yielded promising results.

THYMOSTATIN. Inhibits incorporation of labeled nucleosides in vitro into other cell types, both lymphoid and nonlymphoid, and in vitro and in vivo incorporation of labeled nucleosides into DNA and RNA of lymphocytes.

THYMOPOIETIN. Formerly known as *thymin;* depresses neuromuscular transmission but does not appear to be involved in myasthenia gravis.

THYMOSTERIN. A steroid; inhibits tumor growth and lymphocytopoiesis.

THYMOTOXIN. Closely associated with thymopoietin.

GONADS (TESTES AND OVARIES)

The gonads, both testes and ovaries, serve a double function: (1) the production of reproductive cells (sperm and ova) and (2) the production of hormones. The hormones that induce or stimulate the development of male characteristics are *androgens;* those that stimulate the development of female characteristics are *estrogens* and *progesterone.*

The *structure, blood and lymph supply,* and *nervous innervation* of the testes and ovaries are discussed on pages 264 and 271.

Hormonal Bisexuality. Both sexes secrete both male and female hormones. In mature males, the interstitial cells of the testes produce principally testosterone but also secrete minute amounts of estrogens and progesterone. Sertoli cells are also thought to be a source of estrogens. In mature females, the ovary produces principally estrogens but also small quantities of testosterone. Testosterone is also produced in the liver from ovarian and adrenocortical steroids. In both sexes, the adrenal cortex produces sex steroids including testosterone, estrogens, and progesterone. Gonadotropic hormones produced by the anterior pituitary are identical or nearly so in both sexes.

Hormone of the Testes. The principal hormone produced by the testes is *testosterone,* a potent androgenic substance having the formula $C_{19}H_{32}O_2$. It is a sterol synthesized from cholesterol. Testosterone is produced by the interstitial cells (of Leydig), which occur singly or as small, compact masses in the angular spaces between the seminiferous tubules. The cells are irregular in shape, variable in size, and have large nuclei. They contain the usual cell organelles, and the electron microscope shows a well-developed smooth endoplasmic reticulum characteristic of steroid hormone-producing cells. Present also in the cytoplasm are protein crystals of Reinke, of unknown significance. The cells of Leydig are unique in that they are scattered throughout the testes and form a diffuse type of endocrine gland. They arise from mesenchyme instead of epithelium, as is the case with most endocrine glands. They are prominent in fetal testes due to stimulation by chorionic gonadotropins. They regress following birth but reappear in large numbers at puberty.

FUNCTIONS OF TESTICULAR HORMONES. Testosterone regulates the development of secondary sex characteristics, which appear in normal males at puberty. Among these are growth of hair on the face and in the pubic and axillary regions, enlargement of the larynx and resulting deeper voice, development of male stature and form, development of accessory glands (prostate gland and seminal vessels), and emergence of sexual libido (sex drive).

Testosterone is also essential for the normal development of seminiferous tubules in the testes and the maintenance of spermatogenesis. This is somewhat unusual, since a hormone does not ordinarily act on the organ that produces it. Testosterone also regulates the development of the penis and the scrotum and the development and functioning of all accessory reproductive structures. The development of the epididymis and its functioning in the storage and maturation of spermatozoa are also profoundly influenced.

Testosterone and other androgens produced by the testes exert metabolic effects involving nonreproductive organs, especially the skin, muscle, and bones. Androgens promote protein anabolism and play an important role in the growth changes that occur at puberty. Protein synthesis is increased and amino acid catabolism decreased. Sexual behavior patterns are affected by their secretion. Testosterone is transported in the blood by a testosterone-binding globulin (TBG) present in the plasma. It is catabolized in the liver, principally to 17-ketosteroids (17-KS), which are excreted in the urine.

EFFECTS OF CASTRATION. Removal of the testes (castration) may be necessary owing to disease or injury; in some parts of the world, it is performed as a religious rite. If the testes are removed *before* puberty, secondary sex characteristics fail to develop; hair does not grow on the face, the voice remains high-pitched, and there tends to be excess adipose tissue. Skeletal development is also altered, with late ossification of epiphyses bringing about an increase in the length of the long bones, especially those of the legs. If castration occurs *after* puberty, libido gradually lessens and disappears altogether, and there is a gradual regression of secondary sex characteristics and involution of the accessory sex glands. In the experimental castration of animals, similar changes are noted, but if testicular tissue is implanted or testicular hormones are injected, the changes will not occur.

REGULATION OF SECRETION OF TESTICULAR HORMONES. Both the spermatogenic and the endocrine functions of the testes are under the control of gonadotropic hormones secreted by the pituitary gland. These are the same gonadotropins produced in females, namely, the follicle-stimulating hormone (FSH), which stimulates spermatogenesis, and the luteinizing hormone (LH, or ICSH in the male), which stimulates development of interstitial cells of the testes and secretion of testicular androgens. There is little evidence of direct nervous regulation, but emotional and physical stress have been shown to lower testosterone levels.

Hormones of the Ovaries. The ovaries are the principal source of female sex hormones, which are produced chiefly in two structures, the *ovarian (Graafian) follicle* and the *corpus luteum.* The Graafian follicle is a spherical mass of cells in which the ovum develops; when mature, it is a large vesicle 5 to 10 mm in diameter, containing follicular liquor. The developing follicle is the principal source of estrogens. The corpus luteum is a mass of cells that develops within a ruptured Graafian follicle after ovulation. It is the source of both estrogens and progesterone. A third hormone, *relaxin,* is also produced in pregnant women.

ESTROGENS. These hormones in the female are comparable to androgens in the male in that they are responsible for changes that occur at puberty. They stimulate the growth, development, and functioning of the accessory sex organs (uterus, uterine tubes, vagina) and the development and maintenance of secondary sex characteristics. The latter include body form, especially the distribution of fat, development of the mammary glands, structure of the skeleton, texture and secretory functions of the skin and mucous membranes, distribution and quality of the hair, and pitch and quality of the voice. Estrogens are involved in the cyclic changes that occur in the uterine endometrium, especially those occurring in the first half of the menstrual cycle. They stimulate the secretion by the cervical glands of a watery mucus that facilitates the movement of sperm. Estrogens exert a negative feedback control over FSH secretion by the anterior pituitary, inhibiting its production.

Estrogens also exert many extragenital effects. They play a role in water and electrolyte metabolism and in regulation of the basal metabolic rate.

They promote protein synthesis, especially in the liver, and enzyme activities in various tissues. Estrogens, however, do not play a significant role in sexual desire or libido in the female. This depends on androgens secreted by the adrenal cortex. Estrogens are transported in the blood, bound to certain plasma proteins.

Principal Estrogens. Several estrogenic substances have been identified in the body tissues or fluids, and a number of these have been synthesized. The principal ones are *estradiol, estrone,* and *estriol.* It is believed that estradiol is the primary hormone and that estrone and estriol are derivatives of it. All three have been isolated from the placenta and from the urine of pregnant women. Most of the evidence indicates that estradiol is formed principally by follicles of the ovary, although it may be secreted by the corpus luteum, placenta, adrenal cortex, and testes. In all these structures, estrogens are synthesized from progesterone or testosterone. The liver plays an important role in estrogen metabolism, binding estrogens to blood proteins. The liver also catabolizes them, and the degradation products are excreted in the urine.

Experimental Observations. Much of the experimental data on estrogenic hormones have come from studies of their effects on the estrous cycles of rodents. In immature females or in castrated adult females, the estrous cycle and vaginal changes (determined by examination of vaginal smears) do not occur. Injection of estrogenic substances brings about estrus and correlated changes in the reproductive organs of such females, especially the uterus and the vagina. Estrus is the period of sexual excitement in animals (heat) during which mating occurs.

Chemical Composition of Estrogens. All estrogens are steroids and are similar in structure to the testicular and adrenal cortical hormones, which are also steroids. Estradiol is the most potent of them, being 4 to 8 times more effective than estrone, which, in turn, is about 10 times more effective than estriol. However, in clinical use, the potency of hormones varies greatly, depending on the method of administration and the conditions under which they are administered.

Effects of Menopause on Estrogen Production. The cessation of ovarian function (menopause) brings on a wide range of adjustments of the body, which are described briefly on page 282. Ovarian production of estrogens by the follicles ceases, and the circulating level becomes low. Some estrogens continue to be produced by the adrenal gland, but these are inadequate to maintain normal blood levels. The structural, physiologic, and psychological changes that often occur during and following the menopause are largely consequences of reduced production of estrogens.

Synthetic Estrogens. Estrogens are widely used for therapeutic and contraceptive purposes. Natural estrogens are available from a number of sources, especially the urine of equines, both male and female. However, natural hormones are rapidly degraded in the liver, and their effectiveness is limited. A number of nonsteroidal synthetic estrogenic agents have been developed, and since these are effective when taken orally, they are widely used in contraceptive pills. Among these substances are *mestranal* and *diethylstilbestrol* (DES). The latter has been widely used clinically, but its use is now greatly restricted because of its suspected carcinogenic effects.

PROGESTERONE. This steroid hormone is produced by the corpus luteum, which develops after the rupture of the mature vesicular follicle at the time of ovulation.

Functions of Progesterone. Progesterone supplements the action of the estrogens and is responsible for changes in the uterine mucosa or endometrium during the second half of the menstrual cycle. Under its influence, the glands of the uterus develop and secrete. The endometrium thickens and becomes prepared to receive the fertilized ovum. If no ovum is fertilized, the corpus luteum degenerates; there is a reduction in the amount of progesterone secreted, and in the absence of its influence, the endometrium regresses, portions undergo necrosis, and small hemorrhages occur. The discharge of portions of the endometrium together with noncoagulating blood constitutes *menstruation.* If the ovum is fertilized, the corpus luteum persists. Under the influence of progesterone, the endometrium is sensitized, and when the blastocyst enters the uterus, implantation occurs and the decidual membranes develop. The nourishment and early development of the embryo depend on the continued action of progesterone. During pregnancy, progesterone brings about enlargement of the mammary glands by stimulating the development and differentiation of the secretory alveoli.

Estrogens and progesterone act synergistically. Most of the actions of progesterone require a prior conditioning or priming action by estrogen. The following additional actions of progesterone are of significance: Progesterone has a relaxing effect on uterine muscle, inhibiting contractions that might bring about premature expulsion of the embryo or fetus. Progesterone acts on cervical glands, producing a highly viscous cervical mucus. It reduces motility in the uterine tubes and inhibits the action of prolactin on mammary secretion. Progesterone restores the vaginal epithelium to its normal condition following cornification by estrogen, and it acts on the hypothalamus to bring about a slight elevation of body temperature (0.2° to 0.6°C). As with estrogens, a number of synthetic progestogens (agents that produce progesterone-like effects) are available that are effective orally.

In addition to the ovary, progesterone is produced in the adrenal cortex, placenta, and the testes, and it is an intermediate product in the synthesis of other steroid hormones. Progesterone is transported in the blood bound to a *corticosteroid-binding globulin* (CBG) and to *serum globulin.* It is degraded mainly in the liver, and its metabolites, principally pregnanediol, are excreted in the urine. Its secretion is regulated by the luteinizing hormone (LH) secreted by the anterior pituitary.

RELAXIN. Toward the end of pregnancy, this hormone is produced by the ovaries and can be extracted from the uterus and placenta. It is also present in the blood. It is a polypeptide and acts on tissues that have been primed by estrogen. It relaxes the birth canal and softens the pelvic ligaments, especially those of the pubic symphysis, thus facilitating parturition. Its specific source is unknown.

PLACENTA

During pregnancy, the placenta secretes a number of hormones that include a chorionic gonadotropin, estrogens, and progesterone. *Human chorionic gonadotropin* (*HCG*) is a substance similar to anterior pituitary gonadotropins but differs from them in certain physiologic and chemical aspects. It appears in the blood and urine of women within two weeks after conception, for which reason its presence serves as confirmation of pregnancy when revealed by appropriate tests. Its concentration increases until the second month, after which it declines to a low level that is maintained until after parturition.

Human chorionic gonadotropin is produced by cells that cover the chorionic villi. Following conception, it causes the corpus luteum to persist and continue its secretory activity during the first three months of pregnancy. Following this period, the placenta becomes the principal source of estrogens and progesterone, which support pregnancy and the development of the mammary glands. Therapeutically, human chorionic gonadotropin is used to bring about the descent of the testes in cases of *cryptorchism,* which indicates that it may function in a similar way in the development of the male fetus. When injected into males, it stimulates the interstitial cells, increasing androgen production. In females, the amount of estrogens in the urine increases during pregnancy and then disappears after parturition. This, together with the facts that follicles are not developing in the ovary and that, in a pregnant woman, estrogens are present in the urine even after removal of the ovaries, suggests that there is an extraovarian source of estrogens, namely, the placenta. It is evident, then, that during pregnancy, the placenta takes over or supplements the endocrine functions of the ovary.

Other Hormones of the Placenta. These include *human placental lactogen* (*HPL*) and *human chorionic thyrotropin* (*HCT*).

Human placental lactogen, which is present in maternal blood during pregnancy, is similar chemically to the human growth hormone (HGH) or somatotropin; however, its actions are more closely related to prolactin. It promotes maternal utilization of lipids, conserving glucose for fetal nutrition; it promotes the development of the mammary glands during the second half of pregnancy; and it is luteotrophic, acting to maintain the corpus luteum during pregnancy.

Human chorionic thyrotropin, of placental origin, has been identified. It resembles the pituitary *thyroid-stimulating hormone* and is thought to stimulate development of the fetal thyroid gland.

SKIN

The skin is regarded as an endocrine structure because of its production of vitamin D, the antirachitic vitamin. Vitamin D_3 (cholecalciferol), the effective form of the vitamin in humans, is now classified as a hormone rather than a food supplement because it is produced endogenously, is transported by the circulatory system, and minute amounts exert specific actions on target organs.

Cholecalciferol is produced in the stratum granulosum of the epidermis of the skin by the action of ultraviolet rays on a precursor, 7-dehydrocholesterol. It or its metabolites are essential for the absorption of calcium and phosphorus from the digestive tract and for their utilization in the mineralization of bone matrix. Details of the structure of the skin are given in Chapter 4, Vol. 1. Further details concerning vitamin D are given in Chapter 7, Vol. 1.

VARIOUS TISSUES

Prostaglandins (*PG*) comprise a recently discovered group of lipids that exert hormonelike actions. They were first found in human semen, and their effects on contractility of the uterus were noted. Since then they have been found in many mammalian tissues (brain, spinal cord, kidney, lungs, spleen, thymus, placenta, and other organs). Naturally occurring prostaglandins fall into four groups, designated E, F, A, and B. Prostaglandins are synthesized from arachidonic acid, an essential fatty acid, in reactions catalyzed by enzymes collectively called *prostaglandin synthetase.* Synthesis takes place in nearly all tissues but is especially high in the kidney medulla and seminal vesicles. Prostaglandins are inactivated in the kidney cortex, lungs, and liver.

The exact role of prostaglandins in metabolism is not completely known, but they exert potent effects in nearly every organ system. Some of their actions involve smooth muscle contraction, both stimulation and inhibition, depending on the prostaglandin involved. Their action on blood vessels may raise or lower blood pressure. Prostaglandins induce contractions of the uterus and thus facilitate parturition; they may also be used as an abortifacient in therapeutic abortions. They stimulate gastrointestinal motility and secretion and have a luteolytic effect on the corpus luteum, which reduces progesterone secretion. They are thought to mediate various responses that occur in local and systemic inflammation. Systemic responses such as headache and fever are alleviated by aspirin and other analgesic and antipyretic agents that inhibit prostaglandin synthesis. In the central nervous system, they may have tranquilizing and anticonvulsive effects, and there is evidence that they serve as neurotransmitters. Prostaglandins are thought to exert their effects through the activation of adenyl cyclase, which increases cyclic adenosine monophosphate (cAMP) concentrations. With an increase in cAMP, the reaction of target cells to tropic hormones is greatly enhanced. Research on prostaglandins indicates that they may prove to be extremely useful as therapeutic agents in the treatment of various human disorders and pathologic conditions. They also hold promise of being effective contraceptives. (See Table 6-1.)

DYSFUNCTION OF ENDOCRINE GLANDS

Dysfunction of an endocrine gland is usually manifested as hyposecretion or hypersecretion.

Hyposecretion or hypofunctioning is characterized by an absence of (or a marked diminution in the amount of) hormone secreted. The effects noted are similar to those encountered in the experimental removal of the gland.

Hypersecretion or hyperfunctioning is characterized by excessive secretion of the hormone. The effects noted are similar to those observed following injection of excessive quantities of the hormone in normal animals.

Hyposecretion and hypersecretion may be of structural or functional origin. Structural changes resulting from atrophy, hypertrophy, or hyperplasia of glandular tissue or from the development of cysts or tumors may result in glandular dysfunction. Functional disturbances usually involve disorders in nervous or chemical control of glandular activity.

Table 6-2 summarizes some of the common disorders or pathologic conditions resulting from dysfunction of some of the endocrine glands.

DISORDERS OF THE THYROID GLAND

Simple Goiter. Enlargement of the thyroid is frequently due to inadequate intake of iodine, as indicated by the incidence of this type of goiter and the lack of iodine in the food. Other causes may include inflammation from infection and ingestion of specific goiter-producing substances in food.

Cretinism. This condition of retarded physical and mental growth results from imperfect development or early atrophy of the thyroid. It is characterized by dwarfism, subnormal bodily functions (lowered temperature, lower metabolic rate, slower pulse), thickened skin, and retardation in mental development and sexual maturation.

Myxedema (Gull's Disease). In the adult, hypofunctioning of the thyroid may cause myxedema, a condition characterized by a subnormal metabolic rate, lowered nervous activity, coarsened and rounded features and thickened nostrils, roughened skin, and loss of hair. Lethargy and obesity are common symptoms.

Exophthalmic Goiter (Graves' Disease). This condition is caused by hypertrophy and hyperplasia of thyroid tissue, resulting in excessive production of thyroid hormones. It is characterized by a marked increase in metabolic rate, muscular weakness, rapid heartbeat, shortness of breath, and nervous disturbances such as

TABLE 6-1 A TABULAR SUMMARY OF THE PRINCIPAL ENDOCRINE GLANDS

Glands and Hormones Secreted	*Functions*
PITUITARY GLAND (HYPOPHYSIS)	
A. *Anterior lobe*	
Somatotropin (STH) or growth hormone (GH)	Regulates growth in general; promotes protein synthesis; promotes metabolism in general
Thyrotropin (thyroid-stimulating hormone, TSH)	Stimulates secretion of thyroid hormones
Gonadotropins:	
Follicle-stimulating hormone (FSH)	Stimulates growth of ovarian follicles and secretion of estrogens
Luteinizing hormone (LH) in females	Stimulates development of corpora lutea and secretion of progesterone
Interstitial cell–stimulating hormone (ICSH) in males	Stimulates development of interstitial cells of Leydig and secretion of androgens
Prolactin (PRL) (lactogenic hormone)	Stimulation of mammary glands and initiation of milk secretion
Adrenocorticotrophin (ACTH)	Stimulates secretion of adrenal cortical steroids
B. *Intermediate lobe*	
Melanocyte-stimulating hormone (MSH)	Promotes synthesis of melanin
C. *Posterior lobe*	
Vasopressin (antidiuretic hormone, ADH)	Constricts arterioles; elevates blood pressure; promotes reabsorption of water in kidney
Oxytocin	Stimulates uterine contractions and ejection of milk from mammary glands
HYPOTHALAMUS	
Release factors for all hormones listed under pituitary gland (A and B)	Stimulate the secretion of and release of hormones by the pituitary gland
Inhibiting factors for growth hormone, prolactin, and melanocyte-stimulating hormone	Inhibit the secretion and release of these hormones by the pituitary gland
Vasopressin and oxytocin (These hormones pass by way of axons to the posterior lobe of the pituitary gland, from which they are released.) See functions under pituitary gland (C).	
THYROID GLAND	
Thyroxine	Regulates basal metabolism; is essential for normal growth and development
Calcitonin	Lowers level of blood calcium
PARATHYROID GLAND	
Parathyroid hormone	Regulates the blood level of calcium and phosphorus

TABLE 6-1 (*Continued*)

Glands and Hormones Secreted	*Functions*
ADRENAL GLANDS	
A. *Cortex*	
Glucocorticoids (hydrocortisone, cortisone, corticosterone)	Influence carbohydrate, protein, and lipid metabolism; anti-inflammatory and immunosuppressive
Mineralocorticoids (aldosterone)	Regulate sodium, potassium, and water metabolism
Sex hormones	Supplement action of ovarian and testicular hormones
B. *Medulla* Epinephrine (adrenaline) Norepinephrine (noradrenaline) Dopamine	Through the sympathetic nervous system, stimulate "fight or flight" responses; are involved in emotional adjustments; serve as neurotransmitters
PANCREAS	
Insulin	Facilitates utilization of glucose; promotes glycogen storage; inhibits glycogenolysis; lowers blood glucose
Glucagon	Acts antagonistically to insulin; increases blood glucose
GASTROINTESTINAL TRACT	
Secretin	Stimulates pancreatic secretion; increases flow of bile
Cholecystokinin-pancreozymin	Stimulates contraction of gallbladder and secretion of enzyme-rich pancreatic juice; stimulates intestinal musculature
Gastrin	Stimulates gastric secretion and other digestive actions
PINEAL GLAND	
Melatonin (MLT)	Influences cyclic reproductive activities
THYMUS GLAND Thymosin Thymopoietin (and others)	Stimulate production of lymphocytes and induce their acquisition of immunologic competence
GONADS	
A. *Testis*	
Androgens (testosterone)	Regulate the development and functioning of male accessory sex organs and development of secondary sex characteristics; influence libido
B. *Ovary*	
Estrogens (estradiol)	Regulate the development and functioning of female accessory sex organs and development of secondary sex characteristics

TABLE 6-1 (*Continued*)

Glands and Hormones Secreted	*Functions*
Progesterone	Supplements estrogens in regulation of accessory sex organs; prepares uterus for implantation; stimulates development of alveoli of mammary glands
Relaxin	Relaxes pelvic ligaments
PLACENTA	
Human chorionic gonadotropin (HCG)	Is essential for persistence and development of corpora lutea; in urine, forms basis for pregnancy tests
Estrogens and progesterone	Same as for ovarian estrogens and progesterone
Human placental lactogen	Promotes development of mammary glands
SKIN	
Cholecalciferol (vitamin D_3)	Is essential for absorption of calcium and phosphorus and their utilization in bone formation
VARIOUS TISSUES	
Prostaglandins	Act on the smooth muscle of blood vessels; stimulate uterine contractions; promote luteolysis; many other actions

tremors. Exophthalmos (protrusion of the eyeballs) is a common symptom. The follicles of the thyroid become irregular in form, and the secretory epithelium becomes folded. The gland may increase to the extent of interfering with respiration and deglutition.

Toxic Goiter (Adenomatous Goiter). This condition is attributed to a neoplasm involving the secreting portion of the thyroid. It is characterized by the presence of capsulated or nonencapsulated masses of hyperplastic thyroid tissue. Symptoms are similar to those of exophthalmic goiter except for the absence of exophthalmos and a less marked increase in basal metabolism.

DISORDERS OF THE PITUITARY GLAND

Dwarfism, Pituitary Cachexia, Acromegaly, and Gigantism. Dwarfism and pituitary cachexia (Simmond's disease) are associated with hyposecretion; acromegaly and gigantism with hypersecretion of the anterior lobe of the hypophysis.

Diabetes Insipidus. This condition, associated with dysfunctioning of the posterior lobe (especially the pars nervosa) of the hypophysis, is characterized by excessive loss of body fluids, by a great increase in the amount of urine (which, however, contains no sugar), and by polydipsia (excessive thirst). It is due to a deficiency of vasopressin (antidiuretic hormone, ADH) secreted by cells of the hypothalamus and stored in the posterior lobe.

TABLE 6-2

Gland	*Hyposecretion*	*Hypersecretion*
Thyroid	Simple goiter Cretinism Myxedema (Gull's disease)	Exophthalmic goiter (Graves' disease) Toxic goiter
Pituitary (hypophysis) Anterior lobe	Dwarfism Pituitary cachexia (Simmond's disease)	Acromegaly Gigantism Pituitary basophilism
Posterior lobe	Diabetes insipidus	
Parathyroid	Tetany	Osteitis fibrosa
Adrenal cortex	Addison's disease	Sexual precocity Virilism Cushing's syndrome Adrenogenital syndrome
Pancreas (islets of Langerhans)	Diabetes mellitus	Hyperinsulinism
Gonads (ovaries and testes)	Failure of sexual differentiation and maturation	Sexual precocity

DISORDERS OF THE PARATHYROID GLAND

Tetany. Hyposecretion of the parathyroid glands causes a rapid drop in the concentration of calcium in the blood plasma. The resulting condition, *tetany,* is characterized by intermittent tonic muscle spasms, muscle cramps, and hyperirritability of both the central and the peripheral divisions of the nervous system. Hypoparathyroidism may occur spontaneously or may follow removal of the parathyroid glands. The symptoms are relieved by injection of calcium or parathyroid hormone.

Osteitis Fibrosa. Hyperplasia of parathyroid tissue may cause hypersecretion by the parathyroids, as in the case of tumors. The accompanying increase in blood calcium and withdrawal of calcium from the bones may lead to extensive changes in bone structure and to the deposition of calcium in the soft tissues, especially the kidneys. The condition is characterized by skeletal deformities, diminished neuromuscular irritability, loss of muscle tone, and difficulties of movement.

DISORDERS OF THE ADRENAL GLAND

Addison's Disease (Chronic Adrenal Insufficiency). This disease involves atrophy and degeneration of the adrenal cortex, resulting in extreme muscular debility, fatigue, gastrointestinal disturbances, low blood pressure, nervous impairment, reduced basal metabolism, and, frequently, a remarkable pigmentation of the skin, apparent in normally pigmented areas as well as in such exposed parts as the face and hands.

Sexual Precocity, Virilism, Cushing's Syndrome, and Adrenogenital Syndrome.

Tumors involving cortical tissue may cause hyperfunctioning of the adrenal cortex, accompanied by disturbances in functions of the sex organs and changes in sexual characteristics. In fetal life or early childhood, sexual precocity may result. In females, at later stages of maturity, manifestations include enlargement of the clitoris, growth of pubic and facial hair, deepening of the voice, atrophy of the sex organs, and cessation of menstruation. The cortical changes involved occur most frequently at adolescence and during menopause, with development of male secondary characteristics and the regression of female characteristics. Hyperfunctioning of the adrenal cortex in males is associated with precocious development of male accessory organs and secondary sexual characteristics but not of the gonads.

DISORDERS OF THE PANCREATIC ISLETS OF LANGERHANS

Diabetes Mellitus. This condition, the commonest form of diabetes, is caused by the failure of the islets of Langerhans to secrete adequate amounts of insulin (hypoinsulinism) or by decreased effectiveness of insulin due to the presence of antagonists. The symptoms indicate disturbances in sugar metabolism, especially in the ability of the tissues to utilize glucose. There is a marked rise in the blood sugar level (hyperglycemia). The sugar content of the blood is normally about 100 mg (70–120 mg)/100 ml. When the level rises to 150 mg, the renal threshold is reached, sugar appears in the urine (glycosuria), and the output of urine is greatly increased (polyuria). Symptoms include intense thirst and a marked increase in water intake. Fat metabolism is altered, since the oxidation of fats depends on the oxidation of glucose. Incomplete combustion of fats gives rise to ketone bodies, and acetone appears in the urine (acetonuria). The odor of acetone may be apparent in the breath. Other ketone bodies (beta-hydroxybutyric and acetoacetic acids) are formed in excessive quantities, depleting the alkali reserve and giving rise to acidosis. Tissue proteins are depleted because of the failure of glucose to play its role as a "protein sparer" and the lack of insulin for the promotion of protein anabolism.

The development of generalized acidosis and the reduced capacity of the blood to carry carbon dioxide (air hunger) may bring on diabetic coma and death unless it is treated. Moreover, the high blood sugar level renders diabetics particularly susceptible to infections. Wounds and injuries heal slowly, there is a tendency to develop skin ulcers, circulation is impaired (especially in the extremities), and dental and visual defects are common. Disorders of large blood vessels, such as atherosclerosis, are accelerated, and the walls of capillaries become abnormally thickened. Dysfunction of the brain, spinal cord, and peripheral nerves may occur.

The exact cause of the failure of the pancreatic islets to secrete insulin is not known. Diabetes may occur at any age, but it is much more common in persons over 45. Incidence in females is higher than in males. Genetic factors play a role, since it is often a familial disease. One of the most common conditions associated with diabetes is *obesity,* which imposes an extra demand on the islets to produce more insulin to maintain normal metabolism. Certain drugs, such as adrenocortical steroids and thiazide diuretics, may be precipitating factors by creating a condition of insulin resistance. Stress such as that resulting from severe injury or illness often induces hyperglycemia.

The treatment of diabetes depends on the type and severity of the disease. Many diabetics with a mild form of the disease (*maturity-onset diabetes*) can be treated satisfactorily by diet alone. For overweight persons, weight reduction is an important part of treatment. Diet must be carefully regulated, depending on age, height, weight, activity, severity of the disease, and other factors. Severe cases (*juvenile-onset diabetes*) require, in addition to dietary modifications, the administration of insulin or an oral hypoglycemic agent. Such agents, which include sulfonylureas and biguanides, are effective in lowering the blood sugar level in mild cases or in diabetics free of ketosis and infection, but they cannot replace insulin in severe cases.

The administration of insulin does not cure the disease but merely supplies the hormone that the pancreas fails to manufacture. Since insulin utilization is determined principally by carbohydrate intake and utilization, dosage must be regulated on the basis of diet and activity of the individual; overdosage may produce insulin shock, convulsions, and coma; underdosage may cause diabetic symptoms. If the insulin injected has been combined with protamin and zinc, it is absorbed at a slower rate, fewer injections are needed, and the blood sugar level is kept more constant.

Hyperinsulinism. Oversecretion by the islets of Langerhans causes symptoms that are the opposite of those of diabetes mellitus: hypoglycemia, with extreme hunger, fatigue, muscular weakness, excessive perspiration, and pallor; nervous irritability, anxiety, and neuroses; and, in extreme cases, convulsions and coma. Immediate ingestion or intravenous injection of glucose gives temporary relief.

An overdose of insulin may induce hyperinsulinism (insulin shock). But a chronic condition may be due to tumors involving the islet tissue or to a hypersensitivity of the islets to the blood sugar level. In the latter case, a sudden rise in blood sugar increases the secretion of insulin, which in turn reduces the blood sugar level. The administration of glucose will aggravate the condition. Treatment consists of the surgical removal of tumorous tissue or, in functional hyperinsulinism, modification of the diet, which must be high in proteins and fats (to depress insulin secretion) and low in carbohydrates.

REGULATION OF BLOOD SUGAR. The blood sugar level is determined by a state of equilibrium between the addition of sugar to and the withdrawal of sugar from the blood. The liver plays a dominant role in the homeostatic mechanisms involved. Addition of sugar is by absorption from the intestine or through the conversion of glycogen stores to sugar; withdrawal is by oxidation of sugar or through the conversion of sugar into glycogen or fat. Normally the blood sugar level remains relatively constant throughout a variety of diets or total abstinence from food. Abnormal conditions can be detected by means of the *glucose tolerance test,* in which 100 g of glucose is administered after a night of fasting. In healthy subjects, the blood sugar level will rise, though rarely exceeding 150 mg, and then slowly recede to normal. In diabetics, the blood sugar level will rise more quickly, usually exceeding 150 mg, and tends to remain above normal; glucose may appear in the urine. In cases of hyperinsulinism, the blood sugar level may fall to a subnormal figure (70 mg or lower) about 4 hr after ingestion of glucose.

The concentration of glucose in the blood regulates the secretion of insulin by the islet cells. A sudden rise in blood sugar increases insulin production; as blood sugar falls, insulin production declines. The pancreas is supplied with autonomic nerve endings. Stimulation of parasympathetic fibers increases insulin output; stimulation of sympathetic fibers reduces insulin release. The effective life of insulin within the body is short (1 hr or less). It is constantly degraded in all tissues, especially the liver and kidneys, into its constituent amino acids.

It should be kept in mind that a number of hormones other than insulin affect carbohydrate, fat, and protein metabolism. Among these are glucagon, thyroxin, epinephrine, norepinephrine, and various corticosteroids, pituitary, and sex hormones. Some stimulate insulin secretion; others inhibit it. Some produce insulinlike effects in the tissues; others act antagonistically to insulin.

DISORDERS OF THE GONADS (OVARIES AND TESTES)

Hyposecretion by the Gonads (Hypogonadism). In the *male,* this results from the failure of the interstitial cells of Leydig in the testes to secrete adequate amounts of androgens. When it occurs before puberty, it produces *eunuchoidism,* characterized by failure of development of secondary sexual characteristics. Genital organs remain in an infantile condition, and libido is lacking. In young persons, administration of androgenic hormones may produce favorable results. When testicular failure occurs after puberty, physical changes are less pronounced, and regression occurs slowly.

In the *female,* hypogonadism results from the failure of the ovaries to develop and produce estrogens and progesterone. When it occurs before puberty, secondary sexual characteristics, especially the breasts, fail to develop, the onset of menstruation is delayed or does not occur, sex organs remain infantile, and growth may be retarded.

In adult females, hypogonadism occurs naturally at the menopause when the ovaries cease producing their hormones. The physical and physiologic changes that accompany the menopause are described on page 282. It may also result from surgical removal of the ovaries, inactivation of ovarian tissue by irradiation, the action of certain drugs, or various pathological disorders. Reduced secretion of ovarian hormones profoundly affects changes that occur in the uterine endometrium during the menstrual cycle. Progesterone is essential at the beginning and during the early months of pregnancy, since it is necessary for implantation of the blastocyst and for development of the embryo or fetus. Reduced secretion usually results in abortion or miscarriage.

Hyposecretion by the gonads in both sexes may result from inadequate production of pituitary gonadotropins and growth hormones. This may result from congenital maldevelopment, chromosomal disorders, or unknown causes. Administration of sex hormones is sometimes effective in correcting the condition.

Hypersecretion of Androgens and Estrogens. Excessive secretion of these hormones during early development results in *sexual precocity,* characterized by acceleration of growth and development and the early appearance of secondary sexual characteristics. Epiphyseal fusion of bones occurs, often bringing growth to an end before adult height is reached. Precocious puberty is marked in males by the early development of the external genital organs and early production of spermatozoa; in females by early ovulation, the development of the mammary glands, and the initiation of menstrual cycles. Sexual development before the age of 10 in males and 8 in females is considered precocious. *Hypergonadism* is usually the result of excessive hypothalamic-hypophyseal activity.

7: THE REPRODUCTIVE SYSTEM

The reproductive system comprises the organs concerned with the production of new individuals of the same biologic variety, that is, with the perpetuation of the species. These organs include the primary sex organs (testes and ovaries), which produce the sex cells (spermatozoa and ova), and the accessory sex organs, including a number of structures concerned with bringing the sex cells together and protecting and nourishing the developing embryo.

In addition to its reproductive functions, the reproductive system exerts widespread effects on the entire life of the individual through hormones produced by the sex glands. These hormones influence bodily development and behavior—in fact, the whole psychosomatic makeup of an individual.

THE PRINCIPAL REPRODUCTIVE ORGANS

The primary sex organs are the testes in the male and the ovaries in the female. There are a number of accessory sex organs.

Reproductive Organs in the Male. The principal organs of reproduction in the male are:

1. The two *testes,* which produce germ cells (*spermatozoa*). They are suspended from the body wall by a *spermatic cord* and enclosed in a sac, the *scrotum.*
2. A *duct system* for conveying spermatozoa to the outside. This includes the *epididymus,* two seminal *ducts* (sing., *ductus deferens,* also called *vas deferens*), two *ejaculatory ducts,* and the *urethra.*
3. *Accessory glands* (two seminal vesicles, a prostate gland, and two bulbourethral glands), which contribute to the formation of seminal fluid, the vehicle for spermatozoa.
4. The *penis,* a copulatory organ by which spermatozoa are transferred to the female.

Reproductive Organs in the Female. The principal organs of reproduction in the female are:

1. The two *ovaries,* which produce germ cells (*ova*).
2. Two *uterine tubes,* by which ova are transported from the ovaries toward the uterus. Within these tubes, fertilization of the ovum may occur if spermatozoa are present, in which case the developing *zygote* is transported to the uterus.
3. The *uterus* or *womb,* in which the embryo or fetus develops.

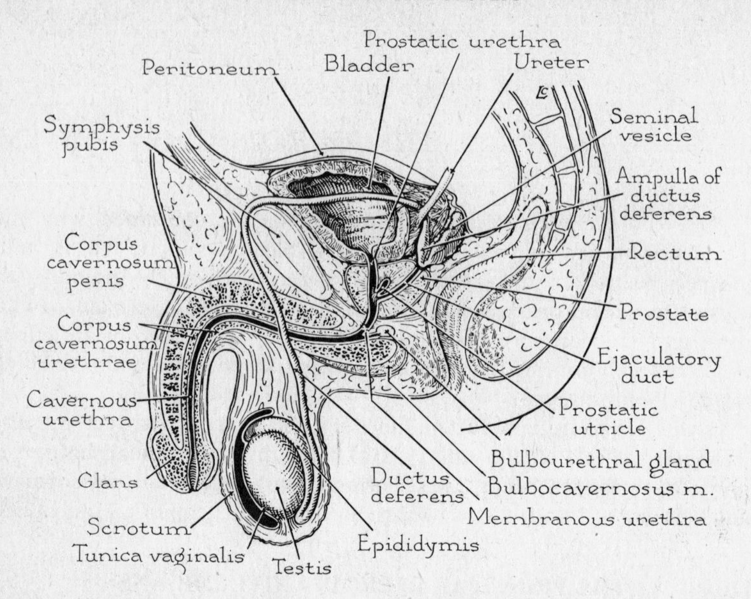

Fig. 7-1. Diagrammatic sagittal section of the male pelvis. (After Turner, *General Endocrinology.*) (Reprinted with permission of W. B. Saunders Co., Philadelphia, from B. G. King and M. J. Showers, *Human Anatomy and Physiology,* 6th ed., 1969.)

4. The *vagina,* which serves as both a copulatory organ in which spermatozoa from the male are deposited and a birth canal through which the fetus passes at birth.
5. The *external genitalia* or *vulva,* composed of the labia majora, labia minora, clitoris, and vestibule.

THE MALE REPRODUCTIVE SYSTEM

The male reproductive system (Fig. 7-1) comprises the following organs: the testes, the duct system, accessory glands, and the penis.

Testes. The *testes,* the primary organs of reproduction in the male, serve a double function. They are the source of *spermatozoa,* male gametes or reproductive cells, and of *androgens,* male hormones. Their function as endocrine organs is discussed on pages 249–250.

The testes lie outside the body cavity in a sac, the *scrotum.* The testes, scrotum, and penis constitute the external genitalia of the male. Each testis is an ovoid body about 4 cm long, 2.5 cm wide, and 2 cm thick. The substance of the testis is divided into *lobules,* each testis containing about 250, separated from each other by partitions of connective tissue called *septa.* Each lobule contains from one to four minute and extremely

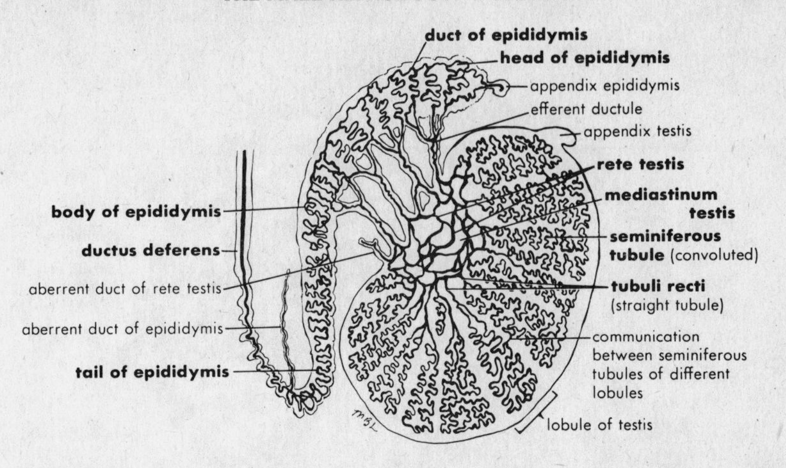

Fig. 7-2. Tubules and ducts of the testis. (Reprinted with permission of Lea & Febiger, Philadelphia, from J. E. Crouch, *Functional Human Anatomy,* 1965.)

convoluted *seminiferous tubules.* These tubules converge at the apex of the lobule and, joining others, form short *straight tubules.* These lead to a plexus, the *rete testis,* from which 15 to 20 *efferent ducts* lead to the epididymis.

COVERINGS OF THE TESTES (Fig. 7-2). Each testis is invested by three membranes. The innermost, the *tunica vasculosa,* a delicate membrane containing blood vessels, covers the surface and continues into the testis, covering the septa. External to it is the *tunica albuginea,* a dense fibrous membrane that is thickened on the posterior border to form an incomplete septum, the *mediastinum testis,* which supports vessels entering and leaving the testis. The outermost membrane is the *tunica vaginalis,* a serous pouch formed from an outpocketing of the peritoneum of the body wall, which was pushed ahead of the testis in its descent. It forms a closed sac enveloping the testis and consists of a *visceral* and a *parietal layer* separated by a serous cavity. Within it, the testis can move freely.

FORMATION OF SPERMATOZOA (Fig. 7-3). Sperm are formed in the seminiferous tubules. Each tubule is lined with epithelium consisting of two types of cells: *spermatogenic cells* and *sustentacular cells* (*cells of Sertoli*). The spermatogenic cells are of three types: *spermatogonia, spermatocytes,* and *spermatids,* which represent the successive stages in development up to the final form of the spermatozoon. The mature spermatozoa lie on the inner surface of the tubule with their heads embedded in Sertoli cells. Sertoli cells provide mechanical support and nutrients for developing spermatozoa. They also phagocytize degenerating sperm and the residual portions of spermatids that are cast off in the maturation of sperm cells. In response to FSH, they synthesize an androgen-binding protein that,

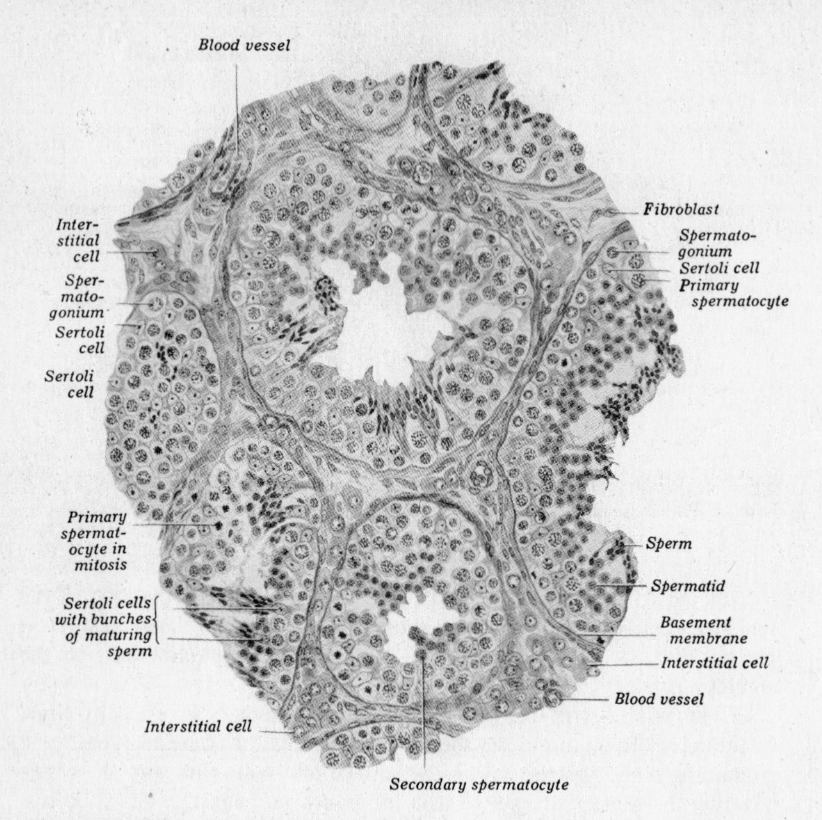

Fig. 7-3. Section of testis showing interstitial cells and stages of spermatogenesis. (Reprinted with permission of W. B. Saunders Co., Philadelphia, from W. Bloom and D. W. Fawcett, *Histology,* 10th ed., 1975.)

within testes tubules, transports androgens to the epididymis, an androgen target organ. They also serve as a blood-testis barrier.

Following maturation, the spermatozoa are discharged into the lumen of the tubule, through which they pass to the straight tubules. They continue through the rete testis and the efferent ducts of the testis to the epididymis, where they are stored.

The entire sequence of events by which spermatogonia give rise to mature spermatozoa is called *spermatogenesis.* Through mitotic division, spermatogonia form two types of cells: stem cells and derivative cells. *Stem cells* divide and perpetuate themselves, giving rise to future spermatogonia and derivative cells. Spermatogonia thus persist indefinitely, enabling sperm production to continue without limit. There is no specific time when the formation of spermatozoa ceases. *Derivative cells* give rise to spermatocytes, which develop successively into spermatids and mature spermatozoa. The details of the process are given in Chapter 8, pages 292–294.

BLOOD AND LYMPH VESSELS OF THE TESTES. Each testis is supplied by a *testicular (spermatic) artery,* which arises from the abdominal aorta near the kidney. It descends, passes through the inguinal canal, and, with other components of the spermatic cord, enters the scrotum, where it becomes tortuous and divides into several branches supplying the testis and epididymis. Capillaries connect with the *testicular (spermatic) vein,* which, on leaving the testis, forms a convoluted *pampiniform plexus* in the spermatic cord. The vessels of the plexus enclose the testicular artery and serve as a *thermoregulatory mechanism* for cooling the blood going to the testis and warming the blood before its reentry into the body cavity. Vessels from the plexus ascend in the spermatic cord and, after passing through the inguinal canal, form a single *testicular (spermatic) vein,* which on the right side enters the vena cava and on the left side enters the left renal vein.

Lymphatic vessels draining the testis and epididymis include several vessels that accompany the testicular veins. They terminate in the lower lumbar lymph nodes.

NERVE SUPPLY OF THE TESTIS. The testis receives nerve fibers from both sympathetic and parasympathetic divisions of the autonomic nervous system. Sympathetic fibers originating principally from the aortic plexus terminate in blood vessels of the testis. In the epididymis, they terminate in blood vessels and in the smooth muscles of the walls of the epididymal ducts. Terminations of parasympathetic fibers are uncertain. Visceral afferent fibers carry sensory impulses from the testis. Injuries, as from a severe blow, or swelling, as in orchitis, cause intense pain.

DESCENT OF THE TESTES; SPERMATIC CORD. In the embryo, each testis originates in a genital ridge of mesodermal tissue located close to the kidney. During the final three months of fetal development, the testes migrate downward through the ventral abdominal wall and take their position in the scrotum. In their descent, each testis carries with it the ductus deferens, blood and lymphatic vessels, and nerves. These structures together with their covering form the *spermatic cord.* The canal through which they pass is the *inguinal canal,* which has an internal opening, the *internal* or *abdominal inguinal ring,* and an external opening, the *subcutaneous inguinal ring.* The covering of the cord consists of two layers of *fascia,* which contain fibers of the *cremaster muscle* that terminate on the surface of the testis. Contraction of this muscle elevates the testis, bringing it close to the abdominal wall.

The development of normal spermatozoa depends on the migration of the testes from the body cavity into the scrotum, where the temperature is about 2°C lower. When the testes remain in the body cavity, a condition called *cryptorchism,* spermatogenesis fails to occur, and fertile spermatozoa are not produced. The production of androgens by the interstitial cells of Leydig, however, is not impaired.

Duct System in the Male. The system of ducts that serves the male reproductive organs includes the epididymis, the ductus (vas) deferens, the ejaculatory duct, and the urethra. All except the urethra are paired.

EPIDIDYMIS. Along the posterior and superior borders of each testis lies an elongated structure, the *epididymis.* It consists of an enlarged portion or *head,* on the superior border of the testis, and a *tail,* lying along the posterior inferior surface. Within the epididymis are 12 to 14 *efferent ducts.* These leave the rete testis as almost straight tubes, but each becomes greatly coiled before emptying into a common duct, the *ductus epididymidis.* This descends in the tail of the epididymis and becomes continuous with the ductus deferens. The ductus epididymis is much coiled and averages 6 to 7 m in length. It is lined with pseudostratified epithelium consisting of small basal cells and tall columnar cells. On the free surfaces, nonmotile processes (*stereocilia*) project. These pass secretions from the cells into the lumen. A circular layer of smooth muscle fibers is present in its wall.

DUCTUS OR VAS DEFERENS. This duct is a continuation of the ductus epididymis. From the tail of epididymis it passes upward within the spermatic cord and enters the body cavity through the inguinal canal, then crosses the brim of the pelvis. It continues posteriorly alongside the lateral wall of the bladder, loops over the distal end of the ureter, and turns abruptly downward. This portion is slightly enlarged, forming an *ampulla.* At this point a lateral sacculation, the *seminal vesicle,* connects with the ductus deferens. Below the ampulla, the duct narrows to form the *ejaculatory duct.*

The ductus deferens is lined with pseudostratified epithelium resting on a lamina propria rich in elastic fibers. Its lining is thrown into folds of moderate height. The tall columnar cells of the epithelium bear stereocilia, except in the region of the ampulla. The middle muscular coat consists of three layers of smooth muscle fibers. The fibers are disposed longitudinally in the inner and outer layers, circularly in the middle layer. The thick muscular coat gives to the ductus deferens a hard, cordlike feel when palpated. The outermost layer of the ductus deferens is the *adventitia,* consisting principally of elastic fibers. The ductus deferens averages about 30 cm in length. The lumen of the duct is small.

EJACULATORY DUCT. This duct is formed by union of the ductus deferens and the duct from the seminal vesicle. It pierces the posterior surface of the prostate gland, passes downward through its substance, and enters the prostatic portion of the urethra, opening on the *colliculus seminalis.*

URETHRA. This common urinary and reproductive duct leads from the bladder to the outside. It consists of three portions: the *prostatic,* the *membranous* (which pierces the pelvic wall), and the *cavernous* (which traverses the penis). Its external orifice is on the tip of the glans penis. (These portions are described in Chapter 1.)

Accessory Reproductive Glands in the Male. These include the seminal vesicles, the prostate gland, and the bulbourethral (Cowper's) glands.

SEMINAL VESICLES. These paired structures lie against the lower posterior surface of the bladder immediately above the prostate gland. Each is a sacculated structure averaging about 5 cm in length and 2 cm in width. A seminal vesicle consists essentially of a tube bearing blind diverticula and folded on itself many times. Its duct joins the ductus deferens to form the ejaculatory duct.

The seminal vesicles are primarily secretory in function. They do not store sperm. They produce a thick, viscid fluid, which is added to the sperm coming from the testes. The secretory product contains prostaglandins; fructose, a sugar that provides energy for spermatozoa; and a proteinaceous material essential for coagulation of semen. It is a yellow, alkaline fluid of sticky consistency.

PROSTATE GLAND. The prostate is a lobular structure that surrounds the urethra close to its origin from the bladder. It consists of glandular and muscular tissue enclosed in a thin, fibrous capsule. The glandular tissue consists of some 30 to 40 compound alveolar glands whose ducts, 16 to 32 in number, open into the prostatic portion of the urethra on each side of the *colliculus seminalis.* The glands are embedded in a stroma of connective and smooth muscle tissue.

The glands of the prostate are of three types: *mucosal glands,* which lie in periurethral tissue; *submucous glands,* arranged in a ring about the mucosal glands; and *prostatic glands proper,* which constitute the bulk of the gland and provide the major portion of its secretion. The product of the prostate gland is a thin, slightly alkaline (pH 7.5), opalescent fluid that is responsible for the odor of semen. The secretion contains citric acid, zinc, magnesium, acid phosphatase, and a number of other enzymes. It may also contain *prostatic concretions,* small spherical or ovoid bodies that are frequently calcified. Larger concretions are commonly found in the gland itself, especially in cysts.

BULBOURETHRAL (COWPER'S) GLANDS. These are two small tubuloalveolar glands, each about the size of a pea, lying alongside the membranous portion of the urethra. Each is drained by a duct 3 to 4 cm in length that empties into the bulb of the cavernous portion of the urethra. Their secretion is a clear, viscid, mucuslike substance, which serves as a lubricant.

The Penis. The penis is a cylindrical organ composed principally of erectile tissue. It consists of a root, a body, and the glans penis. The *root* is the portion by which the penis is attached to the body. At the proximal end of the penis, the two *corpora cavernosa penis* diverge to form the *crura.* Each crus passes abruptly laterally and downward alongside the inferior ramus of each pubis to which it is attached. The proximal portion of the *corpus cavernosum urethrae* (*corpus spongiosum*) is enlarged to form the bulb, which is attached to the lower surface of the urogenital diaphragm. A *suspensory ligament* connects the sheath of the corpora cavernosa penis to the symphysis pubis.

The *body* of the penis consists of three cylindrical masses of erectile tissue, known as the *corpora cavernosa.* Two of these masses, the *corpora cavernosa penis,* lie alongside each other on the dorsum of the penis. The third, the *corpus cavernosum urethrae,* is medially located and is traversed throughout its entire length by the urethra. The corpora cavernosa contain large blood spaces, which upon sexual stimulation become engorged with blood, bringing about erection of the penis. Each of these masses is enclosed by a dense fibrous membrane, the *tunica albuginea.*

The *glans penis,* the cone-shaped distal end of the penis, is a continuation of the corpus cavernosum urethrae. Its rounded posterior border is the *corona glandis,* behind which is a slightly narrower *neck.* At its tip lies the *meatus* or urethral orifice.

INTEGUMENT OF THE PENIS. The penis is covered by skin that is continuous with that of the scrotum. The skin, which is hairless, is loosely attached over the body of the penis, thus permitting free movement. Just behind the corona, the skin forms a fold, the *prepuce,* that extends distally, covering the glans. The extent of this covering varies. In children and in some adults, the glans may be completely covered (when excessive, it is called "redundant foreskin").

The prepuce is attached to the glans on its inferior surface by a median fold, the *frenum.* The skin on the *inner* surface of the prepuce and in the region of the neck behind the glans penis contains modified sebaceous glands (*glands of Tyson*). These glands secrete a substance that, together with shed epithelial cells, forms a cheeselike waste material called *smegma.* This material may accumulate under the prepuce and be a source of irritation.

CIRCUMCISION. This is the process of surgically removing a sufficient portion of the foreskin to permit its free retraction over the glans. It may be performed as a religious rite or for medical reasons. It is resorted to for purposes of cleanliness and to prevent the accumulation of smegma beneath the prepuce. Circumcision is necessary when the opening in the prepuce is so constricted as to prevent retraction of the prepuce, a condition called *phimosis.*

ERECTION OF THE PENIS. The penis may undergo erection even before puberty. It occurs during sexual excitement and is normally involuntary in nature, though it can be voluntarily induced by stimulation of various structures, particularly the sex organs, or through activities of the higher nervous centers. The process is a reflex act during which vasodilatation of the arteries supplying the erectile tissue occurs and the spaces in the corpora cavernosa become filled with blood. Constriction of the veins reduces the outflow of blood. The resultant turgidity of the cavernous bodies causes the penis to become rigid and to assume an erect position.

Ejaculation of Seminal Fluid. Seminal fluid or semen is ejaculated during sexual intercourse (*coitus*) or in states of sexual excitement. It may occur involuntarily at night during sleep (*nocturnal emission*) or may be artificially induced (*masturbation*).

The *semen* is composed of *spermatozoa* and *seminal plasma.* The latter consists of secretions from the epididymides, seminal vesicles, prostate gland, and bulbourethral glands. Seminal plasma comprises about 90 percent of the seminal fluid. While in the seminiferous tubules, the sperm are nonmotile; they are moved along by the flow of gradually increasing quantities of the fluid in which they are suspended. The sperm pass through the straight tubules and rete testis into the efferent ducts of the testes, where cilia propel them to the epididymis. They pass slowly through the epididymis and may be retained there for a long time. Smooth muscles in the tubules are probably the principal factor in their passage through this region. Secretions from the tubule cells nourish the sperm, and their development is completed here.

In the ductus deferens, the spermatozoa are moved rapidly by contractions of the muscles in its walls. Ejaculation is brought about by contractions of the bulbocavernous muscle and smooth muscle of the prostate gland in conjunction with the peristaltic contraction of muscles in the excretory ducts, especially the ductus deferens. An average ejaculation amounts to about 2 to 3 cm^3 and contains from 200 to 400 million spermatozoa. The sperm in the ejaculate are very active, achieving locomotion by lashing movements of their tails. They may retain their motility for several days, but their ability to fertilize an egg is limited to approximately 24 hr. *Capacitation,* the acquisition of the ability to fertilize an ovum, occurs in the vagina or uterus. It involves biochemical, physiologic, and structural changes in the sperm.

THE FEMALE REPRODUCTIVE SYSTEM

The *female reproductive system* (Fig. 7-4) comprises the following organs: the ovaries, uterine tubes, uterus, vagina, and external genitalia. These organs and, in addition, the mammary glands are described in the following pages.

Ovaries. These primary organs of reproduction are paired organs lying close to the lateral walls of the pelvic cavity. Each averages 3 to 5 cm in length and 2 to 3 cm in width and weighs 6 to 8 g. They are held in position by the *mesovarium,* a mesentery that connects with the broad ligament of the uterus; the *ovarian ligament,* which connects with the uterus; and the *suspensory ligament,* which attaches to the pelvic wall. Each ovary is supplied by branches of the ovarian and uterine arteries in the mesovarium. Veins form a pampiniform plexus that drains into the ovarian vein. Nerves and lymph vessels accompany the blood vessels.

The ovaries serve a double function. They are the source of *ova,* female reproductive cells or gametes, and of *female hormones* (estrogens, progesterone, and relaxin).

MICROSCOPIC STRUCTURE OF THE OVARIES. Each ovary has a central

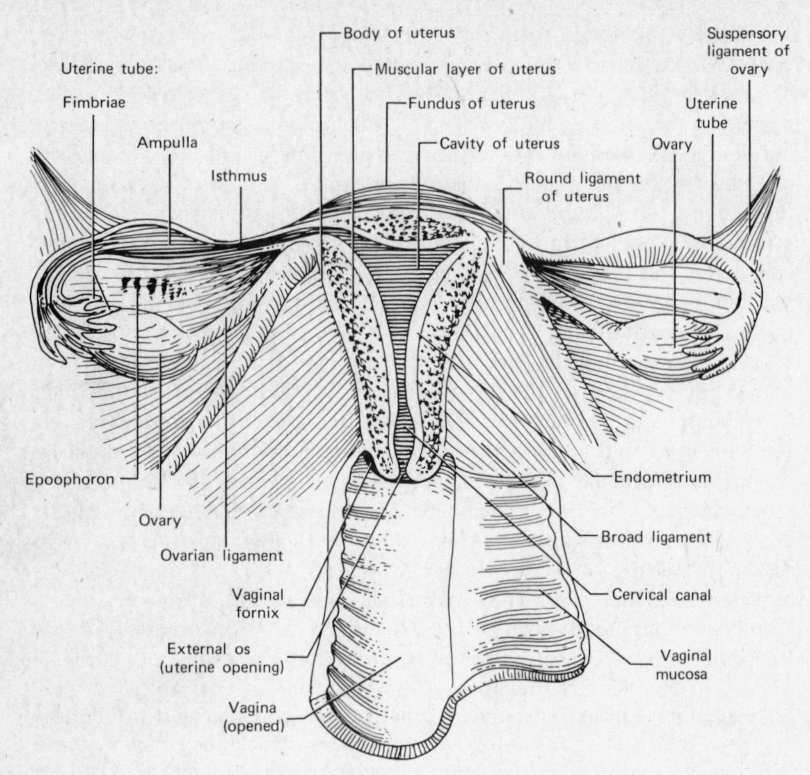

Fig. 7-4. The internal female reproductive organs. (Modified from E. B. Steen, *Laboratory Manual and Study Guide for Anatomy and Physiology,* Brown, 1963.) (Reprinted with permission of John Wiley & Sons, Inc., from E. B. Steen and J. H. Price, *Human Sex and Sexuality,* 1977.)

portion, the *medulla,* and an outer layer, the *cortex.* The *medulla* consists of a stroma of loose connective tissue containing blood vessels, lymphatic vessels, and some smooth muscle fibers. *Interstitial cells* derived from atretic follicles may be present. The *cortex* consists of a connective-tissue stroma in which are located *follicles* in various stages of development. The outermost layer of the cortex is the *germinal epithelium,* a single layer of cuboidal epithelial cells that covers the free surface of the ovary. A dense layer of connective tissue lying immediately beneath the germinal epithelium is the *tunica albuginea.* Follicles may be *primary, growing,* or *mature.* A corpus luteum or corpus albicans may also be present.

DEVELOPMENT OF THE OVUM AND FOLLICLE (Fig. 7-5). A follicle consists of an *ovum* (or more strictly speaking, an *oocyte*) surrounded by a layer or several layers of *follicle cells.* The number of follicles present is large, 200,000 or more in a single ovary in the mature female. They are present

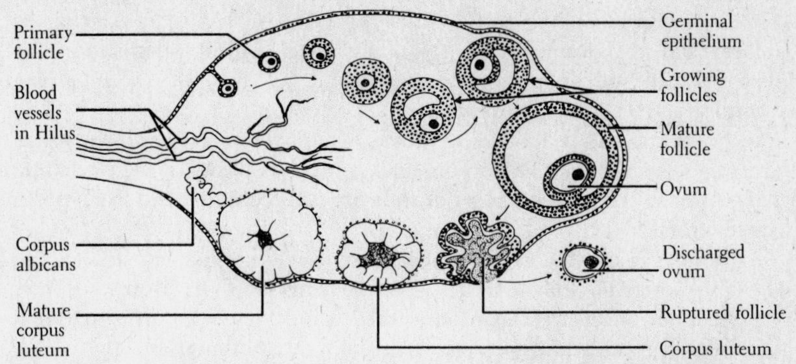

Fig. 7-5. Ovary showing development of follicle and corpus luteum. (From E. B. Steen, *Laboratory Manual and Study Guide for Anatomy and Physiology,* 3d ed., Brown, 1976.)

at birth, from which time they decrease in number until at menopause few, if any, remain. The decrease is brought about by an involutional process, *atresia.*

Primary Follicles. These are by far the most numerous. They lie near the periphery just beneath the tunica albuginea. Each consists of an ovum surrounded by a layer of flattened follicle cells.

Growing Follicles. As the follicle grows, the ovum increases in size, and yolk granules appear. The follicle cells become cuboidal and proliferate, forming several layers around the ovum, from which they now become separated by a clear membrane, the *zona pellucida.* As development proceeds, the follicle increases in size, and irregular spaces filled with follicular fluid appear between the follicle cells. Later, these spaces coalesce and form a single cavity, the *antrum.* The follicle is now *mature* and is known as a *vesicular* or *Graafian follicle.* The ovum occupies a position at the periphery where it is embedded in a mass of cells, the *cumulus oophorus.* The connective tissue surrounding the follicle becomes differentiated into a capsule, the *theca,* which is differentiated into two layers, an inner, highly vascularized *theca interna* and an outer *theca externa.*

Mature Graafian Follicles. The cavity within the follicle increases in size, eventually occupying the entire thickness of the cortex and even protruding from the surface. The human follicle reaches a diameter of 1 to 2 cm and when fully developed occupies about one-fourth of the volume of the ovary. The ovum also increases in size, attaining a diameter of about 0.15 mm. It has a large vesicular nucleus with a conspicuous nucleolus. The tall columnar follicle cells around the ovum are arranged in a radial fashion and form the *corona radiata* (*discus proligerus*). The ovum is attached to the follicle wall by a narrow strand of follicle cells.

At this point the follicle may either (1) undergo involution (retrogression) or (2) rupture, liberating its contained ovum. The latter process is *ovulation.*

HORMONAL CONTROL OF OVARIAN ACTIVITY. The development of the follicle and the corpus luteum is under the control of gonadotropic hormones of the anterior pituitary. These are the *follicle-stimulating hormone* (*FSH*), which stimulates the development of immature follicles into mature Graafian follicles, and the *luteinizing hormone* (*LH*), which acts synergistically with FSH to bring about ovulation. Following ovulation, LH stimulates the development of the corpus luteum and the secretion of progesterone.

Ovulation is the discharge of the ovum from the Graafian follicle. At the point where the follicle bulges from the ovary wall, the stroma becomes very thin and vascular. This region is called the *stigma,* and it is here that the follicular wall undergoes lysis (loosening or thinning) and the ovum, with its corona radiata and follicular fluid, slowly flows from the follicle out into the peritoneal cavity. This emergence is often incorrectly described as an "eruption" or "bursting" through the ovarian wall, but neither term, nor "rupture," provides the best description of the process, which is actually rather a gentle one.

The actual precipitating factor in ovulation is unknown, but the following have been suggested as possible causes: (1) increase in intrafollicular pressure, (2) enzyme activity, (3) action by the fimbria, (4) vascular changes in the stigma, and (5) contraction of smooth muscle fibers within the ovary or follicular wall. Ovulation follows an abrupt rise in the levels of the pituitary hormone, LH and FSH.

In the human female, ovulation occurs on an average once every 28 days from the time of puberty to the menopause, except during periods of pregnancy, when no follicles are developed and ovulation ceases. Ovulation may occur alternately in each ovary, or a single ovary may liberate ova several times in succession. Usually a single ovum is discharged, but two or more from different follicles may be liberated. Ordinarily a follicle contains a single ovum, but it may hold two or more.

Involution or *atresia* is the process in which follicles degenerate and disappear. Of the 400,000 follicles present in the ovaries of a female at birth, only about 400 mature and discharge their ova; the remainder undergo atresia. This may occur at any stage in the development of a follicle, and it may occur at any stage of life of the woman. Atresia of follicles begins during intrauterine development and is most active between the years of birth and puberty. It continues during active sexual life until menopause, at which time follicles cease to develop, though occasionally a few may persist until old age. Follicles undergoing atresia become incorporated into cells that constitute an *interstitial gland* that is thought to synthesize androgens and other steroids.

EVENTS FOLLOWING OVULATION. After ovulation has occurred, the ovum enters the fimbriated end of the uterine tube. If coitus has taken place, sperm are usually present in the tube, and fertilization occurs. The fertilized ovum or *zygote* now begins development, which continues as it passes through the uterine tube into the uterus. Here *nidation* (*implantation*) in the uterine mucosa takes place on about the eighth or ninth day. If no sperm are present, the ovum is not fertilized and fails to develop. It disintegrates in its passage through the uterine tube.

DEVELOPMENT OF THE CORPUS LUTEUM. Upon discharge of the ovum and the liquor folliculi at ovulation, the wall of the follicle collapses, and its inner layer, the *theca interna,* becomes folded. The rupture of capillaries causes a small amount of blood to accumulate in the follicular cavity, forming the *corpus hemorrhagicum.* Cells from the theca interna enlarge, and the number of cell layers increases. *Lutein,* a yellowish lipoidal substance, fills the cytoplasm of these cells. These *lutein cells* form a mass, the *corpus luteum,* which fills the follicular cavity and reaches the peak of its development about 12 to 14 days after ovulation.

If the ovum is not fertilized, this body is known as the *corpus luteum of menstruation,* which, after a period of about 14 days, begins to degenerate. The lutein cells decrease in size and are replaced by connective tissue. After a period of a few weeks, a small white body, the *corpus albicans,* is all that remains where the follicle was located; this body may persist for months.

If the ovum is fertilized, the corpus luteum does not degenerate but persists as the *true corpus luteum.* About the third month of pregnancy, degenerative changes begin, which proceed throughout the remainder of pregnancy, and the corpus luteum slowly regresses.

BLOOD AND NERVE SUPPLY OF THE OVARIES. The ovary receives blood through the ovarian artery and a branch of the uterine artery. Veins emerge from the hilus as tortuous vessels that unite with those from the uterus to form a venous complex, the *pampiniform plexus.* From this plexus, blood leaves by the ovarian or uterine veins. The right ovarian vein empties into the vena cava, the left into the left renal vein; the uterine veins empty into the internal iliac veins.

Efferent lymphatic vessels convey lymph to the lumbar lymph nodes.

Nerves from both divisions of the autonomic nervous system innervate the ovaries.

Uterine Tubes (Fallopian Tubes, Oviducts). These two tubes convey the ova that are discharged by the ovary toward the uterus. Each tube lies in the upper portion of the broad ligament and averages about 12 cm in length and 1 cm in diameter. The proximal end of each tube opens into the uterus, the distal end into the pelvic portion of the peritoneal cavity by an opening, the *ostium.* The wide, expanded portion that surrounds the ostium is the *infundibulum.* Its border has a fringed appearance, bearing many processes called *fimbriae;* one of these, the *fimbria ovarica,* is much longer than the others, reaching nearly to the ovary. The infundibulum lies close to and may partially enclose the ovary, but there is no direct connection between the two; a small peritoneal space always intervenes. The portion of the tube into which the infundibulum opens is the *ampulla.*

Near the uterus the tube becomes straight and narrow to form the *isthmus*, which continues through the wall of the uterus as the *intramural portion.*

MICROSCOPIC STRUCTURE OF THE UTERINE TUBES. The wall of a uterine tube consists of three layers: mucosa, muscular layer, and serosa. The *mucosa* or innermost layer consists of columnar epithelial cells resting on a thin lamina propria of connective tissue. Some of the epithelial cells are ciliated; others are glandular. The cilia beat toward the uterus. The mucosa is thrown into numerous folds that bear secondary and tertiary folds. The folds are numerous in the ampulla, but in the isthmus they are few in number. The *muscular layer* consists of an inner circular and an outer longitudinal layer. These are not clearly separated from each other, for some fibers of the circular layers are directed in a spiral fashion. In the region of the fimbria, the longitudinal layer is lacking. The *serosa* or outer coat consists of connective tissue underlying the outermost layer of *peritoneum.* The connective tissue is continuous with that of the *broad ligament.*

FUNCTION OF THE UTERINE TUBES. After the ovum has left the ovary, it enters the fimbriated end of the uterine tube. At the time of ovulation, vascular changes in the fimbriae cause them to become enlarged and turgid. This, together with contraction of muscular fibers, brings the ostium in close proximity to the ovary. When the ovum is discharged, currents produced by the action of cilia cause the ovum to be drawn into the oviduct. Further transport of the ovum is accomplished primarily by the peristaltic muscular contractions of the oviduct. If the ovum is not fertilized, it disintegrates before reaching the uterus.

However, if spermatozoa are present in the uterine tube, *fertilization* may occur, usually taking place in the fimbriated end. Following fertilization, the *zygote* (fertilized egg) begins development, undergoing early cleavage and transformation into a *blastocyst* as it passes through the tube. The trip to the uterus usually requires 4 or 5 days.

Sometimes the egg following fertilization fails to reach the uterus and develops in an abnormal location. Such a condition, called an *ectopic* or *extrauterine pregnancy,* occurs about once in every 300 pregnancies. Development may occur within the uterine tube (*tubal pregnancy*), within a ruptured follicle of the ovary (*ovarian pregnancy*), or within the body cavity (*abdominal pregnancy*). Tubal pregnancy may result from failure of the tube to transport the developing zygote or from obstruction of the tube as may occur following inflammation. Since the uterine tube is unable to accommodate the developing embryo, rupture invariably occurs, often with serious effects from ensuing hemorrhage. Ectopic pregnancies usually necessitate an abdominal operation to remove the conceptus, and only in exceptional cases does such a pregnancy ever go to full term.

Uterus. This muscular, pear-shaped organ lies in the pelvic cavity between the bladder and the rectum. It averages 6 cm in length, 4.5 cm in width, and 2.5 cm in thickness. It is slightly flattened anteroposteriorly.

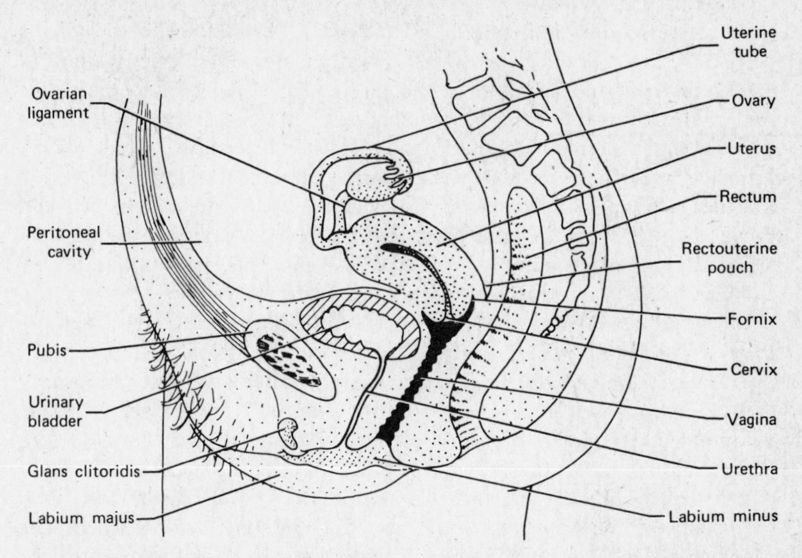

Fig. 7-6. Diagrammatic sagittal section of the female pelvis showing reproductive organs and related structures. (From E. G. Steen, *Laboratory Manual and Study Guide for Anatomy and Physiology,* 3d ed., Brown, 1976.)

During pregnancy it increases in size to accommodate the developing fetus.

REGIONS OF THE UTERUS. The uterus consists of three regions: body, isthmus, and cervix. The *body* forms the expanded upper portion; the *cervix,* the narrower lower portion. These are separated by a slight constriction, the *isthmus.* The superior portion of the body, which lies above the openings of the two uterine tubes, is the *fundus.* A portion of the cervix protrudes slightly into the vagina.

The *cavity* of the uterus is roughly triangular in shape and relatively small, owing to the thickness of the uterine wall. Laterally, it receives the two uterine tubes; inferiorly, it narrows and continues through the cervix as the *cervical canal,* which opens into the vagina. The vaginal opening is called the *external os;* the internal or uterine opening is the *internal os.* On the anterior and posterior inner surfaces of the cervical canal are two longitudinal folds, the *plicae palmatae.*

POSITION AND ATTACHMENT OF THE UTERUS (Figs. 7-4 and 7-6). In an erect female with the bladder empty, the uterus lies nearly horizontal, with the fundus directed forward. Its anterior surface is separated from the bladder by a space lined with peritoneum, the *vesicouterine pouch.* Posterior to the uterus and separating it from the rectum is a space, the *rectouterine pouch* or *pouch of Douglas.* This pouch is the lowermost portion of the peritoneal cavity. It extends inferiorly as far as the vagina.

The uterus is held in position by (1) two *broad ligaments,* winglike structures extending horizontally, that attach to the floor and wall of the pelvis; (2) two *round ligaments,* which extend from the upper portion of the uterus near the openings of the uterine tubes laterally to the pelvic wall (these continue through the inguinal canal and end in the subcutaneous tissue of the labia majora); and (3) two *uterosacral ligaments,* which extend from the upper portion of the cervix to the sacrum. The lower portion of the uterus is relatively fixed in position; the body is relatively free and movable. In addition to these structures, the pelvic diaphragm (levator ani and coccygeus muscles and their fascial coverings) plays an important role in the support of the uterus and other pelvic structures.

MICROSCOPIC STRUCTURE OF THE UTERUS. The wall of the uterus in the region of the fundus measures 1 to 1.5 cm in thickness. It consists of three layers: the mucosa (endometrium), muscular layer (myometrium), and serosa (perimetrium). The *mucosa* consists of a single layer of simple columnar epithelium resting on a connective tissue stroma that is mesenchymal in nature. Irregularly distributed over its surface are groups of ciliated cells. The uterine glands are invaginations of the epithelium. They are usually of the simple tubular type, although some are branched. The tubular portion of a uterine gland follows a tortuous course through the stroma, and the terminal portion may extend into the myometrium (muscle layer). Some of the cells of these glands bear cilia. The *muscular layer* constitutes the major portion of the uterine wall. It consists of bundles of smooth muscle cells arranged in three indistinct layers (inner longitudinal, middle circular, and outer longitudinal). The muscle cells are long, averaging 50 μm or more in length. In a pregnant uterus, however, they hypertrophy, sometimes attaining a length of 500 μm or more. There is also an increase in the *number* of muscle cells as a result of cell division and transformation of other cellular elements into muscle cells. The bundles of muscle cells are bound together by interstitial tissue consisting principally of collagenous fibers but containing embryonic connective tissue cells and other cellular elements. Elastic fibers are abundant, especially between the serosa and the muscular layer. The *serosa,* the outermost layer, consists of fibroelastic tissue. On the entire posterior surface and the upper portion of the anterior surface, a layer of mesothelium continuous with the abdominal peritoneum forms the outer surface.

CYCLIC CHANGES IN THE ENDOMETRIUM—MENSTRUATION. In nonpregnant women from the time of puberty (ages 12 to 14), when the *menarche* or first menstruation occurs, to the *menopause* (ages 45 to 55), when menstruation ceases, the endometrium of the uterus undergoes a series of cyclic changes that occurs on the average every 28 days (range, 21 to 35 days). These cyclic changes of the endometrium, known as the *menstrual cycle,* are correlated with cyclic changes in the ovaries already described. (See Fig. 7-7.)

Stages of the Menstrual Cycle. Menstruation is the periodic discharge from the vagina of blood that contains necrotic tissue elements that have

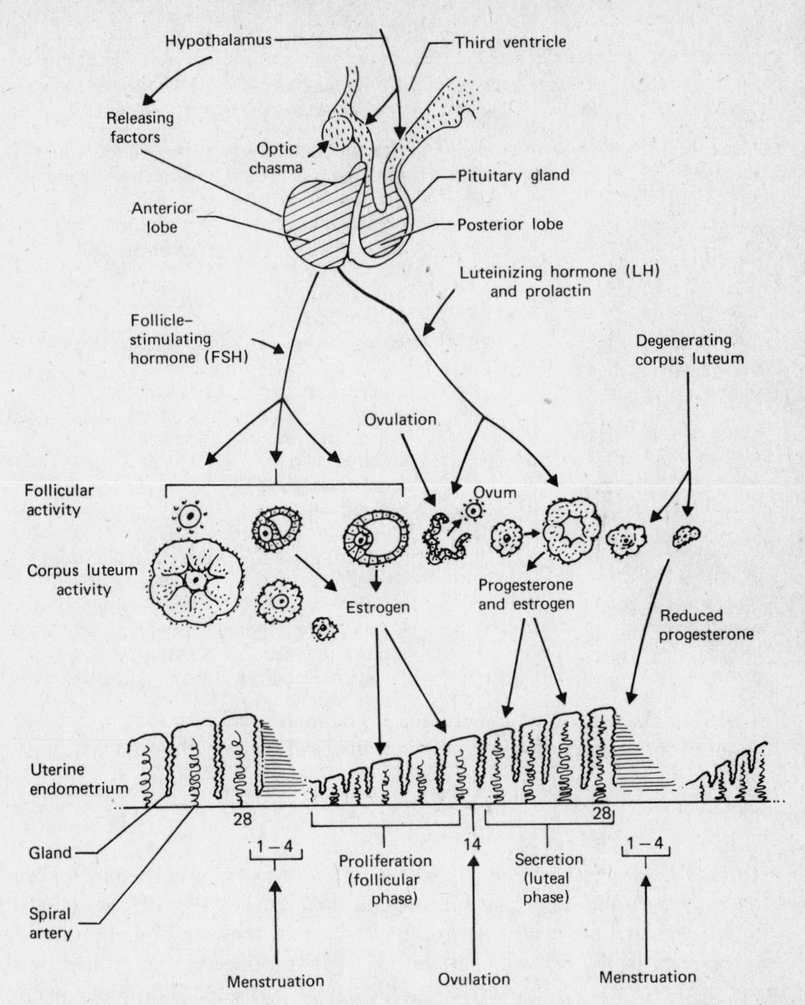

Fig. 7-7. Diagram showing the relationships of pituitary, ovarian, and uterine (endometrial or menstrual) cycles. (Reprinted with permission of John Wiley & Sons, Inc., from E. B. Steen and J. H. Price, *Human Sex and Sexuality,* 1977.)

been sloughed off from the uterine endometrium. Table 7-1 lists the periods of the menstrual cycle and the changes that occur in the uterine mucosa and in the ovaries.

Menstruation is the final event of the cycle, but it is customary to speak of the "menstrual cycle" as beginning with the onset of menstruation; that is, the first day that menstrual discharge appears is considered to be the first day of the cycle.

TABLE 7-1

Period	Duration in Days	Day of Cycle	Changes in Uterine Mucosa	Changes in Ovaries
Menstruation (menses)	5	1–5	Endometrial cells undergo necrosis; desquamation occurs; glands release secretions, blood vessels rupture, menstrual flow occurs	Corpus luteum continues to degenerate New follicle begins development
Proliferative (follicular) phase	9	6–14*	Uterine mucosa is restored Endometrium grows, stroma becomes thicker and more vascular; glands become longer but remain straight	Follicle develops Follicle matures Estrogens are secreted Ovulation marks end of stage
Premenstrual, secretory, progestational, or luteal phase	14	15–28	Endometrium continues to grow; glands become longer, more tortuous and coiled. Cells enter secretory phase, secrete glycogen, mucoid material, and fat	Corpus luteum develops, matures, and begins to retrogress Secretes progesterone and estrogens

* In long cycles of 30 to 35 days, a period of rest intervenes between the first and second stages.

Uterine Changes During the Menstrual Cycle. As shown in the table, the uterine changes that take place during the menstrual cycle are correlated with changes that are occurring in the ovary, since they are brought about and regulated by the effects of certain hormones produced in the endocrine tissue of the ovaries. These are the follicular or estrogenic hormones and the corpus luteum hormone (progesterone).

Estrogenic hormones, of which estradiol is the principal one, are produced by the cells of the developing follicle and the corpus luteum. They exert their effects during the entire menstrual cycle, but primarily during the first half. Under their influence, repair of the endometrium shed during menstruation takes place. The mucosa becomes thicker, glands develop and begin secreting, and the blood supply increases. These changes prepare the uterus for reception of the fertilized egg. Upon ovulation, secretion of the follicular hormones is reduced, but ovarian control is continued through the action of progesterone.

The *corpus luteum hormone* (*progesterone*) is produced by the corpus luteum, a structure that develops within the ruptured follicle. It is responsible for changes that occur in the uterine mucosa during the second half of the menstrual cycle (premenstrual changes). These changes vary, depending on whether or not the ovum is fertilized. If the ovum is *not* fertilized, the mucosa becomes thicker, glands increase in size and become tortuous, cells become filled with glycogen, and the blood supply increases. As the corpus luteum regresses, a reduction in the amount of progesterone and estrogen secreted brings about changes in the surface layer of the uterine endometrium. Coiled arteries supplying the layer constrict, and blood supply to the endometrium is reduced. Cells die; portions of the endometrium undergo necrosis and are sloughed off. Capillaries rupture and blood escapes into the uterine cavity. This material, together with thickened secretions of the uterine glands, constitutes the *menstrual fluid,* which is discharged through the vagina in the process of *menstruation.*

If the ovum is fertilized, the foregoing changes are modified as follows: The mucosa is in a state ready to receive the fertilized ovum; consequently, when the developing blastocyst enters the uterus, it embeds itself (*implantation*) and continues its development. Under this condition, the corpus luteum does not involute but persists as the corpus luteum of pregnancy (*true corpus luteum*). Menstruation does not take place, and further ovulation is inhibited. Progesterone further influences the endometrium, and its continued secretion is necessary for the proper development of the embryonic membranes and the placenta.

Irregularities in the Menstrual Cycle. Though it tends to occur with regularity, the menstrual cycle is subject to considerable variation. Factors that affect physical and mental health may also affect the onset of the menstrual flow. In wasting diseases, the menstrual periods are usually suppressed. Emotional disturbances may bring about delay in menstruation. Except during pregnancy, the absence of menstrual flow is regarded as abnormal; the condition is termed *amenorrhea.* Scanty menstrual flow is called *oligomenorrhea;* painful menstruation, *dysmenorrhea;* excessive flow, *menorrhagia.* Menstrual disorders may result from diseases involving the uterus or ovaries, abnormal growths (cysts, tumors), endocrine dysfunction, or psychogenic disturbances.

Anovulatory Menstruation. Menstruation may occur in the absence of ovulation, that is, before the endometrium has entered the secretory period. Such cycles, which are usually shorter than ordinary cycles, are sterile, for no ovum is produced. They occur most commonly during the years immediately following the onset of puberty and in those just preceding the menopause. The anovulatory nature of these cycles accounts for the relative infertility of women during those periods of life.

"FERTILE" AND "SAFE" PERIODS. Most experimental evidence indicates that ovulation occurs between the fourteenth and sixteenth days of the menstrual cycle; as a consequence, fertilization or conception takes place most frequently about midway between menstrual periods. This is the so-called fertile period. After ovulation, the ovum is thought to maintain its viability for a limited period (not exceeding 1 or 2 days). From this it can be concluded that fertilization is unlikely to occur during the week prior to menstruation. The advent of menstruation indicates that the ovum has not become fertilized, and in the period following menstruation, when the new follicle is being developed, fertilization cannot occur because an ovum has not been liberated from the follicle. Thus, the week preceding and the week following menstruation may be considered relatively "safe" periods, that is, periods in which fertilization is not likely to occur. However, there is so much variation among women that there is no truly "safe period," that is, a period during which contraceptives, if used, can be dispensed with.

MENOPAUSE. This is the period marked by the permanent cessation of menstruation. It may take place at any time between the ages of 35 and 55 but is most common between 45 and 50. Development of ovarian follicles ceases; consequently, menopause marks the end of the childbearing period. Normal libido usually persists.

Certain *anatomical* changes accompany menopause, among them atrophy of the sex organs, including the external genitalia, uterus, uterine tubes, ovaries, and breasts. The vagina becomes conical in shape, and its mucous membrane atrophies. Secretory activity of glands associated with the reproductive organs is reduced. Most of these changes are gradual and extend over a period of years. Pronounced *physiologic* changes may occur. Among these are "hot flashes," sweating, dizziness, rheumatic pains in joints, susceptibility to fatigue, and headaches. There may also be *psychic* changes: excessive irritability, abnormal fears, and a tendency to worry. Sometimes, excessive sexual desire is manifested. Many of these symptoms can be alleviated by the cautious administration of hormones; during menopause, there is marked hormonal imbalance.

Menopause can be induced prematurely (*artificial menopause*) by removal or deactivation of ovarian tissue. This is accomplished either surgically or by irradiation.

The Vagina. The vagina is the canal that extends from the uterus to its external opening in the *vestibule.* It is a highly muscular, dilatable tube, averaging 7.5 cm in length. Its upper end meets the cervix of the uterus at almost a right angle so that its posterior wall is longer then the anterior wall. Because the cervix projects downward into the vagina, a circular groove encircles the lowermost portion of the cervix. This groove is called the *fornix.* Its posterior portion (*posterior fornix*) is deeper than the *anterior fornix.*

THE HYMEN. The external orifice of the vagina lies between the minor labia and is partially closed by a fold of connective tissue, the *hymen,* which is usually ruptured at the first coitus, though this is often caused in other ways. When the hymen completely closes the vaginal orifice, it is said to be an *imperforate hymen.*

LAYERS OF THE VAGINAL WALL. The wall of the vagina consists of three layers: mucosa, muscular layer, and fibrous coat.

Mucosa. This is lined with stratified squamous epithelial cells resting on a thin lamina propria of connective tissue. The lamina propria sends numerous extensions or papillae into the overlying epithelium, especially on the posterior wall of the vagina. It also contains numerous lymphocytes, which migrate into the epithelium. Sometimes the lymphocytes are grouped into small nodules.

The mucosa is thrown into large folds or *rugae* in which blood vessels abound, giving it the structure of erectile tissue. Glands are absent, but the vagina is moistened by mucous secretions of the uterine glands. The pH of the vagina is slightly acid owing to the action of lactic acid–forming bacteria (Döderlein's bacilli), which ferment the glycogen. The pH varies with the phases of the menstrual cycle, being lowest in the late, proliferative phase.

Muscular Coat. This consists of bundles of smooth muscle, the majority of which are directed longitudinally. Circular bundles are interspersed among the longitudinal ones. At the external orifice, a sphincter muscle of striated fibers is present.

Fibrous Coat. This thin layer of connective tissue merges with the loose tissue of surrounding structures. Posterior to the vagina is a sheet of fibroelastic tissue, the *retrovaginal septum,* which separates the vagina from the rectum. Anteriorly, the *urethrovaginal septum* separates the vagina from the urethra.

CHANGES IN THE VAGINA DURING PREGNANCY. During pregnancy, changes in the structure of the vagina correlate with its role in childbirth. The epithelium proliferates, muscle cells hypertrophy, and the connective tissue becomes more loosely arranged. This enables the vagina to stretch to accommodate the fetus. Prolonged and difficult labor may result in development of lesions, as, for example, laceration or fistulas. A *fistula,* an abnormal passageway between two cavities or from a cavity to the outside, may develop from the vagina to the bladder, urethra, or rectum. In cases in which laceration of the perineum is imminent, it is common practice to make an incision on one side through the vulva to direct the tear to the side, a procedure called *episiotomy.*

External Genitalia. The external female sexual organs include the labia majora, labia minora, clitoris, and vestibule. Collectively these constitute the *vulva* (Fig. 7-8).

The *labia majora,* two prominent folds of skin bearing pubic hair, form the lateral borders of the vulva. They unite anteriorly and are continuous with the *mons veneris* (*mons pubis*), a rounded prominence that covers the pubic symphysis. The mons pubis has a dense pad of fat beneath the integument and is thickly covered with pubic hair. Posteriorly, the labia majora unite to form the *posterior commissure.*

The *labia minora* are two longitudinal folds that lie just within the labia

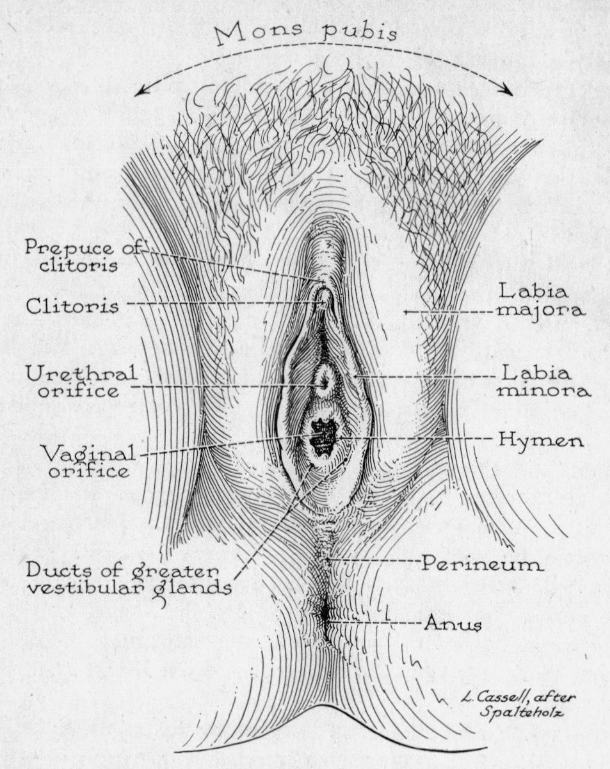

Fig. 7-8. The female external genitalia or vulva. (After Spalteholz.) (Reprinted with permission of W. B. Saunders Co., Philadelphia, from B. G. King and M. J. Showers, *Human Anatomy and Physiology,* 6th ed., 1969.)

majora. They form the lateral borders of the vestibule. Anteriorly, they unite to form the *prepuce* and the *frenulum clitoridis.*

The *clitoris* is a small erectile structure lying beneath the pubic symphysis at the juncture of the labia minora. It consists of a *body,* two *crura,* and a *glans* and is covered by the prepuce. It contains many sensory nerve endings. The clitoris is a homolog of the penis in the male.

The *vestibule* is the region between the labia minora into which the vagina and the urethra open. On each side of the vaginal orifice and just within the labia minora are the openings of the larger *vestibular glands* (*glands of Bartholin*). Smaller vestibular glands open at various points in the vestibule.

The muscles and fascia that form the pelvic floor and enclose the reproductive, urinary, and digestive openings constitute the *perineum.* In obstetrics the term *perineum* is restricted to the region between the vulva and the anus (*obstetrical perineum*).

Mammary Glands. In the strictest anatomical sense, the mammary glands belong to the integumentary system, since structurally they are modified sweat glands. Their development and functioning are, however, closely related to the reproductive system. Mammary glands are present in both sexes, but involution occurs in the male after puberty, and they persist in only a rudimentary state. They undergo conspicuous changes in the female at puberty, during pregnancy, during and after lactation, and after menopause.

STRUCTURE OF THE MAMMARY GLANDS. The mammary glands are contained in the breasts or *mammae* (Fig. 7-9), two rounded bodies that lie on the anterior surface of the thorax. The secreting portion is surrounded by a considerable amount of loose connective tissue in which are variable quantities of adipose tissue. The *nipple,* a cylindrical or conical projection at about the center of each mamma, bears a rounded tip on which open some 15 to 20 *lactiferous ducts.* Surrounding each nipple is a circular pigmented area, the *areola.* The skin over the nipple and areola bears numerous small papillae and contains a considerable amount of pigment, which becomes much darker during pregnancy. During pregnancy, the nipple increases in size, becomes more sensitive, and is more easily erectile. Smooth muscle fibers are present in the skin beneath the areola and the nipple. About the base of the nipple they are disposed in a circular fashion. Also present in the skin of the areola are small *areolar glands of Montgomery* (*apocrine sweat glands*) and large numbers of sebaceous glands, which produce slight elevations.

Each mammary gland is a compound alveolar gland consisting of 15 to 20 lobes of glandular tissue separated from each other by *interlobular septa.* In an inactive gland, the alveoli or secreting elements are small and undeveloped. In fact, their presence has been questioned by those who maintain that only ducts and their branches persist.

When pregnancy occurs, the mammary glands become active and undergo marked changes. After lactation they undergo involution, in which the secreting portions become greatly reduced in size or disappear. After menopause they become still further reduced in size, and in old age they appear as mere fibrous cords. The interstitial tissue also retrogresses, giving the entire structure a shrunken appearance.

HORMONAL CONTROL OF THE MAMMARY GLANDS. The growth and development of the mammary glands and the secretion of milk by them are regulated by endocrine secretions of the ovaries and the hypophysis. Estrogens produced by the follicle cells of the ovary and by the placenta induce early growth and development of the duct system. *Progesterone,* produced by the corpus luteum of the ovary and the placenta, brings about development of the alveoli and induces presecretory changes. *Prolactin,*

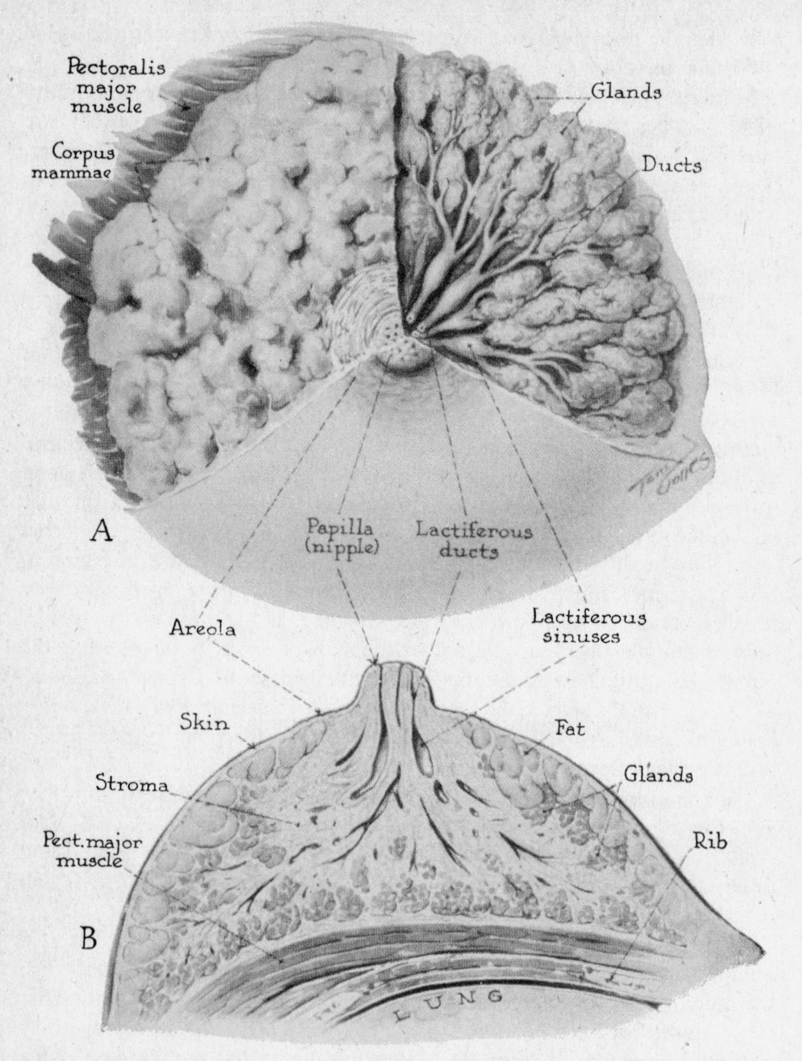

Fig. 7-9. The breast or mamma. (*A*) Dissection of a lactating breast. (*B*) Relation of the breast to the chest wall. (Reprinted by permission of Naturalwear Division, Camp International, Inc. from *Anatomical Studies for Physicians and Surgeons.*)

from the anterior lobe of the pituitary gland (hypophysis cerebri), brings about the active secretion of milk. *Oxytocin,* secreted by the hypothalamus but released by the posterior lobe of the pituitary gland, is responsible for milk letdown and its release.

BLOOD, LYMPH, AND NERVE SUPPLY OF THE MAMMARY GLANDS. The

mammary glands receive their principal blood supply from branches of the internal mammary artery supplemented by branches from the axillary and intercostal arteries. Veins empty into the axillary and internal mammary veins. Lymphatic vessels are numerous; the efferent vessels drain by two principal trunks leading to the axillary nodes. Supplementary drainage leads through the pectoralis major muscle to the mediastinal nodes or the abdominal lymph nodes. The nerve supply to the glands is abundant. Branches from the intercostal nerves and from the supraclavicular nerve innervate the glands and the overlying skin. Branches of the autonomic nervous system innervate muscles of the blood vessels of the nipple, areola, and glandular tissues. The nipple is well supplied with afferent endings.

EMBRYONIC AND LATER DEVELOPMENT OF THE REPRODUCTIVE SYSTEM

In early embryonic development, the male and female reproductive systems follow an almost identical course, and the sexes are indistinguishable. This is known as the *indifferent stage.* In this stage, the gonad cannot be identified as either a testis or an ovary, and the embryo possesses a double set of sex ducts. At about the sixth week of embryonic life, internal changes begin to take place. These result in differentiation of the testis and ovary, and certain tubules and ducts (mesonephric tubules and ducts, Müllerian ducts) are transformed into structures that, in some cases, become functional; in other cases, they remain rudimentary or nonfunctional. The testes and ovaries shift their positions (they descend), and the external genitalia develop.

The development of male structures involves marked changes in basically female embryonic structures and requires a masculinizing hormone (testosterone) acting during fetal development. Androgens produced by fetal testicular cells of Leydig promote the development of male structures and inhibit female structures. The development of female reproductive structures is autonomous; no female hormones are required for their differentiation.

Table 7-2 shows some of the homologies between the male and female sexual organs.

Developmental Anomalies. Occasionally, individuals are seen in whom anatomical characteristics of both sexes are present in greater or lesser degree. When the primary sex organs of both sexes are present in the same person, the condition is *hermaphroditism. True hermaphroditism,* in which *functional* ovaries and testes are present, has rarely been recorded in humans, although this condition is normal in some invertebrates such as worms and mollusks. In the development of secondary sex characteristics, such as beard, mammae, and body form, and in the development of the external genitalia, there is a wide range of variation. This sometimes leads to a condition known as *pseudohermaphroditism,* in which an individual

TABLE 7-2

Adult Structures in the Male	*Indifferent Stage of the Embryo*	*Adult Structures in the Female*
Testis	Gonad on urogenital ridge	Ovary
Efferent ducts, appendix epididymis,* paradidymis*	Mesonephric tubules	Epoophoron,* paroophoron*
Ductus epididymis, ductus deferens, seminal vesicle, ejaculatory duct	Mesonephric or Wolffian duct	Gartner's duct of epoophoron*
Appendix testis*	Müllerian duct	Uterine tubes, uterus, vagina
Urethra, prostate gland, bulbourethral glands	Urogenital sinus	Urethra, vestibule, paraurethral ducts, vestibular glands
Penis	Phallus	Clitoris
Anal surface of penis	Lips of urogenital groove	Labia minora
Scrotum	Labioscrotal swellings	Labia majora

* Vestigial structures.

of one sex may possess some characteristics of the opposite sex. In the male, pseudohermaphroditism may result from failure of the testes to descend (*cryptorchism*) or from retarded development of the penis (*hypospadias*) in which the urethral opening is on the underside of the penis. Such anomalies simulate female structure. In the female, hypertrophy of the clitoris or fusion of the lips of the vulva produces a degree of resemblance to the male. Conditions in which individuals have the appropriate primary sexual organs but have secondary organs and characteristics of the other sex are usually brought on by abnormal functioning of endocrine glands (gonads, adrenals, and pituitary gland).

DISEASES AND DISORDERS OF THE REPRODUCTIVE SYSTEM

In the Male

Anorchidism. Congenital absence of testes.

Cryptorchism. Failure of the testes to descend into the scrotum during embryonic development. It results in sterility; sperm fail to develop if the testes are retained in the body cavity.

Hydrocele. Accumulation of fluid in the tunica vaginalis or testis.

Monorchidism. Presence of only one testis.

Paraphimosis. Condition in which the prepuce is retracted and constricted, forming a tight band about the glans penis.

Phimosis. Inability to retract the prepuce over the glans penis. It may be due to adhesions of the prepuce to the glans or to a narrowing of the preputial orifice. The condition is corrected by circumcision.

Prostatic Hypertrophy. Enlargement of prostate gland. It occurs frequently in men after age 60. When enlarged, the prostate compresses the prostatic portion of the urethra, leading to difficulty in or complete suppression of urination. Enlargement may be due to benign hypertrophy, carcinoma, or fibrosis.

Urethral Stricture. Localized constriction of the urethra. It may be due to spasm of the urethral muscle, inflammation of the urethral lining, or formation of fibrous tissue.

Vesiculitis. Inflammation of a seminal vesicle.

In the Female

Anteversion (Retroversion) of the Uterus. Forward (backward) tilt or displacement of the fundus toward (or away from) the pubis.

Cysts. These are common within or upon the ovary. Among them are *dermoid cysts,* which contain such ectodermal derivatives as skin, hair, and teeth.

Prolapsed Uterus. This is the condition that prevails when the normal supporting structures weaken and permit the uterus to drop from its usual position.

Tumors. The uterus, ovaries, and breasts are especially prone to development of neoplasms (new growths). Although most of these are benign, some are malignant.

Vaginismus. Spasm of the muscles of the vagina or those surrounding its entrance that prevents coitus or results in painful or difficult coitus (*dyspareunia*). Such may be caused by lack of lubrication, structural disorders, inflammatory conditions, or psychogenic factors.

Vaginitis. Inflammation of the vagina, usually due to infectious organisms.

In Both Sexes

Infertility. Inability to produce offspring. In the *male,* this may be due to (1) failure of the testes to produce sperm or production of sperm that lack fertilizing power, (2) *impotency* or inability to perform the sexual act, (3) obstruction of the seminal ducts, (4) disorders in the secretory functions of the accessory sex glands, or (5) failure of the sperm to undergo capacitation.

In the *female,* infertility may be due to (1) malfunctioning of the ovaries, such as failure of the follicles to develop or failure of ovulation to occur; (2) obstructed or incompetent uterine tubes; (3) structural or functional disorders of the uterus; (4) conditions that interfere with sperm survival or transport, such as an excessively acid vagina or thick cervical mucus; or (5) immobilization and agglutination of sperm by antibodies.

Sexually Transmitted Diseases (STDs). These infectious diseases, commonly called *venereal diseases,* are usually acquired through sexual contact with an infected person. They include chancroid, gonorrhea, syphilis, and herpes simplex genitalis.

CHANCROID. *Chancroid* results in the formation of multiple soft lesions (chancres); the infecting agent is a bacterium, *Hemophilus ducreyi.* The lymph nodes are usually involved.

CHLAMYDIAL DISEASES. Chlamydias are nonmotile, obligate, intracellular parasites and are considered to be the organisms responsible for lymphogranuloma venereum, trachoma, and inclusion conjunctivitis. The latter two are eye disorders. Chlamydias have been shown to cause a number of sexually transmitted diseases including nongonococcal urethritis and epididymitis in the male and cervicitis, urethritis, and pelvic inflammatory disease in women. Neonatal conjunctivitis may be transmitted from an infected mother to her newborn.

GONORRHEA. *Gonorrhea* is an infectious inflammatory condition involving the mucous membranes of the reproductive organs. It is caused by a gonococcus, *Neisseria gonorrhoeae.*

HERPES SIMPLEX GENITALIS. *Genital herpes* is a viral disease characterized by the formation of fluid-filled vesicles or cold sores on the skin or mucous membranes of the sex organs. It is usually a self-limited disease, the symptoms of which disappear in a few weeks, but once acquired it tends to recur following physical or emotional stress. It is a common cause of birth defects.

SYPHILIS. *Syphilis,* usually chronic in nature, is caused by a spirochete, *Treponema pallidum.* Symptoms of syphilis appear in three stages. In the *primary stage,* the initial lesion appears 2 to 4 weeks after infection; it develops into a hard chancre. In the *secondary stage,* constitutional symptoms appear 6 to 18 weeks after the exposure to infection and may involve almost any organ of the body; skin manifestations are common. The symptoms of this stage are often mistaken for those of other diseases, a fact that has caused syphilis to be called "the great imitator." In the *tertiary stage,* there is degeneration of structures such as bone, walls of blood vessels, or the brain; any organ may be involved. The symptoms of this stage do not develop in less than 6 months and in some cases not for years. If untreated, the central nervous system usually becomes involved. Tabes dorsalis (locomotor ataxia) results from degeneration of the sensory and motor columns of the spinal cord. General paresis results from involvement of the cerebral cortex.

8: REPRODUCTION AND DEVELOPMENT

Reproduction and development include all processes concerned with the production of a new individual: the formation of germ cells (spermatozoa and ova), transfer of sperm to the female by sexual intercourse (coitus), fertilization of the ovum, segmentation of the zygote, implantation of the blastocyst, and development of the embryo and fetus. It also includes the formation of embryonic membranes, the placenta, parturition or childbirth, and lactation.

CELLULAR BASIS OF HUMAN REPRODUCTION

Individuals begin development as a single cell, a fertilized ovum or *zygote.* By the process of mitosis, this cell divides into two cells, these two into four, the four into eight, and so on until a complex individual consisting of billions of cells results. At first all the cells are nearly identical, but as development proceeds, *differentiation* and *specialization* take place, giving rise to *tissues.* These tissues eventually become associated, forming *organs,* and the organs become grouped together in body *systems.*

Fundamental Cell Types. In the mature organism, cells of the body are of one of two types: (1) *germ* or *reproductive cells,* which have the capacity to produce a new individual, and (2) *somatic* or *body cells,* which constitute the major portion of the body.

Development and Maturation of Germ Cells. The development of female germ cells (ova) is known as oogenesis; of male germ cells (spermatozoa), spermatogenesis.

OOGENESIS (Fig. 8-1). There are two views concerning the origin of functional human ova: (1) they develop within follicles, which arise from the germinal epithelium of the ovary very early in the development of that organ; most of the ova are present before birth and lie dormant until maturity, when a limited number mature and become functional. (2) Functional ova arise from indifferent cells of the germinal epithelium of the mature ovary as they are needed. The first view is the one most generally accepted.

In the process of maturation, the ovum is prepared for fertilization. It increases in size through accumulation of food materials, and nuclear changes occur in which the chromosome number is reduced to one-half the normal number.

In oogenesis there are four stages, as follows:

1. Oogonia. These are generalized cells that develop from the primordial germ cells in the germinal epithelium. They become enclosed in a single layer of follicle cells. From these the follicle develops, and the germ cell enlarges to become a primary oocyte.

2. Primary Oocyte. The follicle matures, and a day or two before ovulation, the primary oocyte undergoes division. In this division, the chromosomes come together in pairs (*synapsis*), and each synaptic pair divides, one chromosome of each pair going to each of the daughter cells. This results in two daughter cells, each with one-half the normal chromosome content. The daughter cells vary greatly in size. One is very large, retaining practically all the cytoplasm of the parent cell; it is called a *secondary oocyte.* The other, which is very small and contains little cytoplasm, is the *first polar body.*

In this division of the primary oocyte, the normal somatic number of chromosomes has been reduced from 46 to 23; accordingly, it is known as the *reduction division.*

3. Secondary Oocyte. In this stage the oocyte is usually liberated from the follicle (*ovulation*). It is drawn into the uterine tube, where, if sperm are present, fertilization will probably occur. The entry of sperm into the oocyte initiates this second division, in the course of which the secondary oocyte divides into a very large cell (the *functional ovum*) and the *second polar body.* Sometimes the first polar body also divides, in which case there may be three polar bodies from one oogonium. The polar bodies are nonfunctional and quickly disintegrate.

4. Mature Ovum. In this stage the ovum is mature and ready for the completion of fertilization, the sperm having already entered the cytoplasm of the oocyte. (The events of fertilization are described on page 297).

The mature human ovum is the largest cell produced by the body, though it is barely visible to the unaided eye, averaging 0.15 mm in diameter. It contains a large spherical nucleus within which are scattered granules of *chromatin* and a large *nucleolus.* The cytoplasm is filled with granules of deutoplasm or *yolk.* Surrounding the cytoplasm is a thick striated membrane, the *zona pellucida.* It is an extracellular membrane composed of mucoproteins synthesized by the follicular cells. Beneath the zona pellucida is a narrow cleft, the *perivitelline space.*

Surrounding the ovum are cells from the follicle, which form the *corona radiata,* from which fine protoplasmic processes extend through the zona pellucida. These carry nutritive materials to the ovum.

SPERMATOGENESIS (Fig. 8-1). Spermatozoa develop within the seminiferous tubules of the testes. In spermatogenesis there are five stages, as follows:

1. Spermatogonia. These are generalized cells located at the periphery of the seminiferous tubule. They increase in number by mitosis. Some remain as stem cells; the other migrate centrally, increase in size, and become primary spermatocytes.

2. Primary Spermatocytes. The chromosomes unite in pairs (*synapsis*), and each spermatocyte divides by meiosis into two *secondary spermatocytes,*

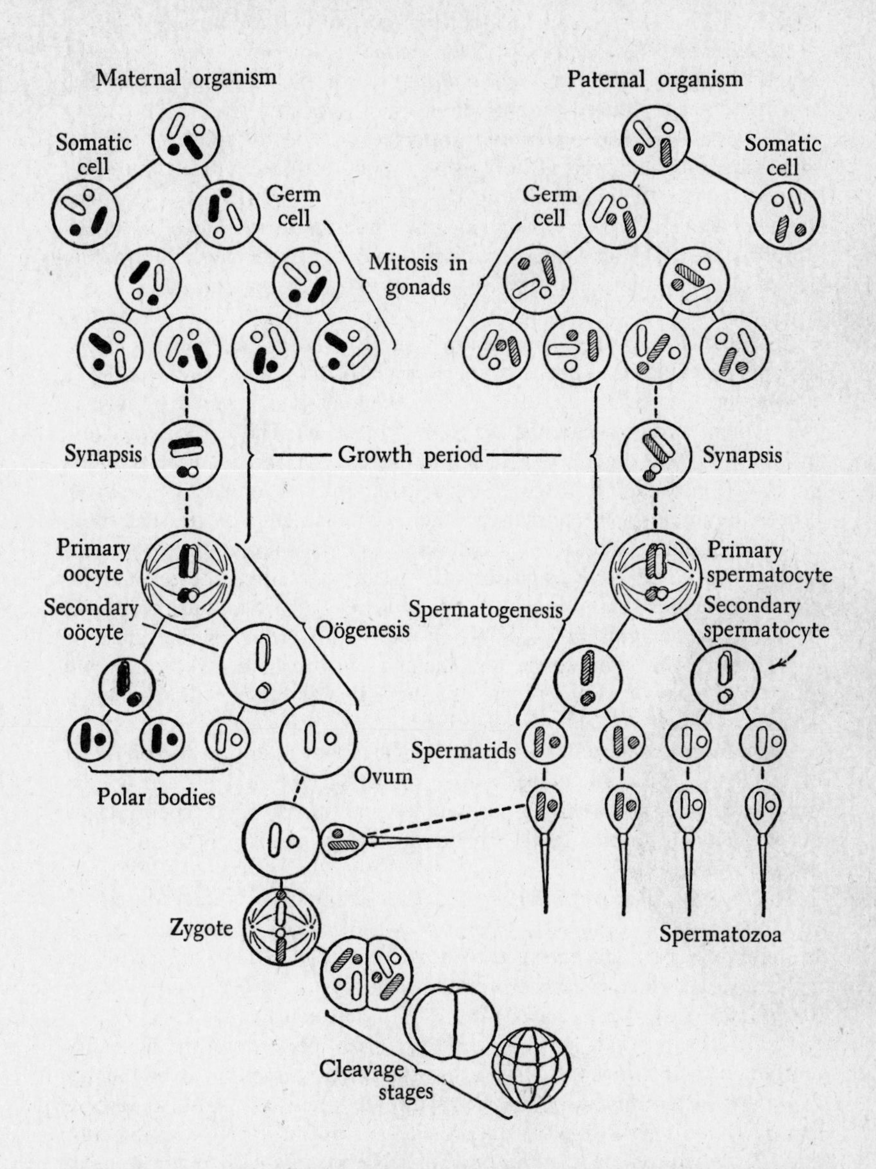

Fig. 8-1. Diagram illustrating gametogenesis (oogenesis in the female, spermatogenesis in the male) and the early stages of embryonic development in animals. Also illustrated is the continuity of chromosomes from the zygote that constitutes each parent, through the germ cells in the gonads of these individuals, to the embryo that is their progeny. (From G. Alexander and D. G. Alexander, *Biology,* 9th ed., 1970, Barnes & Noble Books, A Division of Harper & Row, Publishers.)

each with 23 chromosomes (one-half the normal diploid number).

3. Secondary Spermatocytes. The secondary spermatocytes divide by mitosis to form cells called *spermatids,* two for each spermatocyte. Each spermatid has a reduced number of chromosomes.

4. Spermatids. Each spermatid undergoes a series of complex changes in which a typical nonmotile cell becomes modified into a motile cell. This process, which takes place within the cytoplasm of Sertoli cells, is called *spermiogenesis.* The principal changes include the formation of a nuclear cap from the Golgi apparatus, condensation of nuclear material, formation of a motile flagellum, and the shedding of excess cytoplasm, which forms a *residual body.*

5. Spermatozoa. These are mature male germ cells. Each consists of two regions: an ovoid head and an elongated tail or flagellum. The *head* has a nuclear portion containing the genetic material. It is covered by a *head cap,* which contains flattened *acrosomal material.* The *tail* or *flagellum* includes the following regions: neck, middle piece, principal piece, and end piece. The *neck,* which adjoins the head, contains two centrioles. The *distal centriole* initiates development of the axial filaments of the tail; both centrioles are involved in the formation of the first spindle in the fertilized ovum. The *middle piece* consists of a striated, collarlike sheath of mitochondria surrounding a central *axial filament,* which continues to the tip of the tail. The *principal piece* forms the main portion of the tail. It is enclosed in a thin, fibrillar sheath. The *end piece* is the naked termination of the tail. Cross sections of the tail show that it has the same internal structure as a typical cilium or flagellum.

Spermatozoa average 0.05 mm in length (about one-third the diameter of the ovum). In volume the ovum exceeds the sperm by about 85,000 times. While in the seminiferous tubules and sperm ducts, spermatozoa are not actively motile. In the epididymis, they undergo maturation and acquire motility.

The process of spermatogenesis in the seminiferous tubules is not a continuous one. It occurs in cycles in a wavelike fashion along the tubule. In humans, spermatogenesis extends over a period of four cycles of approximately 16 days each or a total of 64 days.

Chromosomes, Genes, and Twins. The processes of maturation serve a twofold function: (1) to produce cells capable of performing the necessary reproductive functions (that is, in the sperm, the function of finding an ovum and penetrating it, and in the ovum, the function of storing a reserve supply of nutrient substances to provide energy for early development), and (2) to reduce the chromosome number by one-half to the haploid number so that the chromosome number, being restored to the normal diploid number at fertilization, remains constant in each generation.

The chromosomes contain *genes* or *hereditary factors* that are transmitted from one generation to the next. Genes appear to be arranged in a lineal order in each chromosome. The characteristics of an individual are determined by (1) the genes the individual receives from the egg and the sperm and (2) the environment in which the genes undergo development and exert their effects.

The transmission of hereditary traits follows certain general laws or principles, which were established by Mendel in 1865. These include the principles of segregation and independent assortment, which, with modifications, form the basis of modern genetics.

The reduction of the number of chromosomes from the *diploid* number found in somatic cells to the *haploid* number found in reproductive cells results in the production of germ cells that differ not only in the number of chromosomes but in kind as well; that is, they contain chromosomes that are unlike in their genetic (genic) makeup. In the formation of germ cells, the chromosomes are distributed at random, with the result that it is probable that no two eggs or sperm are ever exactly alike in their genic constitution. When two individuals are very much alike, as in the case of *identical twins,* it is assumed that they are *monozygotic* (developed from one zygote). *Fraternal twins,* usually not more alike than ordinary siblings, are assumed to develop from two ova that are fertilized by different sperm and develop at the same time.

Determination of Sex. Chromosomes play a significant role in the determination of the sex of the individual. Before reduction division, the chromosome complement consists of 46 chromosomes, arranged in 23 pairs. In the human female oocyte, these include 22 pairs of *autosomes* (containing genes for body characteristics) and one pair of *sex chromosomes* (designated XX). After the reduction division, each mature ovum contains 22 + X chromosomes. In the male, spermatocytes as well as all the somatic cells contain 22 pairs of autosomes plus 2 sex chromosomes, designated X and Y. In the process of maturation, two kinds of sperm are produced, half containing 22 + X and half 22 + Y. These two types are designated "X sperm" and "Y sperm." An ovum is fertilized by one or the other of these types. If an X-containing sperm unites with the ovum, the resulting zygote will contain 22 pairs of autosomes plus XX chromosomes; this zygote develops into a *female.* If the ovum is fertilized by a sperm containing a Y chromosome, the resulting zygote (44 + XY) develops into a *male.* From this it is seen that the chromosome makeup of the zygote determines the sex of the individual.

SEXUAL INTERCOURSE (COITUS)

After the sex cells or *gametes* are formed, the next essential step in the process of reproduction is to bring them together. If the ovum is not fertilized within a few days after it is discharged from the ovary, it disintegrates. Among mammals, fertilization takes place within the body of the female. Sperm are transferred from the male to the female during sexual intercourse or *coitus.* In most mammals, mating occurs when the female is in a receptive state (*estrus*), which usually occurs at the time of ovulation. Lower animals are said to be "in heat" during this period. In primates, including man, there is no definite period of receptivity, and intercourse may occur at any time during the menstrual cycle. Ovulation takes place about midway between the menstrual periods, and at this time conception is most likely to occur.

Sexual Responses—Orgasm. During sexual intercourse, the erect penis is inserted into the vagina, and mechanisms are brought into play to bring about the discharge or *ejaculation* of semen. During sexual intercourse, a number of generalized body reactions occur. Muscles in various parts of the body contract, especially those in the arms, legs, abdomen, and buttocks. The heart rate increases, resulting in increases in pulse rate and blood pressure. Engorgement of blood vessels occurs, especially in the pelvic region. The rate of breathing increases, with perspiration over much of the body. The nipples become erect, and the skin over much of the body reddens (sex flush).

At the moment of ejaculation, the male experiences a brief period of paroxysmal excitement called an *orgasm.* An orgasm, or possibly repeated orgasms, may occur in the female, but without any associated discharge. Following the orgasm or *climax* of the sexual act, a *resolution phase* or recovery sets in. Vasocongestion decreases, the penis becomes flaccid, and in the male, a refractory period sets in during which another erection and ejaculation cannot occur for a period of minutes or hours.

Sperm Transport. Spermatozoa are deposited in the upper end of the vagina in the region of the opening of the cervix. Human semen coagulates within one minute but liquefies a few minutes later. Cervical mucus secreted under estrogen stimulation at the time of ovulation is thin and watery and favors sperm receptivity. By lashing movements of their tails aided by contractile movements of the vagina and uterus, sperm quickly pass through the cervical canal into the cavity of the uterus, through which they pass to the openings of the uterine tubes. Within the tubes, spermatozoa, by their own movements supplemented by segmental muscular contractions of the uterine tubes, eventually reach the fimbriated region, where fertilization usually occurs.

Sperm Viability. Human sperm degenerate quickly after leaving the male genital tract. Although sperm may survive for several days in the female genital tract, their fertilizing power is generally thought to be limited to one or two days. "Overripe" sperm or ova are commonly thought to result in the formation of abnormal zygotes. Factors that limit the life span and reduce the vitality of sperm include the high body temperature of the female, the acidity of the vagina, the phagocytic activity of leukocytes, especially within the uterus, and the possible production by the female of antibodies that can immobilize and agglutinate sperm. Exhaustion of their limited supply of potential energy due to their activity is also a factor in their rapid decline in vigor.

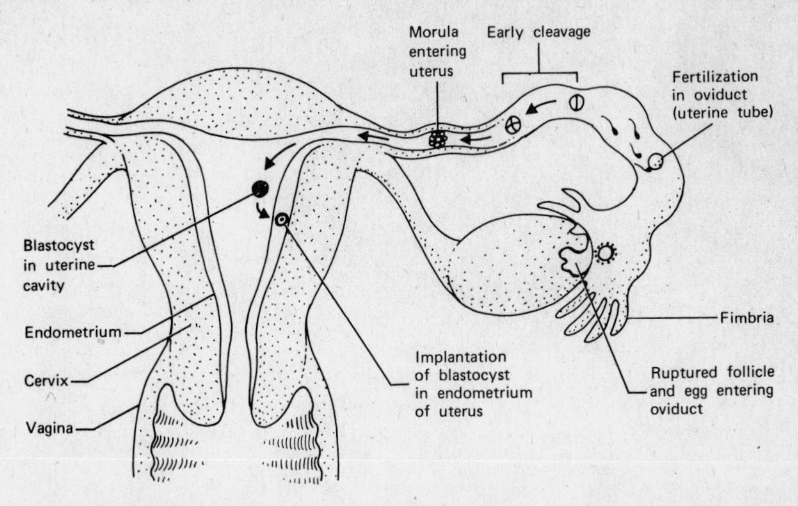

Fig. 8-2. Diagram showing ovulation, fertilization, early developmental stages, and implantation of the blastocyst in the endometrium of the uterus. (Reprinted with permission of John Wiley & Sons, Inc., from E. B. Steen and J. H. Price, *Human Sex and Sexuality,* 1977.)

FERTILIZATION AND EARLY DEVELOPMENT

The next steps in reproduction are fertilization and early development (Figs. 8-2 and 8-3). The latter includes the processes of cleavage, development of the morula and blastocyst, and implantation (nidation).

Fertilization of the Ovum. The penetration of the ovum by a sperm and the fusion of the male and female pronuclei constitute *fertilization.* This accomplishes three ends: (1) It initiates the developmental processes in the ovum, (2) it brings the male genes into the ovum, and (3) it determines the sex of the resulting individual. In some of the lower animals, ova may develop without fertilization, a process known as *parthenogenesis.* In mammals, however, this does not normally occur; in humans, parthenogenetic development of an ovum to maturity has never been demonstrated.

Although fertilization of the human ovum has been accomplished *in vitro,* that is, outside the body, most of the details of fertilization have been worked out from observations on the ova of other mammals: the mouse, rabbit, and certain primates, especially the monkey. Briefly the process is as follows.

In many mammals, the sperm and the egg require certain preparation before fertilization can take place. After being deposited, spermatozoa must remain in the female reproductive tract for a period of time and undergo a series of physiologic changes called *capacitation* before they can penetrate the zona pellucida of an ovum. Although capacitation has not been

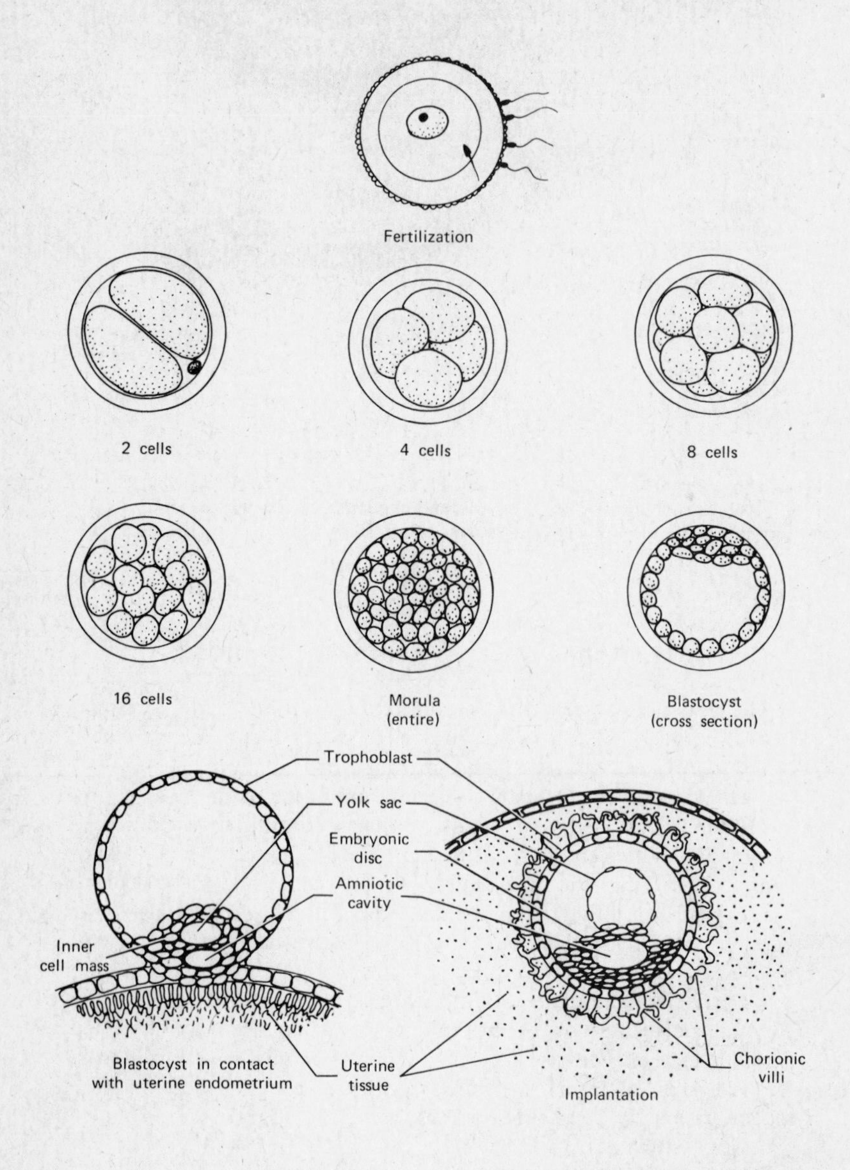

Fig. 8-3. Diagram showing early stages of development and implantation. (Reprinted with permission of John Wiley & Sons, Inc., from E. B. Steen and J. H. Price, *Human Sex and Sexuality,* 1977.)

positively demonstrated in humans, indirect evidence indicates that it is probably essential. On being liberated from the ovary, the ovum is surrounded by a dense layer of cells, the *corona radiata.* Under the influence of bicarbonate ions in the oviduct, hyaluronidase in the semen, and a trypsinlike enzyme in the acrosome, the corona cells are dispersed, enabling spermatozoa more readily to reach the zona pellucida of the ovum.

Spermatozoa reach the ovum by random movements; it is doubtful that chemical substances produced by the ovum have any attractive influence on the sperm. Propelling themselves through the remains of the corona and the zona pellucida, sperm reach the surface membrane of the ovum. Tail movements cease, and the successful sperm is drawn into the protoplasm of the ovum. The head migrates toward the center of the ovum, enlarges, and becomes the *male pronucleus.* The ovum completes the second maturation division, throwing off a second polar body. The *female pronucleus* now approaches the male pronucleus, and the nuclear membranes disappear. The chromatin of the nuclei forms two groups of homologous chromosomes, which arrange themselves in an equatorial plate. Fertilization is now complete, and the ovum undergoes the first mitotic division, producing two daughter cells, each of which receives one-half of each chromosome contributed by the sperm and the ovum.

Cleavage and Formation of the Morula and Blastocyst. Fertilization initiates the process of cell division or *cleavage.* By a series of mitotic divisions, the zygote divides into 2 cells, then 4, 8, 16, 32, 64, and so on until it consists of a spherical mass of cells called a *morula.* While this is occurring, the morula moves slowly through the uterine tube toward the uterus. Within the morula, a cavity begins to develop. This cavity enlarges until the whole comes to consist of a single layer of cells, the *trophoblast,* from which an *inner cell mass* projects centrally. The embryo develops from the inner cell mass; the trophoblast develops into structures concerned with nutrition of the embryo. At this stage the entire structure is called a *blastocyst.*

The inner cell mass gives rise to two cavities within its substance: the *amniotic cavity* and the *yolk sac* (*archenteron*). The amniotic cavity is lined with a single layer of cells, the *ectoderm;* the yolk sac with a single layer, the *entoderm.* These constitute the two *primary germ layers.* Between the two cavities lies the *embryonic disc* from which the embryo proper develops.

At about this stage, a third germ layer, the *mesoderm,* makes its appearance. Its cells occupy the space between the ectoderm and entoderm of the embryonic disc. It also forms a thin layer surrounding the amniotic cavity and yolk sac and the lining of the trophoblast. The trophoblast and its inner lining of mesoderm now constitute the *chorion,* which becomes an important embryonic membrane. The cavity of the chorion is filled with fluid, and the entire structure is called the *chorionic vesicle.* At this

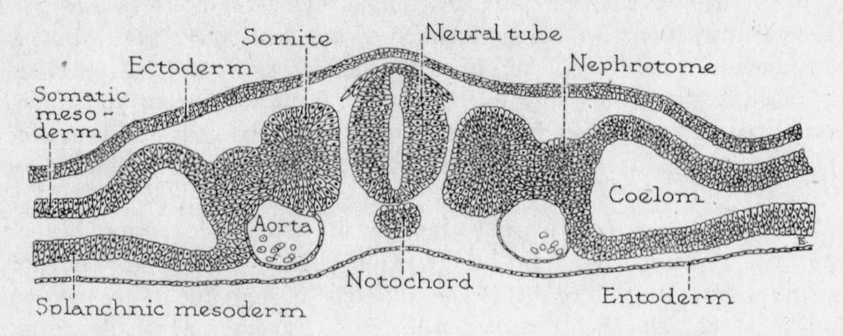

Fig. 8-4. Transverse section of a chick embryo at 48 hr incubation (semidiagrammatic). (Reprinted with permission of W. B. Saunders Co., Philadelphia, from B. G. King and M. J. Showers, *Human Anatomy and Physiology,* 6th ed., 1969.)

stage the developing blastocyst passes from the uterine tube into the uterus, 3 to 4 days after fertilization.

Implantation. The endometrium of the uterus is at this time in the *premenstrual* or *secretory phase* and ready to receive the embryo. About 6 or 7 days after fertilization, the blastocyst, on coming into contact with the endometrium, embeds itself in the uterine mucosa, a process called *implantation* (*nidation*). It is accomplished by the cytolytic or cell-dissolving action of the trophoblast cells through the production of enzymes by which the maternal tissue is dissolved. Now small fingerlike processes, the *chorionic villi,* grow out from the chorion into the endometrium. These develop into complex treelike structures between which are *intervillous spaces* that develop into cavities, the *blood lacunae,* which fill with maternal blood. Blood vessels from the embryo grow out into the villi, and the embryo receives its nourishment through these structures. The uterine tissue at this time completely encloses the developing embryo.

DEVELOPMENT OF THE EMBRYO

The embryonic disc, which is suspended from the chorion by the *body stalk,* now consists of three primary germ layers. On this area a longitudinal streak, known as the *primitive streak,* develops. This marks the primary axis of the embryo. At the anterior end, the cells of the three layers form a mass known as the *primitive knot.* From this mass a rod of mesoderm cells grows forward and fuses with the endoderm beneath. The rod constitutes the *head process,* from which the future notochord develops. The *notochord* is the axial supporting structure of the embryo and the forerunner of the vertebral column. The embryo at this time consists of a flat disc of cells forming the roof of the large yolk cavity below and the floor of the amniotic cavity above. A narrow canal, the *neurenteric canal,*

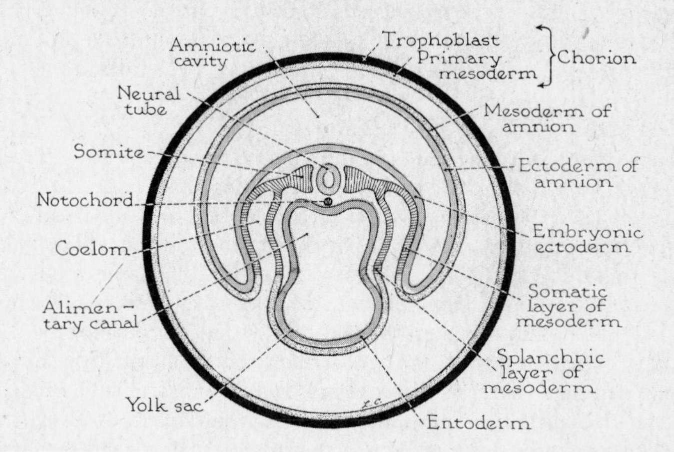

Fig. 8-5. Transverse section of a blastocyst, showing differentiation of the mesoderm and extension of the amnion. (Reprinted with permission of W. B. Saunders Co., Philadelphia, from B. G. King and M. J. Showers, *Human Anatomy and Physiology,* 6th ed., 1969.)

connects the two. The following seven steps in development should be noted:

1. The ectoderm thickens anterior to the primitive knot to form the *neural plate.* On each side *neural folds* arise, enclosing a *neural groove.* These fuse to form a closed *neural tube* that constitutes the primordial brain and spinal cord (Fig. 8-4).
2. The mesoderm on each side of the neural tube thickens and becomes divided into blocklike segments, the *mesodermal segments* or *somites* (Fig. 8-5). Sheets of mesoderm extend laterally and split into a thin outer *somatic* and an inner *splanchnic layer.* The space between the two sheets, the *coelom,* forms the body cavity.
3. The anterior portion of the neural tube grows rapidly anteriorly, also expanding laterally and turning ventrally. By this process the endoderm of the yolk sac is folded in to form a blind pouch, the *foregut,* from which develops the anterior portion of the digestive tract. In a similar way, the *hindgut* is formed in the posterior region; from it develops an evagination, the *allantoic stalk,* which pushes out into the body stalk. The region between the foregut and the hindgut, forming the roof of the yolk sac, constitutes the *midgut.*
4. On either side of the foregut, the mesoderm hollows out to form two tubes lined with endothelium. These tubes meet and form a double-walled structure, which develops into the *heart.* In the embryo, *blood cells* are formed first in masses of mesodermal tissue (*blood islands*) in the walls

of the yolk sac. Later they are formed successively in the body mesenchyme, liver, spleen, thymus, lymph nodes, and finally, at about the fourth month, bone marrow. Blood vessels develop from spaces in the mesoderm.

5. The rapid increase in the size of the heart and the brain and the turning under of the posterior portion of the body constrict the yolk sac so that it becomes connected with the midgut by only a narrow tube, the *yolk stalk.* Downward development of the brain carries with it the surface ectoderm, which, in the region between the brain and the heart, comes into contact with the endoderm of the foregut to form the *oral membrane.* This forms the floor of an external invagination, the *stomodeum,* which gives rise to the mouth parts following rupture of the oral membrane.

6. In the posterior region, the end of the hindgut becomes the *cloaca,* into which open the urinary and genital ducts and the allantoic diverticulum. An external depression, the *proctodeum,* is separated from the hindgut by the *anal* or *cloacal membrane* and a ventral *urogenital sinus.* Evaginations from the foregut give rise to digestive glands (liver and pancreas) and the thyroid gland. The latter loses its connection with the gut and becomes an endocrine gland.

7. Each lateral mass of a somite differentiates into three regions: an upper portion, the *dermatome,* which gives rise to the dermis of the skin; a ventral portion, the *sclerotome,* which gives rise to the axial skeleton; and a lateral portion, the *myotome,* which gives rise to the skeletal muscles. Lateral to each somite is the *nephrotome;* it lies between the mesodermal segment and the somatic and splanchnic layers of mesoderm that enclose the coelom. From the nephrotome develop the urinary organs and their ducts. A rounded mass projects into the dorsal portion of the coelom to form the *urogenital fold,* on the median side of which develops the *genital fold,* which gives rise to the reproductive organs and their ducts.

It can be seen that all embryonic organs comprising the future organs and systems of the body arise from one or another of the three primary germ layers forming the embryonic disc. The principal derivatives of each of the germ layers are given in Table 8-1.

FORMATION OF THE EMBRYONIC MEMBRANES AND PLACENTA

Included among the accessory embryonic structures that have to do with protection, nutrition, respiration, and excretion are the yolk sac, amnion, chorion, allantois, umbilical cord, and placenta.

Yolk Sac. In lower animals (birds and reptiles), this structure is an important organ because it contains yolk, the primary source of food for the embryo. In humans, however, it is small and of little functional importance. Structurally, the endoderm of the yolk sac gives rise to the epithelium of the major portion of the digestive tract. Although (in contrast

TABLE 8-1

Ectoderm	*Mesoderm*	*Endoderm*
General Derivatives		
Nervous system, sense organs, mouth cavity, skin	Muscular, skeletal, circulatory, excretory, and reproductive systems	Digestive and respiratory systems, certain excretory and reproductive ducts
Specific Derivatives		
Epidermis of skin and its derivatives (hair, nails, and sebaceous, sweat, and mammary glands)	All muscle tissues (smooth, skeletal, cardiac)	Lining of alimentary canal (except terminal portions), including pharynx and derivatives (auditory tubes, tympanic cavity, thyroid, parathyroids)
Brain, spinal cord, ganglia, nerves	All connective tissues (bone, cartilage, ligaments, tendons)	Epithelium of digestive glands and their ducts (liver, pancreas, gallbladder, bile duct)
Lens, conjunctiva, retina, external and internal ear	All circulatory organs (heart, blood and lymph vessels, lymphatic organs, blood and blood-forming organs)	Lining of respiratory organs (except nasal cavity), including larynx, trachea, bronchial tree, lungs
Lining of buccal and nasal cavities and parts of the pharynx and the paranasal sinuses	Excretory organs (kidney, ureter, trigone of bladder)	Bladder (except trigone)
Epithelium of the salivary glands and enamel of the teeth	Reproductive organs (testes, ductus deferentes, seminal vesicles, ovaries, uterine tubes, vagina)	Female urethra and vestibular glands
Hypophysis cerebri	Serous membranes (pleura, pericardium, peritoneum)	Male urethra (proximal portion), prostate, and bulbourethral glands
Anus and distal portion of male urethra	Pulp, dentine, and cementum of teeth	
Medulla of adrenal gland	Cortex of the adrenal gland	

to lower animals) no yolk is contained in the yolk sac, an extensive vitelline circulation develops, connected with the embryo through a narrow *yolk stalk.* Blood cells are first formed in the walls of the yolk sac.

This sac becomes incorporated into the umbilical cord, and about the sixth week, the stalk loses its connection with the gut. The yolk sac continues to shrink but may persist up to the time of birth, at which time it is usually found between the amnion and the chorion. In some adults, the proximal end of the yolk stalk persists and forms a blind intestinal pouch called *Meckel's diverticulum.*

Amnion (Fig. 8-5). The amniotic cavity originates as a cavity in the inner cell mass. Its wall, a thin layer of ectoderm to which later a layer of somatic mesoderm is added, develops into the *amnion.* It increases in size until it surrounds the embryo except at the attachment of the umbilical cord. It is filled with *amniotic fluid,* in which the embryo is suspended. At about the third month, the amnion completely fills the chorionic sac and becomes adherent to the chorion, with which it fuses.

The *amniotic fluid* forms a protective envelope that surrounds the embryo, permitting freedom of movement and allowing for growth. At parturition, uterine pressure forces the amniotic sac into the cavity of the cervix, where it serves as a fluid wedge for dilatation of the cervix. The amnion usually ruptures during childbirth, with resultant loss of the fluid (the "flow of waters"). Amniotic fluid serves to moisten, lubricate, and disinfect the birth canal or vagina. If the amnion fails to burst, however, the head of the newborn may be presented covered by the amnion; this covering is the so-called *caul* or veil.

Amniotic fluid obtained by *amniocentesis* (puncture of the amniotic sac) is utilized in the diagnosis of various metabolic disorders, when it is suspected that a child might be born with an inherited defect. By karyotyping (chromosome analysis), disorders such as Down's syndrome and many others can be diagnosed. The sex of the fetus can be determined. Various abnormalities of fetal and placental function can also be detected. The presence of alpha fetoprotein indicates a neurological defect and, in an Rh-sensitized pregnancy, the amount of bilirubin increases.

While immersed within the fluid of the amniotic sac, the surface of the embryo is protected by a fatty deposit, the *vernix caseosa.*

Chorion (Fig. 8-5). The *chorion* is formed from the ectoderm of the trophoblast, to which is added an inner layer of mesoderm. Together these form the wall of the primitive blastocyst. Projections develop from it to form the *primary villi,* which in turn give rise to *secondary villi.* These profusely branched structures extend into the uterine mucosa. In them develop branches of the umbilical blood vessels. The development of the chorion is closely related to the development of the placenta. (See page 306.)

Although the chorion is an enveloping and protective structure, its primary function is the nutrition of the embryo.

Allantois. The *allantois* arises as a small diverticulum or outpocketing of the hindgut. It grows into the body stalk as a small, narrow tube. Blood vessels accompany it in its development and vascularize the chorion. After about the fourth week, the allantois ceases to grow and persists only as a rudimentary threadlike structure in the umbilical cord. Functionally, it is of little importance.

In egg-laying vertebrates such as birds and reptiles, the allantois develops into a large vesicular structure that completely envelops the embryo. It is highly vascularized and forms the lining of the shell, where it functions as a respiratory organ. It also has excretory and nutritive functions.

Umbilical Cord. As the embryo develops and the amniotic sac enlarges,

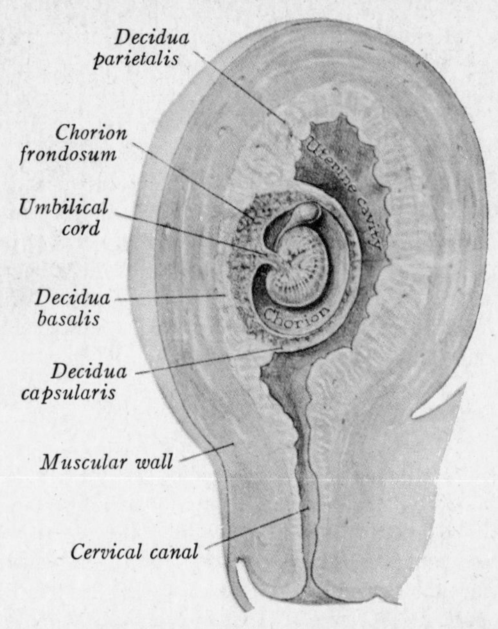

Fig. 8-6. Gravid human uterus of 5 weeks, hemisected to show the decidual relations to an implanted chorionic sac. (Reprinted with permission of W. B. Saunders Co., Philadelphia, from L. B. Arey, *Developmental Anatomy,* rev. 7th ed., 1974.)

the ventral unenclosed area of the embryo is reduced in size. This region, the *umbilicus,* is the point of junction between the embryonic and the extraembryonic tissues. Further development produces a tubular structure, the *umbilical cord,* which is actually a continuation of the body wall of the fetus, connecting distally with the fetal portion of the placenta. This cord encloses the yolk stalk and a portion of the embryonic coelom surrounding it, the allantois, and the allantoic vessels; the last of these become the *umbilical arteries* and *vein.*

At birth the umbilical cord is a spirally twisted cord about 60 cm long and 20 mm in diameter. It consists of an outer epithelial covering of ectoderm enclosing a loose type of embryonic connective tissue (mucous tissue or *Wharton's jelly*). Embedded in this tissue are the yolk stalk (and, in the early embryo, its vitelline vessels), the allantois, two umbilical arteries, and one umbilical vein.

Up to the sixth week, the embryonic gut protrudes into the coelom of the cord to form a temporary umbilical hernia. As development proceeds, the gut is gradually withdrawn, and the cavity of the cord disappears. The yolk stalk disappears, and the allantois persists as a solid strand of tissue.

Decidual Membranes (Fig. 8-6). The mucosa of the maternal uterus comes into so intimate a relationship with the embryonic chorion that at birth there is an extensive sloughing of the uterine lining. The endometrium of a pregnant uterus is known as the *decidua.*

When the blastocyst implants in the uterine mucosa, it becomes completely embedded within the endometrium. As the embryo increases in size, it begins to form an elevation on the mucosal surface. Further growth causes it to protrude into the uterine cavity. At this time, three regions of the uterine mucosa can be differentiated: (1) *decidua parietalis* (the nonplacental lining of the uterus), the portion of the mucosa that has no direct connection with the embryo; (2) *decidua capsularis,* the portion that forms a thin covering on the free surface of the chorion; and (3) *decidua basalis,* the portion between the chorion and the muscular wall of the uterus.

Placenta. Following implantation, villi develop over the entire surface of the chorionic vesicle, but after the vesicle begins to protrude into the uterine cavity, the villi on the surface facing the uterine cavity (i.e., that covered by the decidua capsularis) begin to atrophy, leaving a smooth surface called the *chorion laeve.* The villi in the region of the decidua basalis undergo extensive development and form the *chorion frondosum,* which constitutes the fetal part of the placenta, the decidua basalis forming the maternal part.

Umbilical arteries carry blood from the embryo to the villi of the chorion frondosum. The villi are surrounded by maternal blood, which occupies the *intervillous spaces* or *lacunae.* The blood of the fetus and that of the mother circulate in separate channels. Blood bathes the villi in a way similar to the way water in soil bathes the roots of plants. Substances that can diffuse through cellular membranes pass from one to the other; consequently, some substances of the mother's blood diffuse into the blood of the fetus. Among these are nutritive materials, oxygen, hormones, and antibodies. Coincidentally, substances in the blood of the fetus diffuse into the mother's blood, among which are waste products, such as urea, carbon dioxide, and water. There is, however, *no mixing of fetal and maternal blood.* Red blood cells, being 0.007 mm in diameter, are far too large to pass through the placental filter.

The *mature placenta* (Fig. 8-7) is a large discoid structure averaging 15 to 18 cm in diameter and about 2.5 cm in thickness. In the uterus it is concave; the umbilical vessels enter on the concave side. At its margin is a membrane formed from several structures, the decidua parietalis, decidua capsularis, chorion laeve, and amnion. The uterine surface of the mature placenta is divided into irregular regions or lobules called *cotyledons.*

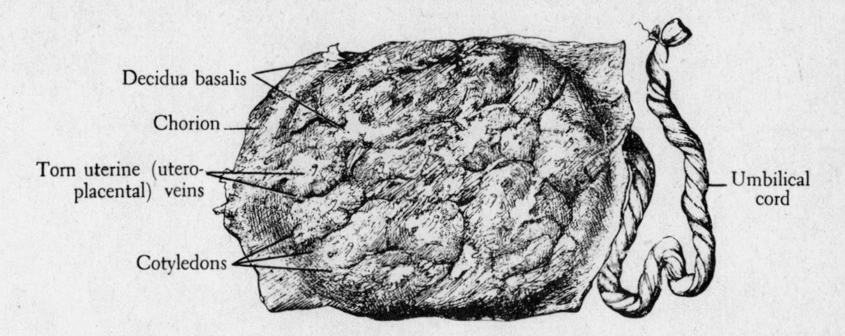

Fig. 8-7. Uterine surface of the placenta at full term. (Reprinted with permission of The Macmillan Company from C. V. Toldt and A. D. Rosa, *Atlas of Human Anatomy for Students and Physicians,* 1948.)

PREGNANCY, PARTURITION, AND LACTATION

The final steps in reproduction are pregnancy, parturition, and lactation. *Pregnancy* is the state of the female from the time of fertilization of the ovum to childbirth. *Parturition* is the process of giving birth to the child, beginning normally with the first uterine contractions that lead to the expulsion of the newborn. *Lactation* is the process of secreting milk for the nourishment of the newborn.

Pregnancy. The state of pregnancy, also called *gestation,* is firmly established after implantation. The corpus luteum persists and becomes the *corpus luteum of pregnancy;* follicle formation is inhibited, and menstruation ceases. The length of the period of pregnancy varies greatly, with extremes of 240 and 300 days. Average duration from time of conception is 9½ lunar months (38 weeks or 266 days). Counting from the first day of the last menstrual period, the average length is 280 days. The development of a fetus within the uterus constitutes *intrauterine* or *prenatal life.*

The periods of intrauterine development (Fig. 8-8) are as follows:

PERIOD OF THE OVUM. This, the pre-embryonic or germinal period, extends from the time of fertilization to the time of implantation, a period of 8 to 10 days. During this period, segmentation occurs: The developing zygote becomes differentiated into embryonic and extraembryonic portions, and the primary germ layers are established.

PERIOD OF THE EMBRYO. This extends from the end of the first period to the end of the second month, or approximately 48 to 50 days. By the end of this period, the embryo has begun to assume a "human" appearance, and the rudiments of all the organ systems have developed. Some of the changes taking place during the embryonic period are as follows:

1. *Change in general body form.* The body frame becomes straighter

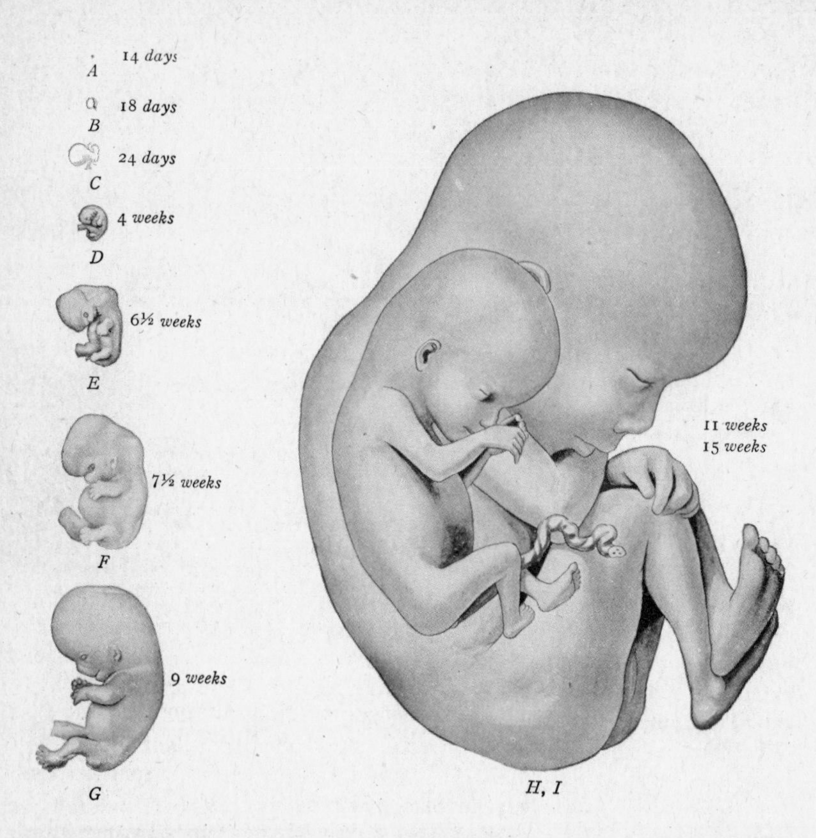

Fig. 8-8. Human embryos and fetuses at various stages of development. (Reprinted with permission of W. B. Saunders Co., Philadelphia, from L. B. Arey, *Developmental Anatomy,* rev. 7th ed., 1974.)

and most nearly erect. The dorsal convexity prominent in the early embryo is lost.

2. *Development of the face and neck* takes place between the fifth and eighth weeks and is closely related to the development of the pharynx and its associated *branchial arches,* elevated regions that appear on the lateral surfaces of the face and neck. The branchial arches, separated by *branchial grooves,* correspond to the gill arches characteristic of such aquatic forms as fishes. In these lower forms, they support gills that serve as respiratory organs, the water flowing through the *clefts* separating the arches. In humans, five arches separated by four external branchial grooves appear. On the inside of the pharynx, in positions corresponding to the four grooves, are outpocketings, the *pharyngeal pouches.* These are lined with

endoderm. Through modifications of these arches, the principal parts of the neck and face develop. The first branchial groove constitutes the primordium of the external ear. The first pharyngeal pouch differentiates into the auditory tube and the tympanic cavity of the middle ear. Other structures derived from the pharynx include the tongue, tonsils, thyroid and parathyroid glands, epiglottis, larynx, trachea, and lungs.

3. *Development and retrogression of the tail.* A tail develops, becoming prominent about the sixth week, after which it regresses and disappears.

4. *Alterations in the ventral portion of the body.* In early embryos, the heart and liver form a prominent ventral bulge. This structural characteristic continues until about the eighth week, when the gut begins to occupy the major portion of the abdominal region.

5. *The umbilical cord develops,* incorporating within itself the yolk stalk and the body stalk.

6. *The external genitalia appear,* representing the "indifferent stage" in sexual development.

7. *Limb buds appear,* and digits become well differentiated by the end of this period.

8. *Bone begins to appear;* centers of ossification develop.

9. *The nervous and muscular systems develop* to the extent that spontaneous movements can be initiated. Sense organs (eye, ear, and olfactory pit) make their appearance.

PERIOD OF THE FETUS. This extends from the beginning of the third month to parturition. Some of the changes that occur during this time are as follows:

In the third month: The head, which is very large in early embryos, becomes relatively smaller; limbs become longer; nails appear; eyelids fuse. The intestine is withdrawn from the umbilical herniation into the body cavity. The external genitalia differentiate to the extent that sex can be distinguished. Centers of bone formation become numerous.

In the fourth month: The fetus appears distinctly human. Hair appears on head and body. Bones become distinctly indicated. Sense organs (eye, ear, nose) assume typical forms.

In the fifth month: The hair coat (*lanugo*) is present. Blood formation in the bone marrow begins.

In the sixth and seventh months: The body form becomes better proportioned; the skin is wrinkled (Fig. 8-9).

In the eighth month: The fetus reaches the age of *viability;* that is, it is capable of surviving should it be born prematurely. (With the aid of incubators, however, fetuses born with as little as 5 months of intrauterine life and weighing only 500 g have survived.) In males, testes descend into the scrotum. Subcutaneous fat is deposited, and the body becomes plumper and smoother.

In the ninth month: Growth continues; additional fat accumulates. The lanugo is shed. Limbs become plumper, and nails extend to the tips of the

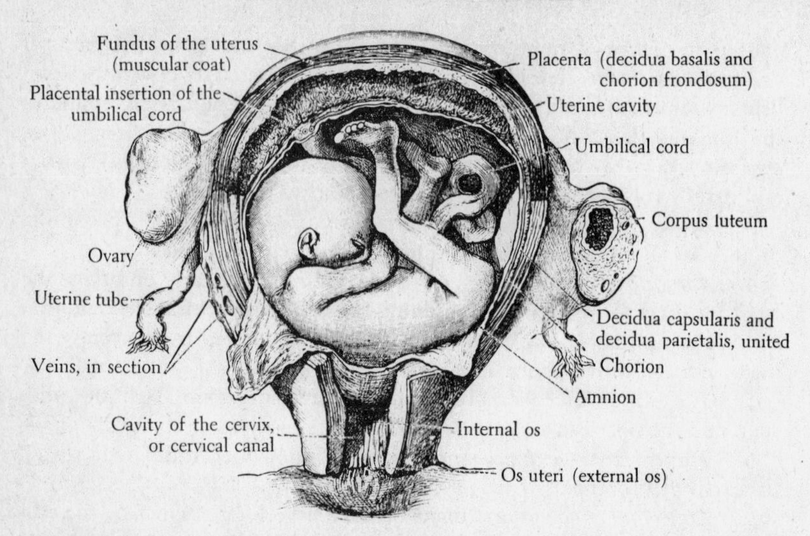

Fig. 8-9. Uterus, sixth month of pregnancy, opened by removal of the posterior wall. Fetus, with membranes and placenta, in transverse section. (Reprinted with permission of The Macmillan Company from C. V. Toldt and A. D. Rosa, *Atlas of Human Anatomy for Students and Physicians,* 1948).

digits. The embryo is now at *full term* and weighs, on the average, slightly more than 3 kg.

Pregnancy Tests. Various tests are employed to determine if a woman who has missed a menstrual period (the first sign of pregnancy) is pregnant or not. All tests are based on the presence or absence of *human chorionic gonadotrophin* (*HCG*), a hormone produced by the placenta and excreted in the urine. Tests commonly employed are of three types: biologic, immunologic, and radioimmunoassay.

In *biologic tests,* the urine of the subject being tested is injected into a laboratory animal (frog, toad, mouse, rat, or rabbit). If HCG is present in the urine, it will bring about the release of eggs or sperm from the gonads or produce observable vascular changes in the gonads of the test animal. These tests are not positive until 3 or 4 weeks after conception and have an accuracy of about 95 percent.

Biologic tests have been largely replaced by immunologic and radioimmunoassay tests, which are less difficult to perform, less expensive, and more accurate and provide results much earlier in the gestation period. *Immunologic tests* include latex agglutination or latex agglutination inhibition tests that can be performed in 2 min or a hemagglutination inhibition test that requires 2 hr. These tests are 95 to 98 percent accurate and have a high degree of sensitivity.

The *radioimmunoassay test* is based on competitive inhibition for binding sites on antibodies. It is extremely sensitive and can give a positive result as early as 10 days after conception. By this method, the presence of HCG in a sample of the subject's blood can be detected. It is 98 to 99 percent accurate.

The early diagnosis of pregnancy enables a pregnant woman to avoid possible complications arising from taking certain medications and avoiding unnecessary X-rays, which have deleterious effects on a developing embryo.

Maternal Changes During Pregnancy. While the foregoing changes are taking place in the embryo and fetus, correlated changes are occurring in the mother (Fig. 8-9). During the period of the ovum, the chorionic vesicle containing the embryo becomes embedded within the uterine mucosa. During the period of the embryo, the vesicle increases in size and fills the cavity of the uterus. By this time, the placenta has become established. As the fetus grows, the uterus increases in size and in weight. The *nonpregnant uterus* is about 6.5 cm in length, weighs about 50 g, and has a capacity of 2 to 5 cm^3. The *pregnant uterus at full term* is about 50 cm in length, weighs about 1000 g, and has a capacity of 5000 to 7000 cm^3.

With increased size, uterine muscles hypertrophy. The muscle cells become longer and increase in number. The blood supply to the uterus is also greatly increased.

Until about the fourth month, the uterus with its contained fetus occupies space only within the pelvic cavity. After this time, an increase in size causes it to push upward into the abdominal cavity, where it brings about considerable displacement of the viscera. The body wall becomes distended, and the skin on the face and neck may acquire a yellowish brown pigment, the "mask of pregnancy." Mammary glands enlarge with the development of their secretory tissue. Posture changes come about, and the pubic symphysis and sacroiliac joints soften. Changes in bone and teeth may occur, especially if there is a calcium deficiency. Kidney function is increased, and disorders of urination often arise owing to pressure on the bladder.

Parturition (Childbirth). About 280 days after the onset of the last menstrual period before pregnancy, the fetus is at *full term* (sometimes described merely as "at term") and ready to be expelled from the body of the mother. The process of expulsion is called *parturition, labor,* or *delivery.*

DURATION OF LABOR. The average duration of labor in a primiparous woman (pregnant with first child) among whites is $16\frac{1}{2}$ hr, among blacks $17\frac{2}{3}$ hr. Among multiparous women (those who have previously borne one or more children in as many pregnancies) these figures are, respectively, 11 and $12\frac{1}{2}$ hr.

Parturition is divided into three stages:

First Stage of Labor. Contractions of the uterus bring on the first "labor pains." Repeated contractions force the amniotic sac into the cervix, bringing about its dilatation, which permits the head of the fetus to enter the cervical canal and the upper end of the vagina. This usually takes about 6 hr; in some mothers the time is much shorter and in others longer. The amnion usually ruptures at this point, the escaping amniotic fluid serving to moisten and lubricate the birth canal.

Second Stage of Labor. This period extends from the time of complete dilatation of the cervix to and including the expulsion of the fetus. It may last several hours. Uterine contractions and concomitant pains occur at much shorter intervals and are more severe. Contractions of the abdominal muscles assist the uterus in expelling the fetus. Following delivery of the child, the umbilical cord is ligated near the umbilicus, then severed. The uterus continues to contract, expelling the remaining portion of the amniotic fluid.

Third Stage of Labor. After a lull, uterine contractions resume, and the *placenta* ("afterbirth") is expelled. This may occur within a short time after delivery of the newborn (20 to 30 min later) or not for some hours. The uterus then continues to contract, serving to close the ends of ruptured blood vessels and to arrest the flow of blood. The processes of this stage are accelerated if the baby is put to suck at the mother's breast immediately after birth.

CESAREAN SECTION. This is removal of a fetus from the uterus through an incision made in the abdominal wall and uterus. It is resorted to (1) in cases of a contracted pelvis or a tumor blocking the birth canal, (2) when a patient has had a previous cesarean section, (3) in certain cases of toxemia of pregnancy, (4) in certain cases of placenta previa or premature separation of the placenta, and (5) in infections of the vagina, especially active herpes genitalis.

TEST-TUBE BABIES. A procedure called *embryo transfer* has been developed in recent years that enables a woman whose uterine tubes have been damaged or removed to become pregnant and give birth. The technique involves treatment of a woman to produce superovulation, removal of the ova to a container outside the body, the addition of spermatozoa (*in vitro fertilization*), the implantation of one or two developing blastocysts into the woman's uterus, and subsequent completion of the normal process of pregnancy.

NATURAL CHILDBIRTH. The birth of a baby is usually accompanied by pain, sometimes slight, at other times more severe. For the relief of pain, various analgesics and anesthetics have been employed. However, recent studies have shown that the use of anesthetic agents may be hazardous to both mother and fetus, anesthetics being a potential cause of maternal death and frequently the cause of respiratory dysfunction and possible brain damage in newborn babies.

As a consequence, various programs have been developed to encourage natural childbirth. These are designed to lessen fear, apprehension, and tension, which are thought to be primary factors in the production of pain. The emphasis is on educational and psychological preparation. Body exercises, controlled breathing, and various relaxation techniques are used.

One method employs the principle of conditioned responses by which a patient blocks out painful sensations by substituting another stimulus at the appropriate time. Education prepares the prospective parent both physically and emotionally for the event of childbirth and encourages the development of attitudes that recognize childbirth as a normal function.

NONVIOLENT CHILDBIRTH. Emphasis is now being placed on lessening the trauma experienced by the emergence of a baby from the quiet, warm, moist, dark environment of the womb into a new world of activity, bright lights, noise, a sudden drop in temperature, a dry atmosphere, and rough handling. It is believed that a newborn is much more conscious of and sensitive to its environment than previously thought, that sights and sounds register with the newborn and have a profound influence on its later life. More attention is now being paid to the Leboyer "birth without violence" procedures designed to lessen the pain, confusion, and fear that are the usual experiences of a newborn infant.

Lactation. The production of milk by the mammary glands for nourishment of the young is called *lactation.* Following implantation, collapsed alveoli of the mammary glands enlarge and expand. New branches grow out from the ducts and develop into new alveoli. As the secreting tissue increases, the interstitial tissue becomes compressed, and fatty tissue disappears. Toward the end of pregnancy, secretion begins and *colostrum* appears. This is a watery fluid that differs from milk in the small amount of fat it contains and in the presence of *colostrum corpuscles* (believed to be phagocytic cells distended with fat particles). Extremely high in protein content, colostrum is believed to contain substances that give the child immunity to the most common infectious diseases during the first few months of life. Colostrum disappears on the second or third day after delivery and is replaced by true milk.

HUMAN MILK VERSUS COW'S MILK. Milk is an emulsion of fat particles suspended in a watery fluid containing proteins, sugar (especially lactose), and inorganic salts. Human milk and cow's milk are compared in the following table:

Substance	*Human Milk*	*Cow's Milk*
Water	88.3%	87.3%
Inorganic salts	0.2%	0.7%
Protein	1.5%	3.8%
Fat	4.0%	4.0%
Sugars	6.0%	4.5%
Reaction	Alkaline	Acid

Human milk is considered superior to cow's milk or other foods in the following respects:

1. It is cleaner and freer from bacteria.
2. Its composition varies during the first few weeks of lactation, in accordance with the needs of the infant.

3. Its protein is principally soluble lactalbumin instead of the relatively insoluble caseinogen of cow's milk. The curd formed in the stomach from human milk is therefore less dense and is more readily digested.
4. The fat is in a more finely emulsified form, and there is a smaller proportion of fatty acids.
5. The percentage of lactose is higher.
6. Immune bodies from the mother's blood induce a greater degree of immunity in the infant.
7. Breast-fed babies, in contrast to bottle-fed babies, suffer less frequently from disturbances of the gastrointestinal tract (diarrhea, for example), and they recover more quickly from postnatal birth-weight loss.

The secretion of milk (*lactogenesis*) and its release are primarily under the control of pituitary hormones. *Prolactin,* secreted by the anterior pituitary and acting synergistically with adrenal cortical hormones, plays a primary role in the initiation and maintenance of milk secretion. However, other hormones, especially ACTH, STH, and TSH, influence milk production. Estrogen and progesterone have an inhibiting effect. Milk ejection depends on *oxytocin,* produced by the hypothalamus but released by the posterior lobe of the pituitary in response to milk removal by a sucking infant.

The average yield of milk in the human nursing mother is from 600 to 800 g per day, but three to four times these amounts may be yielded over a long period without detrimental effects. Because milk is being constantly secreted in the period of lactation, repeated suckling by the infant or artificial removal is necessary. If the milk is not removed, secretory activities cease, and the mammary glands undergo involution. Milk will continue to be secreted as long as the infant suckles, usually a period of 8 to 9 months. The gradual substitution of other foods for breast milk and the cessation of breast feeding constitute *weaning.*

Sucking by the infant aids in the involution of the uterus, gives the mother greater pleasure in her offspring, and plays an important role in the psychological development of the infant.

POSTNATAL DEVELOPMENT AND LATER STAGES OF LIFE

The life of the individual following birth is termed *postnatal* or *extrauterine development.* This usually applies to life through the end of infancy and may even refer to early childhood. The following is a chronological analysis of the periods of life, including "postnatal development":

1. *Period of the newborn.* Also called the *neonatal period,* this extends from birth to about the end of the first month.
2. *Infancy.* This extends from the end of the first month to about the

end of the first year or to the time when an erect posture is assumed (on the average, about the sixteenth month).

3. *Childhood.* This extends from infancy to puberty. *Early childhood,* the "milk-tooth period," lasts from years 1 to 6; *middle childhood,* the "permanent-tooth period," from years 6 to 10; *late childhood* or the "prepubertal period," from age 10 to puberty (13 to 15 in males, 12 to 14 in females).

4. *Puberty.* This stage of life is commonly identified in females with the first menstruation (*menarche*), about the middle of the twelfth year, in males toward the end of the fourteenth year. Puberty should not, however, be identified with a specific event; it should be regarded as a period of gradual development extending through two-thirds of the period of adolescence. During this period, the sex organs become functional and secondary sex characteristics develop.

5. *Adolescence.* This period extends from puberty (which is extremely variable) to about the twenty-first year in females, the twenty-fourth year in males.

6. *Maturity. Early maturity* lasts from the end of adolescent period to about age 35; *later maturity* from then until age 65.

7. *Terminal age.* The period of life following later maturity.

ABNORMAL DEVELOPMENT OF THE FETUS

A considerable degree of variation occurs in the development of the human body. Whenever the organism as a whole or an individual part deviates from the normal range of variation, resulting in malformation, the condition is referred to as an *anomaly.* If the fetus as a whole is grossly malformed, it is called a *monster.* The study of abnormal development constitutes the science of *teratology.*

Nature of Fetal Anomalies. Abnormalities may be functional or structural. Among functional abnormalities are such conditions as abnormal protein metabolism, photophobia (abnormal sensitivity to light), color blindness, and hemophilia. Structural anomalies fall into the following general types:

1. *Absence of an organ or a part* (e.g., absence of a finger, an arm, or an external ear).

2. *Developmental arrest,* in which only partial development occurs (e.g., dwarfism, infantile uterus).

3. *Developmental excess,* in which growth or development is exaggerated (e.g., extra digits, hypertrophy of the clitoris).

4. *Failure to atrophy,* in which an embryonic structure that normally atrophies persists (e.g., a tail, a pupillary membrane, the hyaloid artery in the vitreous body of the eye, Meckel's diverticulum of the ileum).

5. *Failure of fusion,* in which paired embryonic parts that normally fuse fail to do so (e.g., double uterus, cleft palate).

6. *Fusion,* in which parts normally paired are united into a single structure (e.g., horseshoe kidney).

7. *Splitting,* in which parts normally single are paired or split (e.g., a ureter).

8. *Failure to subdivide* (e.g., fused digits).

9. *Persistence of embryonic ducts or openings* that normally close (e.g., a patent ductus arteriosus, persistent foramen ovale, cervical fistula).

10. *Stenosis* or abnormal narrowing of a duct or opening (e.g., pyloric stenosis, aortic stenosis).

11. *Abnormal migration,* in which the normal shifting of an embryonic structure fails to occur (e.g., cryptorchism) or migration of structures to abnormal positions, (e.g., parathyroids occurring in the thorax).

12. *Misplacement,* in which organs develop in abnormal positions (e.g., palatine teeth in the roof of the mouth) or transposition of viscera, (e.g., organs that normally occur on one side of the body appear on the opposite side).

Causes of Abnormalities. Abnormalities are either inherited or acquired.

INHERITED ABNORMALITIES. These are the result of the action of the genes transmitted through the egg or sperm. Anomalies of this type tend to recur generation after generation in a more or less definite pattern following the laws of heredity. New characteristics may make their appearance as a result of mutations (spontaneous changes in the genes) or changes in chromosome number. *Down's syndrome* or *mongolism* is due to an extra chromosome. The primary cause in many cases is not known. Such changes have been produced experimentally in animals and plants by exposing the germ cells to X-rays or radioactive substances.

ACQUIRED ABNORMALITIES. These are due to some environmental factor or influence that acts on the developing individual or a part of the organism. In the process of development, each organ or other structure passes through a *critical period* at which time the growth rate is accelerated and differentiation processes are brought into play; the organs and tissues are markedly susceptible to any abnormal influences brought to bear upon them. In general, most abnormalities have their origin in the embryonic period (from the second to the eighth week), for it is then that all the organ systems arise from their respective germ layers and the fundamental structure of each organ is established. Some of the environmental factors that may modify embryonic development are mechanical factors, radiation, chemicals, and disease and disease organisms.

Mechanical Factors. Experimentally, forces such as pressure or constriction may modify the development of ova outside the body, but the human embryo developing within the uterus seems to be remarkably free from the effects of such influences. Blows to the abdomen, unless of unusual severity, are unlikely to injure the fetus. However, a deficiency of *amniotic fluid* may result in distorted forms, such as compression babies. Uterine disease or faulty placental attachment may cause anomalies. Extreme shortness of the umbilical cord or abnormal attachment of the cord to the placenta may result in abnormal development.

Radiation. Tissues in which cells are multiplying rapidly are peculiarly susceptible to radiation. Such tissues include the spermatogenic cells of the testes and the hemopoietic cells of the bone marrow. In a developing embryo, sublethal doses of X-rays or exposure to radioactive substances have been shown to produce anomalies.

Chemicals. Chemical substances in the mother's blood can be transmitted through the placenta to the fetus. Chemical substances known to influence fetal development include drugs, vitamins, hormones, and antibiotics.

Drugs are taken for a variety of reasons—to relieve pain, to combat infection, to stimulate or inhibit the nervous system, to facilitate various metabolic processes, to effect birth control, or for other reasons. A pregnant woman may be able to utilize these substances effectively; an embryo or fetus may not. Consequently, the use of certain drugs may result in serious anatomic or physiologic disorders. Some drugs are definitely *teratogenic,* that is, capable of causing a physical defect in a developing organism. It is recommended that a prospective mother *take no drugs or medications* of any kind except under the direction of a physician. This includes analgesics (including aspirin), antibiotics, sleeping pills, weight-reducing pills, appetite suppressors, stimulants, tranquilizers, antidepressants, anticoagulants, and hormones, especially corticosteroids. A woman should avoid these drugs at all periods when she could become pregnant, since the damage to the embryo may occur before she knows she is pregnant. It is also recommended that pregnant mothers abstain from smoking and drinking, since both tobacco and alcohol may have detrimental effects on the fetus. Recent studies have shown that caffeine in excessive amounts can cause birth defects.

Experimental work on laboratory animals has shown that a deficiency in *vitamin A* may cause eye defects, diaphragmatic hernia, and abnormal kidney development; *riboflavin* deficiency, skeletal defects and cleft palate; *vitamin* B_{12} deficiency, cranial abnormalities; and *folic acid* deficiency, a number of defects. In all experiments, the fetal death rate was above normal. Adequate vitamin intake is of special importance to pregnant women, and greater daily allowances of all vitamins, especially those of the vitamin B complex, are recommended.

There is evidence that *hormones* produced by the mother's endocrine glands, especially the adrenal gland, have a considerable influence on the development and well-being of the fetus. As the secretion of most hormones is either directly or indirectly under the influence of the nervous system, the mental state of a prospective mother may play an important role in the normal development of the fetus.

An Rh-negative woman bearing an Rh-positive fetus, if she has been previously sensitized, may develop *antibodies* to the Rh factor. These, on entering the fetus through the placenta, will tend to destroy the blood cells of the developing fetus, resulting in death or serious illness of the fetus (erythroblastosis fetalis).

Disease and Disease Organisms. The *syphilis* organism may cross the placental barrier and infect the embryo, resulting in abnormalities of the bone or teeth. The number of stillbirths among syphilitic mothers is much higher than among nonsyphilitic mothers. *Rubella* (*German measles*), a viral disease, if acquired by the mother during the early months of pregnancy, usually results in abnormalities. In connection with this disease, the incidence of *congenital cataract* is usually high, especially when the disease occurs during the second or third month of pregnancy; heart lesions, microcephaly, deaf-mutism, and other defects have been noted. *Herpes genitalis,* another viral infection, is also a common cause of birth defects.

MULTIPLE PREGNANCIES

In a multiple pregnancy, more than one fetus develops within the uterus. The frequency of twins is approximately one in 90 births, triplets one in 90^2 and quadruplets one in 90^3. Births exceeding four in number, e.g. quintuplets, have increased in number in recent years. This has resulted from overstimulation of the ovary by human gonadotropins administered to women to correct ovulatory failure.

A twin is one of two individuals developed within the uterus at the same time as a result of a single impregnation. Twins are of two types: *diovular* and *monovular.*

Diovular Twins. These are also called *dizygotic, fraternal,* or *"false"* *twins.* They result from the fertilization of two separate ova. They may be of the same or opposite sexes. Extraembryonic membranes develop separately for each, though secondary fusion may occur. About 75 percent of all cases of twinning belong to this type.

Monovular Twins. These are also called *monozygotic, similar, duplicate,* or *identical twins,* since both develop from a single fertilized ovum that at an early stage in development divides into two masses, each of which develops into a complete individual. This may occur as a result of (1) the development of two inner cell masses within the blastocyst or (2) the development of two organization centers within a single cell mass (or the division of a single organization center by budding or fission) to form two embryonic axes, each capable of developing into a complete embryo.

Anomalies. Malformations occur frequently among identical twins. The twins may develop unequally, resulting in individuals of unequal size. They may be *conjoined* and form a "double monster." The degree of fusion may be slight, involving only superficial tissues, or it may be extensive, involving visceral and skeletal parts. Sometimes the separation of conjoined twins after birth is possible; in other cases, they remain united through life as "Siamese twins."

In cases involving the splitting of the embryonic axis, the fission may be complete, with all parts of the body duplicated, or it may be partial, with some parts double, others single. The latter process accounts for the formation of two-headed babies or of babies with a single head and neck and a double trunk.

BIRTH CONTROL

Birth control, also called *fertility control,* includes any measures employed to limit the number of offspring. In a broad sense it includes abstinence, contraception, contraimplantation, sterilization, and abortion. All in some way prevent either the production of germ cells, their union to form a zygote, the development of a fertilized ovum, or the birth of a viable baby.

Contraception literally means "prevention of conception." However, the term is now generally used as a synonym for *birth control* and includes any measures used to prevent conception or the development of an embryo or fetus.

Contraceptive measures are resorted to (1) when there is a likelihood that the offspring might be defective as a result of hereditary factors or congenital disease; (2) when a pregnant woman has been subjected to conditions such as disease, drugs, or excessive radiation, which might cause defective development of a fetus; (3) for the prevention of unwanted pregnancies; (4) for the limitation of the number of offspring for social, economic, health, or other reasons; (5) for the optimum spacing of pregnancies; and (6) as a means of limiting world population, which, if unchecked, will within the near future result in overpopulation, with catastrophic results.

Birth control or contraception includes any measures that (1) prevent the germ cells (ova or sperm) from being produced or liberated from the ovary or testis; (2) prevent the sperm from being deposited within the vagina or, if deposited, kill or inactivate them or prevent their union with the egg; (3) prevent the fertilized ovum from implanting in the uterine endometrium, or if it implants, prevent it from developing; (4) limit intercourse to a period when the ovum is not present to be fertilized. Some measures are simple and can be self-employed; others are more complicated and necessitate medical supervision. Sterilization and abortion procedures require surgical intervention.

Methods. Each method of birth control has certain advantages and disadvantages, and the effectiveness of each depends on a number of factors. The following are the principal procedures employed, with a brief explanation of each.

WITHDRAWAL OR COITUS INTERRUPTUS. This involves withdrawal of the penis during intercourse just prior to ejaculation so that theoretically no spermatozoa are deposited within the vagina.

CONDOM OR RUBBER. This is a thin sheath of latex or a similar material that is placed over the erect penis prior to its insertion into the vagina. Semen, when discharged, is collected within the condom, and consequently no spermatozoa are deposited within the vagina. The condom also serves a prophylactic function in that it prevents the male from acquiring a venereal infection or transferring one to the female.

DIAPHRAGM AND CERVICAL CAP. The diaphragm is a dish-shaped device; the cervical cap is thimble-shaped. Both are made of soft rubber and designed to fit over the cervix of the uterus. When properly inserted, they prevent sperm from entering the uterus. Their effectiveness is increased when a spermicidal cream is used with them.

CHEMICAL CONTRACEPTIVES. These are substances that, when placed within the vagina, immobilize or kill spermatozoa. When combined with a gelatinous or oily base, they also form a physical barrier that acts to impede the movement of spermatozoa. They include jellies, creams, and foams that are applied in the form of suppositories or pessaries inserted into the vagina or from aerosol or other types of containers with special applicators inserted into the vagina. These preparations are readily available in most pharmacies and do not require a doctor's prescription.

"SAFE PERIOD" OR RHYTHM METHOD. This method is based on the assumption that the ovary releases an ovum about the fourteenth day of the average menstrual cycle and that intercourse must take place either a day or two before or after ovulation for fertilization to occur. Intercourse at other times—during the week following or the week preceding menstruation—is considered as occurring in a relatively safe period, during which conception is unlikely to occur. The safe period can be monitored by noting changes in body temperature and cervical mucus. At ovulation, body temperature rises about 0.6°C and cervical mucus becomes clear and watery.

INTRAUTERINE DEVICE (IUD). An intrauterine contraceptive device is inserted through the cervix into the uterus. Its presence as a foreign body acts in some way to prevent a blastocyst from implanting in the uterine endometrium (*contraimplantation*). Some IUDs release copper or progesterone, which increases their effectiveness.

ORAL CONTRACEPTIVES ("THE PILL"). Oral contraceptives are synthetic estrogens and progestogens that inhibit the production of the gonadotropic hormones FSH and LH by the pituitary gland. In their absence, ovarian follicles fail to mature and liberate ova. Oral contraceptives are taken for a period of 20 days, then discontinued until the end of the normal menstrual period, which should occur within about 5 days. Pills are of two types, *combination* and *sequential*. In the combination pill, estrogen and progesterone are combined in various proportions. In the sequential pill, the first 15 pills contain estrogen, the remaining 5 progesterone. Contraceptive pills are available only by prescription and should be taken only under medical supervision.

DOUCHING. This is the process of washing out the vagina by injection of a solution containing a germicidal substance.

Advantages, Disadvantages, and Effectiveness. Each contraceptive method presents certain advantages and disadvantages, and the effectiveness of each may vary from person to person. Information on how to make the most effective use of each can be obtained from a physician, health center, sex counselor, or therapist or from most texts dealing with human sex and sexuality.

EFFECTIVENESS. It is difficult to determine the exact effectiveness of the various contraceptive methods because of the many variables involved; the following table gives the *approximate* effectiveness of the common methods employed.

	Estimated Effectiveness (percent)
Highly effective	
The Pill	99
Intrauterine device (IUD)	95
Condom	90
Diaphragm or cervical cap, with foam	90
Moderately effective	
Chemical barriers alone	80
Rhythm method	
For regular women only	85
For all women	70
Low effectiveness	
Withdrawal	40–50
Douche	30–40

Sterilization. This is the process by which a person is rendered incapable of reproducing. In the *female,* it is resorted to when anatomic or physiologic disorders make childbearing difficult or impossible, when genetically defective offspring are likely to be produced, or as a method of permanent birth control. It is accomplished by *ovariectomy* (removal of the ovaries), *hysterectomy* (removal of the uterus), or *tubal ligation* (tying off or removal of a portion of each uterine tube). In the *male,* it may be accomplished by castration or vasectomy. *Castration* or *orchiectomy* is removal of the testes. This is rarely resorted to, for the testes serve not only for the production of germ cells but also for the production of male hormones essential for the maintenance of male libido. *Vasectomy* consists of tying off or removal of a portion of each sperm duct or ductus deferens. It is employed primarily as a contraceptive procedure. In most cases, sterilization measures in both females and males are 100 percent effective, and only rarely can they be reversed.

Abortion. Abortion is the expulsion of an embryo or nonviable fetus. When occurring during the first 12 weeks of gestation, it is called an *abortion;* from this time to viability (28 weeks), a *miscarriage;* from the period of viability to full term, a *premature delivery.* However, the term *abortion* is now generally applied to all categories.

Abortions may occur naturally or may be induced. *Natural abortions* occur commonly as a result of abnormal development, endocrine disturbances, or maternal disease. *Induced abortions* are brought about by artificial means. A *therapeutic abortion* is performed when the life or health of the mother is threatened by disease or complications of pregnancy or when the developing fetus is known to be defective, as determined by *amniocentesis,* in which loose cells from the amniotic fluid are sampled and examined for chromosome abnormalities. Abortions are sometimes utilized as a method of birth control, especially in cases of unwanted pregnancy.

Methods involved in inducing an abortion include dilatation and curettage, uterine aspiration, and intra-amniotic injection.

DILATATION AND CURETTAGE (D & C). In this procedure, the cervix is dilated and the endometrium scraped by a spoonlike instrument, a *curette.* When performed during the first 3 months of pregnancy, it is relatively safe and without serious physical aftereffects.

UTERINE ASPIRATION OR VACUUM CURETTAGE. A suction curette (*vacurette*) is inserted into the uterus and, by suction, the conceptus (embryo and placental material) is withdrawn. When performed within a week or two following a missed menstrual period, it is called *menstrual extraction.*

INTRA-AMNIOTIC INJECTION. This is also known as the *saline method* or "*salting out.*" The amniotic sac is punctured (*amniocentesis*) and the amniotic fluid withdrawn, followed by the injection of a strong saline or glucose solution. Expulsion of the fetus follows a day or two later. This method is employed only in the late stages of pregnancy and is considered a hazardous procedure.

SELECTED REFERENCES

GENERAL ANATOMY (COMPLETE TEXTS)

ANSON, B. J. *Morris' Human Anatomy,* 12th ed. New York: McGraw-Hill, 1966.

BASMAJIAN, J. V. *Grant's Method of Anatomy,* 10th ed. Baltimore: Williams & Wilkins, 1980.

CUNNINGHAM, D. J. *Textbook of Human Anatomy,* 12th ed. New York: Oxford University Press, 1981.

GARDNER, D., et al. *Anatomy,* 4th ed. Philadelphia: Saunders, 1975.

GOSS, C. M. *Gray's Anatomy of the Human Body,* 29th ed. Philadelphia: Lea & Febiger, 1973.

HAMILTON, W. J. *Textbook of Human Anatomy,* 2d ed. Philadelphia: Lippincott, 1976.

LOCKHART, R. D., et al. *Anatomy of the Human Body,* 2d ed. Winchester, Md.: Faber & Faber, 1981.

WOODBURNE, R. T. *Essentials of Human Anatomy,* 7th ed. New York: Oxford University Press, 1983.

ANATOMY AND PHYSIOLOGY (SHORTER TEXTS)

ANTHONY, C. P., and THIBEDEAU, G. A. *Textbook of Anatomy and Physiology,* 11th ed. St. Louis: Mosby, 1983.

BASMAJIAN, J. V. *Primary Anatomy,* 8th ed. Baltimore: Williams & Wilkins, 1982.

CRAFTS, R. C. *Textbook of Human Anatomy,* 2d ed. New York: Wiley, 1979.

CROUCH, J. E., and MCCLINTIC, J. R. *Human Anatomy and Physiology,* 2d ed. New York: Wiley, 1976.

GARDNER, E., et al. *Anatomy.* Philadelphia: Saunders, 1975.

JACOB, S. W., et al. *Structure and Function in Man,* 5th ed. Philadelphia: Saunders, 1982.

KIMBER, D. C., et al. *Textbook of Anatomy and Physiology,* 16th ed. New York: Macmillan, 1972.

MONTGOMERY, R. L. *Basic Anatomy for the Allied Health Professions.* Baltimore: Urban & Schwarzenberg, 1980.

TORTORA, G. J., and ANAGNOSTAKOS, N. P. *Principles of Anatomy and Physiology,* 3d ed. New York: Harper & Row, 1981.

PHYSIOLOGY

BROMBECK, J. R., ed. *Best and Taylor's Physiological Basis of Medical Practice,* 9th ed. Baltimore: Williams & Wilkins, 1973.

FOX, S. I. *Human Physiology.* Dubuque, Ia.: Brown, 1981.

GANONG, W. F. *Review of Medical Physiology,* 11th ed. Los Altos, Calif.: Lange Medical Publications, 1983.

GUYTON, A. C. *Textbook of Medical Physiology,* 6th ed. Philadelphia: Saunders, 1981.

PRICE, S. A., and WILSON, L. M. *Pathophysiology,* 2d ed. New York: McGraw-Hill, 1982.

RUCH, T. C., and PATTEN, H. D., eds. *Physiology and Biophysics,* 20th ed., vols. 1–3. Philadelphia: Saunders, 1974.

VANDER, A. J., et al. *Human Physiology,* 3d ed. New York: McGraw-Hill, 1980.

HISTOLOGY

BLOOM, W., and FAWCETT, D. W. *Textbook of Histology,* 10th ed. Philadelphia: Saunders, 1975.

COPENHAVEN, W. M., et al. *Bailey's Textbook of Histology,* 17th ed. Baltimore: Williams & Wilkins, 1975.

DIFIORE, M. S. H. *Atlas of Human Anatomy,* 5th ed. Philadelphia: Lea & Febiger, 1981.

HAM, A. W., and CORMACK, D. H. *Histology,* 8th ed. Philadelphia: Lippincott, 1979.

LEESON, C. R., and LEESON, T. S. *Histology,* 4th ed. Philadelphia: Saunders, 1981.

WEISS, L., and GREEP, R. O. *Histology,* 4th ed. New York: McGraw-Hill, 1981.

EMBRYOLOGY

BALINSKY, B. I. *An Introduction to Embryology,* 5th ed. Philadelphia: Saunders, 1981.

CARLSON, B. M. *Patten's Foundations of Embryology,* 4th ed. New York: McGraw-Hill, 1981.

KARP, G., and MERRILL, W. J. *Development.* New York: McGraw-Hill, 1981.

TUCHMANN-DAPLEASIS, H., and HAEGEL, P. *Illustrated Human Embryology,* vols. 1–3. New York: Springer-Verlag, 1974.

NEUROANATOMY AND NEUROLOGY

AFIFI, A. K., and BERGMAN, R. A. *Basic Neuroscience.* Baltimore: Urban & Schwarzenberg, 1980.

BANNISTER, R., ed. *Brain's Clinical Neurology,* 5th ed. New York: Oxford University Press, 1978.

CHAUSID, J. G. *Neuroanatomy and Functional Neurology,* 18th ed. Los Altos, Calif.: Lange Medical Publications, 1982.

GILROY, J., and MEYER, J. S. *Medical Neurology,* 3d ed. New York: Macmillan, 1979.

RANSOM, S. W., and CLARK, S. L. *Anatomy of the Nervous System,* 10th ed. Philadelphia: Saunders, 1959.

KINESIOLOGY

BASMAJIAN, J. V. *Muscles Alive,* 4th ed. Baltimore: Williams & Wilkins, 1978.

KENDAL, H. O., et al. *Muscles, Testing and Function,* 2d ed. Baltimore: Williams & Wilkins, 1971.

LEHMKUHL, L. D., and SMITH, L. K. *Brunnstrom's Clinical Kinesiology,* 4th ed. Philadelphia: Davis, 1983.

RASCH, P. J., and BURKE, R. K. *Kinesiology and Applied Anatomy,* 6th ed. Philadelphia: Lea & Febiger, 1978.

WELLS, K. F., and LUTTEENS, K. *Kinesiology.* Philadelphia: Saunders, 1976.

ENDOCRINOLOGY

DILLON, R. S. *Handbook of Endocrinology,* 2d ed. Philadelphia: Lea & Febiger, 1980.

MARTIN, C. M. *Textbook of Endocrine Physiology.* Baltimore: Williams & Wilkins, 1976.

TURNER, C. D., and AGNARA, J. T. *General Endocrinology,* 6th ed. Philadelphia: Saunders, 1976.

VILLEE, D. E. *Human Endocrinology.* Philadelphia: Saunders, 1975.

WILLIAMS, R. H., ed. *Textbook of Endocrinology,* 6th ed. Philadelphia: Saunders, 1981.

ATLASES

ANSON, B. J. *An Atlas of Human Anatomy,* 2d ed. Philadelphia: Saunders, 1963.

FROHSE, F., et al. *Atlas of Human Anatomy,* 6th ed. New York: Barnes & Noble Books, 1961.

LANGMAN, J., and WOERDEMAN, M. W. *Atlas of Medical Anatomy.* Philadelphia: Saunders, 1982.

SOBOTTA, J. *Atlas of Human Anatomy,* vols. 1–3, New York: Macmillan (Hafner Press), 1974.

DICTIONARIES

BLAKISTON'S *Gould Medical Dictionary,* 4th ed. New York: McGraw-Hill, 1979.

DORLAND'S *Illustrated Medical Dictionary,* 26th ed. Philadelphia: Saunders, 1981.

DOX, I., et al. *Melloni's Illustrated Medical Dictionary.* Baltimore: Williams & Wilkins, 1979.

MILLER, B. F., and KEANE, C. B. *Encyclopedia and Dictionary of Medicine, Nursing, and Allied Health,* 3d ed. Philadelphia: Saunders, 1983.

STEDMAN'S *Medical Dictionary,* 24th ed. Baltimore: Williams & Wilkins, 1981.

STEEN, E. B. *Dictionary of Medical Abbreviations,* 5th ed. Philadelphia: Saunders, 1984.

TABER'S *Cyclopedic Medical Dictionary,* 14th ed. Philadelphia: Davis, 1981.

INDEX